Ecological Stoichiometry

Ecological Stoichiometry

The Biology of Elements from

Molecules to the Biosphere

Robert W. Sterner and James J. Elser

PRINCETON UNIVERSITY PRESS · PRINCETON AND OXFORD

Copyright © 2002 by Princeton University Press
Published by Princeton University Press, 41 William Street,
Princeton, New Jersey 08540
In the United Kingdom: Princeton University Press,
3 Market Place, Woodstock,
Oxfordshire OX20 1SY

Library of Congress Cataloging-in-Publication Data

Sterner, Robert Warner.
Ecological stoichiometry : the biology of elements from molecules to the biosphere /
Robert W. Sterner and James J. Elser.
p. cm.
Includes bibliographical references (p.).
ISBN 0-691-07490-9 (alk. paper)—ISBN 0-691-07491-7 (pbk. : alk. paper)
1. Biochemistry.   2. Stoichiometry.   3. Molecular ecology.   I. Elser, James J., 1959–   II.
Title.

QH345 .S74 2002
572—dc21   2002072262

British Library Cataloging-in-Publication Data is available

This book has been composed in Caledonia

Printed on acid-free paper. ∞

www.pupress.princeton.edu

Printed in the United States of America

2   3   4   5   6   7   8   9   10

ISBN-13: 978-0-691-07491-7 (pbk.)

ISBN-10: 0-691-07491-7 (pbk.)

CREDITS

Figure 2.1: Reprinted with permission from A. J. Lotka, *Elements of Physical Biology*,
Williams and Wilkins, 1925

Figure 2.3B: Reprinted with the permission of Scribner, a Division of Simon & Schuster from
The Double Helix by James D. Watson. Copyright © 1968 Elizabeth L. Watson, as Trustee under
Agreement with James D. Watson dated November 2, 1971;
copyright renewed © 1996 James D. Watson.

Figures 2.5A, B: Reprinted from *Molecular Biology of the Cell* (1999, volume 10, pp. 2054, 2055),
with permission by the American Society for Cell Biology.

Contents

Figures

Tables

Foreword

Stoichiometry has had a long and honorable history in ecology. Redfield's pioneering analysis in 1958 is a beautiful and widely known example of an explicitly stoichiometric analysis; Reiners' (1986) too-little-noted discussion of stoichiometry as a major underlying principle in ecosystem-level ecology is another example. Earlier work by Sterner and Elser, their students and colleagues, and increasingly by others, has built substantially on these beginnings. Stoichiometric concepts have been applied implicitly even more widely, for even longer. They underlie discussions of light-nitrogen and light-phosphorus interactions in terrestrial and aquatic ecosystems, of critical ratios of carbon to nitrogen during decomposition/mineralization, and of the nutrient use efficiency of individuals and ecosystems.

That background is reviewed clearly and thoughtfully here. However, I believe the application of stoichiometry to ecology has been changed fundamentally by the publication of this book—that there will be a BS + E and an AS + E (before and after Sterner and Elser) in this field. From simple and clear assumptions and observations, Sterner and Elser here build a powerful overall framework for understanding ecology. Their analysis makes use of physical principles like the conservation of mass and the dissipation of energy, and of biological principles like the importance of trade-offs in energy metabolism and growth at the biochemical and individual levels. The implications of these principles are developed carefully, and applied to the dynamics of individuals, populations, communities, ecosystems, and even of human alteration to the Earth system.

I believe that this book represents a significant milestone in the history of ecology—comparable perhaps to Lindeman's development of trophic dynamics. While that represented a young ecologist's brilliant (and sadly final) insight, this is a mature work, blending insight and synthesis. The framework developed in this book is integrated vertically, from organelle-level biochemistry to ecosystems; horizontally across organisms from bacteria to plants and animals, and across systems from freshwater and marine to grassland and forest; and conceptually across ways of looking at the world from theory, pattern, and process to experiment.

Love it or argue with it—and I do both—most ecologists will be influenced by the framework developed in this book. I know that the ways I develop my in-process book on the relationships between internal (plant-consumer-microbial) and external (input-output) nutrient cycling will change; certainly it will be more explicitly stoichiometric than it would have been otherwise; probably it will be conceptually richer as well. I will be surprised if any ecologist whose work encompasses resources on any level of biological organization does not find him-or herself similarly affected by this book. My teaching has become more explicitly stoichiometric in recent years—and that too will be accelerated by Sterner and Elser. There are points to question here, and many more to test (many of them identified directly in the book)—and if we are both lucky and good, this questioning and testing will advance our field beyond the level achieved in this book. I can't wait to get on with it.

PETER VITOUSEK
Professor of Biological Sciences
Stanford University

Preface

Organisms are made of more than one thing. This obvious statement forms the basis of the book you are about to read. As obvious as it is, this observation has been surprisingly neglected in much ecological and evolutionary thinking. In *Ecological Stoichiometry* we hope to demonstrate that this fundamental feature of life has profound implications for ecological dynamics, as well as for other areas of biology. Up until a few years ago, the words "ecological" and "stoichiometry" had rarely cohabited in the same sentence; this book explains why they often now do. Here we will describe this way of thinking and organizing certain kinds of information about living systems. In doing so, we provide some glimpses into important structuring forces in nature from genes to ecosystems.

The subject of ecological stoichiometry touches on many diverse topics: molecular and cell biology, life history evolution, nutrition, secondary production, nutrient cycling, food-web structure and stability, global change, and others. In the years we have been studying it, we have come to appreciate the many ways that ecological stoichiometry helps us understand patterns and processes in the natural world. In our "home turf" of freshwater plankton ecology, the basic ideas in ecological stoichiometry are increasingly well known, and increasingly scrutinized. This has led to refinement and improvement. That this has happened is gratifying and exciting. We hope that by putting all the diverse material about plankton in one spot, we will make that literature more accessible to workers in other fields.

However, we can see no particular reason why ecological stoichiometry should be more helpful in studying freshwater plankton than other suites of species in the living world. One of our major goals for this book therefore is to make the core concepts of ecological stoichiometry accessible and interesting to workers in other fields. Additionally, we sought to incorporate data from many systems (species, habitats, etc.) to make our points. This exercise often challenged us to stretch intellectually beyond our normal literature comfort zone, but we think the outcome was worth the effort. So, as far as scope goes, this book is—as nearly as we could make it—about all the ecosystems on Earth.

Another big stretch was necessary for connecting organismal stoichiometry to cellular and molecular functioning within an evolutionary framework. This meant assimilating a large bolus of molecular biological detail brand new to us. For example, the detailed literature on rDNA gene expression was initially daunting but, as the patterns within these data crystallized, and the possibility of connecting the structure of the genome to the processes in ecosystems seemed like an attainable goal, zeal replaced trepidation. We present the tentative outcome in Chapter 4.

One of the key observations that drew us into studying ecological stoichiometry was that even closely related animal species of similar size and built on a common body architecture can differ in the content of major elements such as N and P. Awareness of these differences was (and in places still is) surprisingly low. Perhaps the limited and diffuse data on elemental composition in different species contribute to this lack of awareness. This situation may have created a circular chain of scientific inaction: limited data result in little awareness and little ability to perceive patterns of interest. Meanwhile, low awareness and lack of interesting patterns result in little new data being generated and little effort at compiling those already available. One goal for this book is to help break that cycle. We emphasize repeatedly the importance of basic data on elemental composition of different living species, and how surprising some of the ecological and evolutionary implications of that composition turn out to be.

The chapters took shape over a four-year period during our separate sabbaticals and in between grant proposals, a ruptured Achilles tendon, and other important life and professional events. "Ecological stoichiometry" the subject and the book grew in tandem during that time and as we finish editing, new interesting works continue to appear. But it is time to stop writing this book and get it to you! Lead authors on the individual chapters were RWS for Chapters 1, 5, 7, and 8 and JJE for Chapters 2, 3, 4, 6, and 9. In the closing writing stages, it became necessary to determine order of authorship for the official bibliographic entry, and a coin flip in a hotel bar in Albuquerque, New Mexico, won by RWS, decided it.

ACKNOWLEDGMENTS

We extend abundant and sincere thanks to the people who read and helped improve portions of the book: Bill DeMott, John Downing, Jim Grover, Dag Hessen, Sarah Hobbie, Sue Kilham, Irakli Loladze, Kim Schulz, Val Smith, Jotaro Urabe, Mike Vanni, Manuel Villar-Argaiz, and Larry Weider. Don DeAngelis and Dave Karl read the entire manuscript under a deadline and provided pages of helpful and supportive feedback. We also thank Peter Vitousek for writing the Foreword. Tom Andersen (Oslo) provided help with plotting the nullclines of his model in Chapter 7, and Thomas Anderson (Southhampton) had many helpful comments on the minimal model in Chapter 5. Claudia Neuhauser lent a hand with the solution to the differential equation homeostasis model in Chapter 1. We also extend a special thank-you to the graduate students from our labs and others who read sections and drafts along the way, and who helped improve the book in numerous ways: Jessie Clasen, Dean Dobberfuhl, Rebecca Forman, Linda Gudex, Heather Hendrixson, Paul Frost, Amy Galford, Neil MacKay, Andrea Plevan, John Schampel, Tanya Smutka, and Jill Welter. The participants of several graduate seminars at UMN and ASU over the past few years also offered useful feedback. We gratefully acknowledge the people who kept our labs running smoothly during these few years and who also helped produce many of the data discussed; these include Sandra Brovold, Anne Dovciak, Jim Hood, Marcia Kyle, and Judy Olson. Joan Sterner did the final figure production.

And now for some personal words and thank-yous.

RWS: I've been attracted to approaches using resource ratios since my early days in graduate school at the University of Minnesota, where I had the good fortune to be in prime position to observe Dave Tilman write his first monograph on resource competition theory (Tilman 1982). His enthusiasm for an idea that offered new insights was contagious. I hope that with the writing of this book Jim and I might be doing some of the same things for another cohort of ecologists. The fact that I was at Minnesota in graduate school at all is due to one person, Mike Lynch, whom I encountered at the University of Illinois when I was in my undergraduate degree program. My luck was that Mike more than tolerated my inquisitiveness, and with his encouragement I began to learn to read the scientific literature. Taking one more step backward in time, it was my brother-in-law Terry Liersaph who first taught me the names of the birds and offered some scientific explanations for the things I saw happening around me in the natural world. Fast forwarding now to postgraduate work, Winfried Lampert and Ulrich Sommer brought me to the Max Planck Institute in Plön, Germany, where I learned entirely new lessons, not the least of which was how properly to do laboratory-based experimental limnology. I also want to give my personal thanks to some other senior colleagues who offered support and insight at key times, sometimes with just a few perfectly placed words and often with much more: Thomas Chrzanowski, Susan and the late Peter Kilham, David Young, and Joe Shapiro. I am grateful to Jim Elser for his unflagging enthusiasm and dedication to this subject.

I thank my family for their love and support. Mom and Dad Sterner took me camping when I was a boy, which somehow inevitably led to this project. I want to thank my wife Joan, daughter Kathryn, and son Nick for their understanding and patience. The book is done; let's have some fun! I acknowledge financial support during the writing from the Max Planck Society, the University of Minnesota, and the National Science Foundation. My part of this project is dedicated to Mom, Dad, Joan, Katie, and Nick.

JJE: For me this all began in the summer in 1985 when, alongside my wife Monica (a fellow limnologist at the time), I happened to be measuring phytoplankton nutrient limitation status (N vs. P) while investigating cascading trophic interactions as a member of a research team led by Steve Carpenter and Jim Kitchell. Much to everyone's surprise, we found that food-web structure, by shifting zooplankton community dominance from copepods and *Bosmina* to *Daphnia*, could shift the phytoplankton from limitation by N to limitation by P (Fig. 6.1). At the time we published that paper (Elser et al. 1988), we had no idea of the particular mechanisms, proximate or otherwise, that could produce such an effect. Much of my part of the following book is the result of my extended and increasingly stubborn attempt to get to the bottom of this, during which time I had the great fortune to join forces with Bob Sterner and others in this pursuit. At the moment the digging has brought me to strange sections of the library devoted to biochemistry, genetics, cell biology, genomics, and evolution, has led me to show my first image of an electrophoresis gel at a professional meeting, and has motivated me to expend capital equipment funds on a PCR cycler and other arcane items of the molecular trade. It has been a long strange journey for a food-web ecologist.

I thank my family also for accompanying me on this journey, especially Monica who was there at the very start of this long mission. Timothy and Stephen Elser assisted with sampling and with various helpful smiles and bear hugs along the way. Institutional support from Arizona State University is gratefully acknowledged, especially the supportive environment encouraged by my department chair, Jim Collins. University of Minnesota–Duluth and Kyoto University provided peaceful sabbatical homes where early chapters were drafted. Finally, I am grateful to the National Science Foundation for funding of my recent projects and for financial support of my time at Kyoto University. For my part, I dedicate this book to my father John and my late mother Rosanne.

Together: Our first meetings occurred at several different gatherings of scientific societies where we shared our newest results with each other. Quickly it became clear that we often thought exactly on the same plane (this went well beyond science: we shared identical vehicles, years of marriage, dates of birth of children, and other things—very spooky). A close working relationship formed quite naturally, helped in no small measure by the appearance of the Internet. As the two of us found ourselves thinking similar thoughts about the elements comprising the bodies of our precious zooplankton, we learned that other limnologists of our generation had arrived simultaneously at more or less the same set of questions. Key contributions leading to the coalescence of ecological stoichiometry in the plankton came from a set of then thirty-something limnologists from around the world, including, from Norway, Dag Hessen and Tom Andersen, from Japan, Jotaro Urabe, and from the United States of America, Mike Vanni. As with most things in science, there is nothing really new under the sun. All of us "stood on the shoulders of giants" who came before us. We think here of Lotka, Hutchinson, and Redfield of past generations and, among the alive and kicking, J. Goldman, R. Hecky, D. Karl, S. Kilham, Y. Olsen, W. Reiners, G. Rhee, V. Smith, U. Sommer, D. Tilman, and P. Vitousek.

We close these personal remarks by expressing our hope that, in combining these ideas, patterns, and theories in a single place, a deeper appreciation of the unity of biological systems will arise, freshened by a better understanding of the fundamental chemical properties of the living world.

Ecological Stoichiometry

1

Stoichiometry and Homeostasis

There is no science that claims the ecologists' leftovers.—Slobodkin (1988)

What are some of the most powerful explanatory ideas in all of science? Here are a few of our favorites: natural selection, the periodic table of elements, conservation of matter and energy, positive and negative feedback, the central dogma of molecular biology, and the ecosystem. This book you have just opened involves all of these. It is a book about how chemical elements come together to form evolved, living species in ecosystems. It is a book that takes very seriously the constraints of matter and energy. These are among the most powerful forces in nature, and a good understanding of physical and chemical barriers is one of the most helpful paths in understanding the things that are actually achieved. It is a book about biology and chemistry and to a lesser extent about physics and geology. As Slobodkin's statement above points out, ecologists often make serious use of the work from a great many disciplines. This is a book about many things, but it is organized around a single conceptual framework: **ecological stoichiometry** (see p. 42 for definitions of words in bold). We will soon elaborate in some detail on the history and meaning of the term "**stoichiometry**". To the uninitiated, it refers to patterns of mass balance in chemical conversions of different types of matter, which often have definite compositions. Most scientists run into it in beginning chemistry class when they learn to balance chemical reactions.

What is it you notice when you observe an ecosystem such as a grassland, lake, or stream? All biologists perceive things in their own way and that way is much simpler than the reality in front of them. Some have a conception based on the names of the species that they see. Some focus their minds on what they know or infer about the recent or distant history of the site. Some focus on the animals, others on the plants, and still others on the soil and microbes. Some concentrate on the structure, some dwell on the function. Some see constancy while others see dynamic change. Some might wonder how it's possible that the information necessary to build all of the organisms present could possibly be stored and processed by tiny molecules. All of these views are "right" in their own way. As Levins and Lewontin (1980) put it, "The problem ... is to understand the proper domain of each abstraction rather than becoming its prisoner." The human mind is incapable of grasping at once the enormous complexity of the entirety of natural systems, and it seeks simpler abstractions as a path to knowing. The abstraction we will follow is this: organisms can be thought of as complex evolved chemical substances that interact with each other and the abiotic world in a way that resembles a complex, composite, chemical reaction. Ecological interactions invariably involve chemical rearrangements. Like any other "normal" chemical rearrangement at the surface of the Earth, when organisms interact, mass must be conserved and elements are neither created nor destroyed (we will ignore the other more exotic things that can happen, especially at high energy or in radioactive substances, which a nuclear physicist could explain). There is stoichiometry in ecology, just as there is in organic synthesis in a test tube.

The stoichiometric approach considers whole organisms as if they were single abstract molecules. While this is, of course, not strictly true, note that if we wish to identify and understand those properties of living things that are truly biological, we must first identify those features that arise simply from the chemical nature of life. The science of chemistry has persisted despite, and indeed profited from, a deep understanding of the physics of the atom. Likewise, a rich and predictive science of biology will continue to develop as we better understand the chemical nature of life and how its consequences are expressed at all levels of biological organization. Some might find our approach to be a radically reductionist agenda, a charge to which we join Wilson (1998) in pleading "guilty." Wilson's "consilience" concept regards the fragmentation of knowledge and resulting isolation as artifacts of scholarship rather than reflections of how the world is really organized. We have been excited to see stoichiometric reasoning correctly predict very macroscopic phenomena using its very microscopic principles (for example, in the studies of herbivores in light gradients that we will cover in Chapter 7). This book starts with the basic physical chemistry of the elements and progresses in a linear fashion from atoms to ecosystems with an utter disregard for where chemistry ends and biology starts, or where evolution starts and ecology ends, and so forth. Deep philosophizing aside, all organisms must obey the principles of conservation of energy and mass, and it is time we put these principles to new ecological uses.

In this first chapter, we will explain the book's scope and define a set of core concepts. As long as elements are neither created nor destroyed, any multiple-element system will follow many of the rules we discuss here. However, a driving mechanism behind many of the patterns is the difference in stoichiometric variability among different organisms, species, nutritional modes, trophic levels, etc. The chemical composition of different ecological players is constrained to different degrees, and it is this difference in variation, in addition to differences in mean values, that has interesting effects. The chapter continues with some consideration of closely related concepts, including the idea of "yield" as used in ecology and agriculture, an extended discussion of the "Redfield ratio" in terms of organismal C:N:P ratios, and some pragmatic notes about how ratios and growth rates are expressed. Finally, to foreshadow the wide-ranging scope of the materials to follow, we describe a logical structure for stoichiometric analysis first articulated by Reiners (1986), from which we have drawn great inspiration.

SCOPE

To begin to define this book's scope, let us start by considering what portion of chemistry and physics we cover. Our focus is almost entirely on the elements. All of life requires the macroelements C, H, O, N, P, etc., as well as a set of trace elements including Fe, Mg, and others (Fig. 1.1). These provide for a diverse set of functions including structure (such as C, H, and O in cellulose fibers), oxidation-reduction (such as Fe as a cofactor in enzymes), and others (see Chapter 2). The patterns of abundance of elements in living things are major components of ecological stoichiometry. Take humans as an example. The concentrations of at least 22 elements in humans have been determined; these range in total amount in a typical live individual from 35 kg (O) to 1 mg (Co) (Heymsfield et al. 1991; Williams and Fraústo da Silva 1996). From information on the quantities of individual elements, we can calculate the stoichiometric formula for a living human being to be

$$H_{375,000,000} \ O_{132,000,000} \ C_{85,700,000} \ N_{6,430,000} \ Ca_{1,500,000} \ P_{1,020,000} \ S_{206,000}$$
$$Na_{183,000} \ K_{177,000} \ Cl_{127,000} Mg_{40,000} \ Si_{38,600} \ Fe_{2,680} \ Zn_{2,110} \ Cu_{76} I_{14} \ Mn_{13}$$
$$F_{13} \ Cr_7 \ Se_4 \ Mo_3 \ Co_1$$

That is, there are about 375 million H atoms for every Co atom in your body (the formula is based on "wet weight"). This formula combines all compounds in a human being into a single abstract "molecule." This formula sets the value of the scarcest substance (cobalt, mass in humans ≈ 1 mg) equal to a stoichiometric coefficient of 1 and shows relative amounts, not absolute ones. Most stoichiometric analyses are in fact concerned with relative abundances. The large stoichiometric coefficients for H and O, the most numerous atoms in our bodies, arise partly from the fact that life is aqueous and partly from the fact that H and O are key components of many (or all, in the case of H) organic molecules. Our main purpose in introducing this formula for the "human molecule" is to stimulate you to

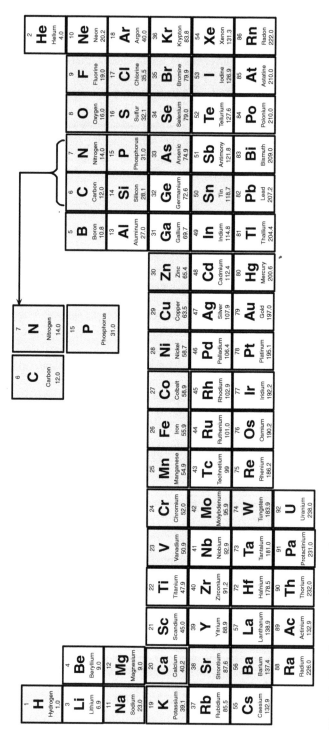

Fig. 1.1. Distribution in the periodic table of the elements known or believed to be essential for bacteria, plants, or animals (shaded) (from Williams and Fraústo da Silva 1996). Ecological stoichiometry is much more advanced for the three elements C, N, and P than for the others, but its principles should be transportable throughout the periodic table.

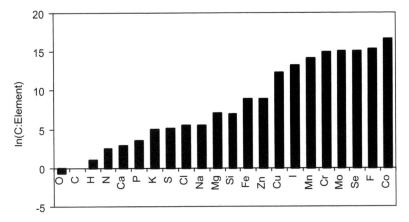

Fig. 1.2. Abundance of 22 elements in humans, expressed relative to carbon (mass:mass) and natural-log-transformed for clarity. High values indicate substances in low relative abundance. Data obtained from two sources, with averages taken when multiple values were available (Heymsfield et al. 1991; Williams and Fraústo da Silva 1996). For a historical presentation of similar information, see Fig. 2.1.

begin to think about how every human being represents the coming together of atoms in proportions that are, if not constant, at least bounded and obeying some rules. Humans must obtain these elements in sufficient but not (for some) oversufficient quantities from an environment that may or may not have similar proportions of these elements. Human physiology must take complex chemical resources containing multiple elements arranged in myriad different molecules, absorb some, metabolize some, rearrange many, and excrete or otherwise release a great deal. This book is concerned both with the way that organisms do this and with its consequences for ecological dynamics.

We will often express element content relative to carbon. A high C:element ratio means that element is in low quantity relative to C. Figure 1.2 shows the distribution of the matter of a human among component elements. The total mass of elements for this human is 64.0 kg, and we can account for 99% of the mass with the elements up to and including K. Although it is customary to speak of "macro" and "micro" elements to contrast those that are in high versus low abundance, there is no clear breakpoint between the two. It should not be assumed that elements in low abundance are unimportant; frequently, in fact, the opposite is the case. Ecological stoichiometry takes information like that illustrated in this figure, combines it with similar information on potentially interacting species, including the resources that are consumed, and makes testable predictions about such things as, "How fast will the organism grow? What nutrients will become limiting in an ecosystem? What accounts for the foraging decisions consumers make? How does the environment constrain the evolution of particular life histories?"

Why do we focus so on the elements? To a certain extent we do this because we are limnologists trained to take an ecosystem perspective and focus on pools and fluxes of energy and matter in the environment. But there are grander reasons to focus on elements. They provide a framework for easily moving between levels of biological organization, as we can calculate the elemental composition and estimate the fluxes of chemical elements to and from a wide variety of biological entities, ranging from organelles and cells to watersheds and indeed the biosphere. In this book we will present data across this entire range and in doing so we hope at least to raise the possibility that stoichiometric analysis may help increase the degree of conceptual and intellectual consilience in modern biology.

A more pragmatic reason for focusing on the elements is that they (at least the nonradioactive ones) are immutable and thus we can put the law of mass conservation to use. For stoichiometry to work, the chemical substances of interest must not be created or destroyed; otherwise we could not rely on conservation rules. This reason implies that stoichiometric analysis might also be applied to certain biochemicals (for a recent application, see Anderson and Pond 2000). For example, some contaminants may follow stoichiometric principles, since organisms do not synthesize them, and they are also highly resistant to biological breakdown. If these substances also showed variability in their concentrations in key ecosystem components, principles of ecological stoichiometry developed for elements might be useful in understanding the distribution of contaminants in ecosystems. With active regulation of contaminant concentration by certain members of an ecosystem, the comparison between elements and contaminants would be very apt. Other biochemicals are essential in consumer diets. "Essential" can be taken to mean they cannot be synthesized at rates necessary to sustain growth and survival, and thus they must be obtained from the food. If catabolism also is negligible, again we might be able to use the same principles discussed here to understand those nutritive biochemicals. There likely are some good applications of ecological stoichiometry to biochemicals; however, in general the data to do so are not yet widely available. The majority of this book is about elements, reflecting the state of the art but perhaps not the ultimate applicability of the ideas.

A focus on elements raises the question of how safe it is to ignore the biochemical arrangement of elements in organisms. Although stoichiometry represents organisms as single abstract molecules, they are of course wondrous collections of countless individual biochemicals and even most biologically active substances are composed of multiple elements. We recognize that an animal's N budget is primarily determined by the physiology and biochemistry of amino acids and proteins, not zillions of independent atoms of N. The biochemical nature of the precise arrangement of elements into compounds and molecules affects how organisms utilize those elements. A different and more detailed approach, based on proportions of different chemical bonds (C—C vs. C—N, etc.), has also been considered (Hunt et al. 1983). Our point in this work is not that organisms *are* bags of independent elements. Rather we wish to see how far this abstraction will take us; we are confident it will take us further than any approach using a single currency (e.g., total mass, energy, C only, P only, etc.). Deevey (1970) considered the "architecture" of assembling elements into compounds and put it this way: "A listing of elements and compounds does not reveal that architecture. There is a big difference between a finished house and a pile of building materials. Nevertheless, a list is a useful point of departure. If it is made with care, it can protect ecologists from the kind of mistakes that architects sometimes make, such as forgetting the plumbing."

We sometimes hear the criticism that we wish to treat organisms as if they were "just" chemical reactions. Our response is, "What do you mean by 'just'?" We are not only reasoning by analogy. Organisms *are* chemical entities and are produced, maintained, and propagated by chemical reactions, albeit in the form of highly complex coupled networks, which are the product of evolution. Statements about nature that are so well accepted as to be referred to as "laws" are exceedingly rare. The **law of conservation of matter** is as close as we get in science to a "fixed point" in nature and we should make good use of it as our understanding of the living world grows.

During the years that the two of us have participated in the development of this field, we have alternately heard that ecological stoichiometry is either too complicated to understand or too simple to be true. With respect to the former, we hope this volume makes things clearer. With respect to the latter, we acknowledge that ecological stoichiometry is unlikely to be a "final theory" (is anything?) and that it does not and cannot explain everything of interest in the ecological and evolutionary worlds. However, bear in mind that simple mechanisms can produce surprisingly complex outcomes, evolution by natural selection being the shining example. Hence, we take inspiration from the words of the physicist Percy Bridgman in 1927 (as quoted by Ferris 2001), who commented on the importance of "conviction" in scientific discovery: "Whatever may be one's opinion as to the simplicity of either the laws or the material structures of Nature, there can be no question that the possessors of such conviction have a real advantage in the race for physical discovery. Doubtless there are many simple connections still to be discovered, and [s/he] who has a strong conviction of the existence of these simple connections is much more likely to find them than [s/he] who is not at all sure that they are there."

STOICHIOMETRY AND HOMEOSTASIS

Let us turn now to a more complete explanation of stoichiometry and several key related concepts. In its common chemical usage, the term "stoichiometry" refers to patterns of proportions of elements in reactants and products of chemical reactions. Stoichiometry can be defined as a branch of chemistry that deals with the application of the **law of definite proportions** and conservation of mass. Lotka (1925) wrote: "we may employ the term Stoichiometry to denote that branch of the science which concerns itself with the material transformations, with the relations between the masses of the components." The word "stoichiometry" comes from the Greek root "stoicheion" for element. "Stoichiometry" thus means "measuring elements." During the earliest days of chemistry, it was discovered that some substances reacted in constrained ways; that is, they combined only in set proportions. This was a key observation in the discovery of the elements, suggesting that something indivisible was combining so that proportions could always be reduced to integers.

Consider a familiar example. On each side of the chemical reaction for photosynthesis,

$$6CO_2 + 12H_2O \rightarrow C_6H_{12}O_6 + 6H_2O + 6O_2, \qquad (1.1)$$

there are six atoms of C, 24 atoms of O, and 24 atoms of H. The reaction stoichiometry indicates that one diatomic molecule of oxygen (O_2) must be produced for every molecule of carbon dioxide that is used. The most important thing about stoichiometry is that it constrains allowable states. Stoichiometry says you can't combine things in arbitrary proportions; you can't, for example, change the proportion of water and dioxygen produced as a result of making glucose. In this reaction, six molecules of each must be produced for every molecule of glucose generated. Stoichiometry also controls the quantitative relations between reactants and products. If you simply added one CO_2, for example, it would be in excess and would not react (and the yield of glucose from carbon dioxide would be diminished, as we explain below). Were it not for the constraints of stoichiometry, chemistry would be a highly chaotic subject. Chemical stoichiometry is essentially a very large set of constraints that greatly limit the combinations of chemical elements and how those combinations interact with each other. As we will see as we go further, ecological stoichiometry sets similar limits to interacting biological systems. These limits to composition combine with those imposed by the conservation of matter. Establishing these constraints and examining how they impinge on ecological processes is what makes ecological stoichiometry potentially a very powerful heuristic and predictive tool.

In the context of ecological stoichiometry, conservation of energy presents somewhat of a semantic problem. The term "stoichiometry" normally refers only to the conservation of matter. The conservation of energy is typically considered separately (in thermodynamics). However, matter and energy are inextricably linked in biological systems as they are indeed in all systems. For example, it is possible to rewrite all mass fluxes in organisms in comparable energy terms (Kooijman 1995). In this book, we will often want to be able to combine energy fluxes and matter processing when considering solar energy and nutrient fluxes to plants. When this happens, we adopt a broad definition of stoichiometry, one that includes energy transformations and conservation as well as conservation of matter. Whether or not this is really stoichiometry is just a semantic nuisance.

The term "stoichiometry" has also been used to refer to the quantitative relationship between constituents in a chemical substance, i.e., the quantitative relationship between two or more substances making up a composite substance, such as in our formula for the chemical makeup of a human given above. In this sense of the word, this chemical formula represents "the stoichiometry of a human." An interesting but archaic term is "stoichiology," which refers to that part of the science of physiology that treats the elements composing animal tissues (*Webster's New Universal Unabridged Dictionary*, 2nd ed., 1983). Actually, it could be said that much of this book is about "stoichiology," not "stoichiometry," but we will not attempt to reverse any etymological tides here; we're perfectly happy with "stoichiometry."

Although many compounds are formed with elements in strict and definite proportions, not all are. In some circles, the former compounds are called stoichiometric and the latter are called nonstoichiometric (Rao 1985; Williams and Fraústo da Silva 1996). Covalently bonded substances such as $C_6H_{12}O_6$ or compounds characterized by ionic bonds and crystalline lattices such as NaCl are stoichiometric. According to this terminology, in stoichiometric substances, the precise pattern of bonding between atoms means that elements must occur in fixed proportions in the material. However, not all substances are like this. Compounds lacking fixed proportions of elements include certain salts, minerals, and alloys. One notable example is semiconductors, some of which are allowing these words to be written (and edited and reedited!) on a personal computer. Another is below your feet. The silicates of Earth are composed of a principal Si-O unit that builds itself in countless ways with Na^+, Mg^{2+}, Al^{3+}, and Ca^{2+} into a vast variety of strings, planes, and three-dimensional structures, making this major portion of the Earth's crust nonstoichiometric. In substances like semiconductors and siliceous minerals, proportions of chemical elements are not fixed. The proportions that are found are largely a function of the proportions of elements available when the substance is made, much as the shade of light blue paint one gets depends on the proportion of white and blue paints being mixed.

To our eyes, the hallmark of the nonliving world (e.g., rocks, air, substances dissolved in water) is stoichiometric variability. Indeed, the stoichiometric composition of certain portions of the abiotic world is fundamentally limitless. The most extreme stoichiometric composition of any material would be 100% of a single element. The characteristics of some of the pure elements are very familiar. Carbon can be found in pure form as graphite, diamond, or even the more exotic C_{60} buckminsterfullerene. As a colorless, odorless gas, pure nitrogen is the major single component of today's atmosphere. Pure phosphorus is a waxy solid, colorless, transparent, and spontaneously combustible. Pure iron and pure gold are both well known to all, the latter at least by reputation. Combinations of elements in inorganic matter do not entirely lack constraints; some combinations of elements spontaneously react, forming products that may have definite proportions. Some refuse to coexist for other reasons. But many common elements are able to combine in (approximately) limitless combinations. For example, common igneous rocks can contain from near zero to more than 50% MgO (Cox 1995). The chemical evolution of Earth has included massive changes in atmospheric chemistry, from a highly reduced state containing 10 atm CO_2 along with N_2 and CH_4 and virtually no O_2 to today's air with 10^{-3} atm CO_2 and N_2 and O_2 as major components; in other words, there has been a major shift in composition from carbon to oxygen (Schopf 1982). The chemistry of aqueous solutions is critical to understanding life, and there are some important purely inorganic controls on the chemistry of aqueous solutions. Ionic strength, which responds to the balance between precipitation and evaporation, has a major influence on the composition of surface waters (Gibbs 1970; Gorham et al. 1983; Kilham 1990). Phosphorus availability in high-oxygen waters is strongly dependent on concentrations of metals such as Fe and Mn (Mortimer 1941, 1942), oxidized precipitates of which bind with P and greatly lower its availability (we will discuss this relationship again near the end of the book). Metals variously speciate and react depending upon pH and oxidation-reduction potential, and so forth. These are some of the purely chemical constraints on the stoichiometry of the abiotic world. However, as we will see as we go, those constraints are so loose, at least in comparison to the stricter ones in living biological systems, that we view the inorganic world as having great stoichiometric variability. In the final chapter of this book we present some summary data comparing the stoichiometry of abiotic and biotic systems.

Let us look in more detail at some of the major components of the nonliving natural world, because there are some interesting and important stoichiometric patterns of elements there. In a progression from the solar system, to the whole Earth, to the Earth's crust, to seawater (Table 1.1), there are numerous gigantic shifts in chemical content of 10^3, 10^6, and even 10^9 parts per mass. The two lightest elements, H and He, are the most abundant elements in the solar system (the mass of which is almost entirely made up by the sun) with oxygen also making up about 1% of total mass. Stoichiometric shifts associated with the formation of the Earth resulted in much greater concentrations of Li and much lower concentrations of He and Ne. Other elements in greater concentration in the Earth than in the solar system as a whole include Be, Mg, Al, Si, S, Ca, Sn, V, and Fe. Stoichiometric shifts in the formation of the crust from the whole Earth are somewhat less extreme, although there is a very large drop-off in H and increases in the concentrations of Fe, B, Ne, Na, and K. Many very large and very important differences in the chemical content of seawater as compared to Earth's crust include massively lower concentrations of Be, Al, Sn, Ti, Cr, Mn, and especially Fe in the sea than in the crust. The inorganic chemistry of the elements, and many very intriguing patterns in abundance as related to fundamental aspects of chemistry, have been engagingly considered by Cox (1989, 1995) and by Williams and Fraústo da Silva (Williams 1981; Williams and Fraústo da Silva 1996). In particular, the latter duo's major opus of 1996, with its stated purpose "to show the relationship of every kind of material around us, living and non-living, to the properties of the chemical elements of the periodic table," is required reading for all those interested in ecological stochimetry.

Now, what of the transition to living things? As Williams (1981) wrote, "Evolution through natural selection implies that there must be a drive within biology to readjust the given accidental abundances of the Earth's crust so as to optimize biological chemistry." Table 1.1 also shows some major stoichiometric jumps in going from modern seawater to the human body. Putting aside the thorny issue of how much the chemistry of today's ocean resembles the setting for the evolution of life many years ago, we can see that, at a coarse level, the recipe for this representative living thing differs dramatically from the bulk of Earth's surficial water today. The human body has much greater concentrations of N, P, and Fe (three elements very commonly regarded as limiting to living systems—no coincidence there), and a much lower concentration of, for example, Ar.

At a deeper level, we must consider whether organisms are like immensely complex covalently bonded molecules, with elements appearing in fixed proportions. If organisms were like covalent molecules, mass flux through them would be highly constrained in the same way that C, H, and O are constrained to combine only in one way in the formation of glucose.

TABLE 1.1

Approximate concentrations of the first 26 elements in the solar system, in the whole Earth, in the Earth's crust, in seawater, and in humans. Concentrations are divided into categories differing by multiple orders of magnitude. As parts by mass they are $A > 10^{-2}$ ($> 1\%$), $10^{-2} > B > 10^{-6}$ (between 1 ppm and 1%), $10^{-6} > C > 10^{-9}$ (between 1 ppm and 1 ppb), and $D < 10^{-9}$ (< 1ppb). Less-than and greater-than symbols are used in the table to indicate changes in categories for individual elements between columns (hence large fractionations). Several elements for humans are reported as "0" as these are not detectable. Graphical depictions of similar data for crust and for humans are given in Figures 1.2 and 2.1. Data from Cox (1995).

Atomic Number	Symbol	Atomic Mass	Solar System		Whole Earth		Earth's Crust		Seawater		Human Body
1	H	1	A	>	B	>>	D		A		A
2	He	4	A	>>	D		D		D	>	0
3	Li	7	D	<<	B		B	<	C		C
4	Be	9	D	<	C	<	B	>>	D		D
5	B	11	C		C	<	B		B		B
6	C	12	B		B		B		B	<	A
7	N	14	B		B		B	<	C	<<	A
8	O	16	A		A		A		A		A
9	F	19	B		B		B		B		B
10	Ne	20	B	>>	D	<	C	>	D	>	0
11	Na	23	B		B	<	A		A	>	B
12	Mg	24	B	<	A		A	>	B		B
13	Al	27	B	<	A		A	>>	D	<	C
14	Si	28	B	<	A		A	>	B		B
15	P	31	B		B		B	>	C	<<	A
16	S	32	B	<	A	>	B		B		B
17	Cl	35	B		B		B	<	A	>	B
18	Ar	40	B	>	C		C		C	>>	0
19	K	39	B		B	<	A	>	B		B
20	Ca	40	B	<	A		A	>	B	<	A
21	Sc	45	C	<	B		B	>>	D	>	0
22	Ti	48	B		B		B	>>	D	<	C
23	V	51	C	<	B		B	>	C		C
24	Cr	52	B		B		B	>>	D	<	C
25	Mn	55	B		B		B	>>	D	<	C
26	Fe	56	B	<	A		A	>>	D	<<	B

Reiners (1986) suggested that organisms lacking major support structures (he called such creatures "protoplasmic life") were like that: "Protoplasmic life has a common stoichiometry of chemical elements in particular proportions." Or are organisms more like Earth's crust, with a variable chemistry formed from differing proportions of a limited set of constituents and with **nutrient content** largely determined by the proportions of elements that they are exposed to? Herbert (1961) wrote, "There are few characteristics of micro-organisms which are so directly and so markedly affected by the environment as their chemical composition. So much is this the case that it is virtually meaningless to speak of the chemical composition of a microorganism without at the same time specifying the environmental conditions that produce it." Similarly, in writing about insect nutrition, Mattson and Scriber (1987) wrote, "Food has profoundly influenced the evolution of animals because in so many respects an animal is what it eats."

The contrast in these stated views is partly a matter of perspective, including whether one is considering physiological, ecological, or evolutionary time scales. Any difference looks small from a great distance or relative to even bigger differences. To resolve these differences in points of view, we will consider how consumer stoichiometry varies with resource stoichiometry, both in terms of theoretical expectations and in a series of empirical examples. First, we will describe the general features of plots of consumer versus resource stoichiometry; then, we will look at some mathematical models that capture key aspects of these patterns.

One situation is both easily analyzed and readily understood. If a consumer's nutrient content passively reflected the content of the resources it consumed—in other words, if a consumer truly *was* what it ate—then all points in a plot of consumer stoichiometry versus resource stoichiometry would lie on a line with slope 1 and intercept zero (Fig. 1.3A, dotted line) (for all such plots, we use the same scale for consumer and resources). For this to occur, species must assimilate and retain the nutrients being plotted in identical proportions to their relative abundance in the food. (We will see later in the book that the relevant term describing this assimilation and retention is the gross growth efficiency.) This "you are what you eat" model provides one simple way that stoichiometries of consumer and resource may be related: they may be equal.

Let us consider **nonhomeostatic** elements within organisms more deeply. These at first seem to violate the idea that **homeostasis** is the essence of life. It is self-evident that no organism can be totally nonhomeostatic and still be alive. There are limits to what combinations of chemical elements can function as a living cell (in its absurd limit, cells cannot take on the composition of pure C, N, P, or any other single element). We will develop those ideas from a molecular and cellular standpoint in the next chapter. However, it is clear already that the stoichiometry of any living system must be bounded. This is one aspect of a complete understanding of the determinants of consumer stoichiometry.

Returning to our plot, departures from the 1:1 line would be caused by differential nutrient processing of the two elements. If the stoichiometry of the consumer was some constant of proportionality multiplied by the stoichiometry of the resource, a family of lines with constant slope and zero intercept would be obtained (Fig. 1.3A, solid lines). We will call this the constant proportional model. Although the two hypothetical consumers represented by the solid lines in Figure 1.3A are not what they eat, they still are passive consumers that do not adjust their stoichiometric balance in response to the stoichiometry of the resources they can consume. In one case, their stoichiometry is a constant multiple higher than the resource stoichiometry, and in the other case, their stoichiometry is a constant multiple lower than the resources. They may generally be more variable than the food they eat (Fig. 1.3A, upper solid line) or less variable than the food they eat (Fig. 1.3A, lower solid line), but they are not regulating their stoichiometry in response to their food. They could be following a simple rule such as, "always retain A percent of all resource 1 ingested and B percent of all resource 2 ingested." Written out in this way, it is apparent that in these sorts of patterns, there is no feedback between the resource's elemental composition and that of the consumer.

Negative feedback (opposite to the direction of perturbation) between an internal condition (such as consumer stoichiometry) and an external condition (resource stoichiometry) in biological systems is a **homeostatic** regulation. We have already noted the use of the terms "stoichiometric" and "nonstoichiometric" compounds to represent different degrees of constancy of element proportions in chemistry. However, in biology "homeostasis" is a more widely accepted term. "Homeostasis" is the resistance to change of the internal milieu of an organism compared to its external world. It has been said that homeostasis is the essence of life. Organisms regulate many of their properties, including water balance, pH, and others. In ecological stoichiometry, homeostasis is observable in the patterns of variation in nutrient elements in organisms relative to their external world, including the resources they eat. Kooijman (1995) defined homeostasis in a stoichiometric context as follows: "The term homeostasis is used to indicate the ability of most organisms to keep the chemical composition of their body constant, despite changes in the chemical composition of the environment, including their food." An organism's stoichiometry is the pattern that results from different degrees of homeostasis operating on chemical composition. Homeostasis generates different degrees of variation in chemical substances in living things. Because in ecological stoichiometry, homeostasis refers to changes in matter in living things, it is often used when discussing growth. However, as a general biological concept, homeostasis may occur with or without growth. We will see that differences in the strength of homeostasis have numerous ecological consequences. Homeostasis is a major reason for this book's existence. Without it, ecological stoichiometry would be a dull subject.

Another easily understood case would be if a consumer's nutrient content were independent of the chemical composition of its food. In our now familiar plot of consumer versus resource stoichiometry, any horizontal line segment above, below, or intersecting with the 1:1 line would represent this situation. We refer to these situations as "**strict homeostasis**"

(Fig. 1.3B, solid lines). A related term that will come up frequently later is **balanced growth**. These two terms describe a **stoichiometric equilibrium** where the proportions of substances in an organism do not change. Strict homeostasis means that consumer stoichiometry does not vary with resource stoichiometry. However, strict homeostasis does not necessarily mean that a group of individuals of such a species from nature, including diverse ages, life stages, sexes, etc., will exhibit zero variation in their chemical content. Individual organisms often show differences in stoichiometry during their life cycles. Young organisms may have different composition from older ones, reproductive organisms may be different from nonreproductive, males may be different from females, etc. For example, a plot of body C:P versus size for a species might show a trend, but if that allometric trend is "hard wired" (not a function of food ingested), such a plot could be consistent with a strict homeostasis of C:P in that species. Intraspecific variation in chemical content alone does not disprove the presence of strict homeostasis.

Also, please be warned that the word "balance" will be used in several other ways in this book. One usage is in the sense of balancing a budget: for example, in the water cycle or nutrient cycles, mass conservation means that input equals output in an equilibrium system, and this is a "balancing." Another usage is in the general sense of where nutrient or energy and nutrient ratios lie relative to some reference point: for example, inorganic N:P ratios in the environment that happen to be similar to those found in the tissues of living things can be considered balanced. Finally, when we discuss interacting stoichiometric systems (such as consumers and their resources), we will consider two things having similar stoichiometry to be "balanced," and two things having dissimilar stoichiometry to be "imbalanced." In that context, **elemental imbalance** is a measure of the dissimilarity in relative supply of an element between an organism and its resources.

Figure 1.3 therefore presents two extremes: the complete absence of stoichiometric regulation by a consumer (panel A) and a regulation so thorough that the consumer's state is independent of the resource state (panel B). These are limits. As we shall soon see, nature often dwells in the space between these end points. Therefore, we need to consider what to do with imperfect homeostatic regulation. What if there is a negative feedback between resource and consumer stoichiometry, but the outcome is not a truly fixed elemental composition? It is important to define homeostasis as precisely as possible because it is a core concept in ecological stoichiometry.

Differing levels of homeostatic regulation can be analyzed both graphically and with equations. First, graphically: Consider an arbitrary point x,y in a plot of consumer versus resource stoichiometry. Homeostatic regulation of nutrient content at x,y can be diagnosed as a slope lower than the

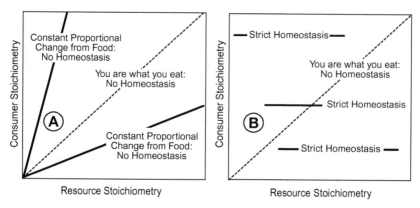

Fig. 1.3. Generalized stoichiometric patterns relating consumer stoichiometry to resource stoichiometry. Horizontal and vertical axes are any single stoichiometric measure, such as N content or C:P ratio. A. Points on the 1:1 line (slope 1, intercept 0) represent identical stoichiometry in consumer and resources. This dashed line represents a consumer with stoichiometry that always matches the stoichiometry of its resources. This is the "you are what you eat" model. The solid lines represent consumers that perform constant differential nutrient retention. These represent the "constant proportional model." B. Strict homeostasis is defined as any horizontal line segment (slope 0, intercept > 0).

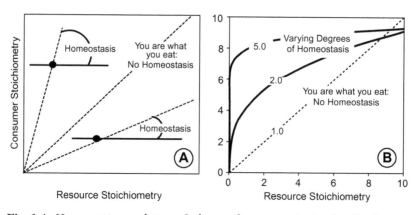

Fig. 1.4. Homeostatic regulation of elemental content. A. Graphically, homeostasis at a point x,y can be defined as a slope between 0 and y/x. B. Degrees of homeostatic regulation based on models with constant coefficient of regulation [H, Eq. (1.3)]. The three curves use values for the coefficients of c and H of 1 and 1, 2 and 3, and 5 and 6, and they are labeled by their value of H. The line marked "1.0" represents no homeostatic regulation. Increasing values of H mean increased regulatory strength.

slope from a constant proportional response shown in Figure 1.3A. In other words, homeostasis is a slope at x,y less than y/x, down to an expected lower limit of zero (Fig. 1.4A). Thus, the arcs in Figure 1.4 indicate expected ranges of homeostatic regulation of consumer stoichiometry. An important thing to understand from this plot is that the slope alone cannot diagnose homeostasis (except where the slope is truly zero). Notice that slopes may be fairly steep and still be consistent with homeostatic negative feedback. Similarly, even fairly shallow slopes may be consistent with an absence of homeostasis.

We now can formalize this concept of stoichiometric homeostasis. First, note that the model of constant proportionality (Fig. 1.3A) can be rewritten as $dy/dx = y/x$, where y = consumer stoichiometry and x = resource stoichiometry (measured on identical scales). By "stoichiometry" here we mean any sort of ratio of substances or masses (e.g., percent P, N:P, etc.). Thus, we use the general symbols y and x here to represent the vertical and horizontal variables. Now, regulation as conceptualized graphically in Figure 1.4A would be given by

$$\frac{dy}{dx} = \frac{1}{H}\frac{y}{x},$$

(1.2)

where H (eta) is a regulation coefficient greater than 1. For x large enough (note it is in the denominator of the last term), as the regulation coefficient approaches infinity, the slope of consumer versus resource stoichiometry approaches zero (i.e., regulation approaches strict homeostasis). Equation (1.2) can be rearranged to show that 1/H relates the proportional change in y (dy/y) to the proportional change in x (dx/x). Equation (1.2) can be integrated to give

$$y = cx^{\frac{1}{H}},$$

(1.3)

where c is a constant. In Figure 1.4B, we have plotted several realizations of Equation (1.3). The three functions in this plot differ in their values of c and H. The latter parameter accounts for the "bendiness." A surprising conclusion from this plot is that, at extremely low values on the resource stoichiometry axis, even a seemingly high sensitivity of consumer stoichiometry to resource stoichiometry may be consistent with homeostatic regulation (note the steep slopes at the left portion of the graph).

A perhaps more useful way to study homeostasis is to linearize Equation (1.3) using logarithms:

$$\log(y) = \log(c) + \frac{\log(x)}{H}.$$

(1.4)

Equation (1.4) indicates that we can easily diagnose homeostatic regulation of consumer stoichiometry by plotting the logarithms of consumer versus resource stoichiometry. On such a plot, slopes (1/H) between 1 and 0 indicate a negative feedback between resources and consumers, and hence indicate homeostatic adjustment of nutrient content because they indicate a lower proportional change in consumer stoichiometry than in resource stoichiometry.

Now that we are equipped to think more precisely about these patterns, let us consider several examples showing how consumer and resource stoichiometry are related. Variation in stoichiometry in living things is in some ways similar to the classical ecological terminology of "regulators" and "conformers." The former have internal conditions (e.g., temperature) little changed by external conditions, while in the latter, inside matches outside. Let us see how closely some real species come to matching these ideal expectations. We will now consider as tests of the degree of homeostasis a set of studies that took a single standardized set of individuals of common age, stage, etc., and fed them different diets experimentally. Studies using collections of miscellaneous individuals of differing ages, etc., might well be interesting for other purposes, but they do not necessarily test homeostasis.

A clear example of nonhomeostasis of a chemical parameter comes from the algae studied by Rhee (1978), who cultured *Scenedesmus* (a Chlorophyte to which we will often refer) at a range of nitrate concentrations, holding phosphate concentration constant, and thus producing a range of N:P resource ratios. He grew his cultures to equilibrium in chemostats (for more on the operation and theory of chemostats, see Chapter 3). Rhee's *Scenedesmus* had a cellular N:P that almost perfectly matched the N:P in its environment (Fig. 1.5). Note that the "you are what you eat" model (Fig. 1.3) holds for this situation on both linear (Fig. 1.5A) and logarithmic (Fig. 1.5B) axes, as it should given Equations (1.3) and (1.4) and letting H = 1. Within the range studied, the alga shows no evidence for homeostasis of N:P. However, does this plot indicate that we might be able to grow *Scenedesmus* of any arbitrary N:P, just by expanding the range of the chemostat conditions? A moment's reflection proves that the answer must be "no." Although Rhee's results do not demonstrate the bounds to cellular N:P, they surely must exist; living algae could never be made of pure N or P nor could they have been grown in the complete absence of P (or N). Our interpretation is that within the range of intermediate N:P he studied, algal homeostasis is essentially absent. We predict that homeostasis would be observed at more extreme values of N:P in the medium. Vascular plants also show similar patterns to these algae. Shaver and Mellilo (1984) reported a wide range of plant N:P ratios in several species of marsh plants as a function of N:P supplied to the plants. We will consider variability in autotrophic C:N:P stoichiometry in much greater detail in Chapter 3 (an analogous plot to Rhee's algal study, but for vascular plants, is shown in Fig. 3.3B).

The next example shows a strict homeostasis. Goldman et al. (1987b) grew natural assemblages of marine bacteria on substrates of differing C and N sources (amino acids, glucose, and NH_4^+), spanning a range of C:N from 1.5 to 10. They determined a number of physiological features we will consider further in Chapter 5. Here, we consider the C:N of the bacterial biomass. Figure 1.6 shows the bacterial C:N as a function of substrate C:N. It is clear that bacterial C:N is homeostatically regulated here. The slope is not statistically different from zero on linear (Fig. 1.6A) or logarithmic (Fig. 1.6B) axes. Note that the range of substrate C:N examined was both higher and lower than the C:N in the bacteria themselves (unweighted mean of 5.6), so that it is reasonable to surmise that bacterial C:N would be homeostatic over nearly any other substrate C:N that could be examined. From these two examples, it is tempting to think of algae as conformers and bacteria as regulators. However, homeostatic regulation is a function of both the species and the resource. As we will see later (Fig. 5.8), bacterial N:P does vary with medium N:P, similar to the patterns we just described for algae. However, Figure 1.6 shows that C:N in this bacterium is homeostatic. Our purpose in giving the two examples in Figures 1.5 and 1.6 is to show two extreme examples, an absence of homeostasis and a strict homeostasis, not to directly compare identical resources in two different species.

Now, we will look at several more complex examples. Levi and Cowling (1969) raised the wood-consuming fungus *Polyporus versicolor* in a synthetic medium containing varying amounts of C and N provided as the sugar glucose and the amino acid asparagine. They sampled the fungus and determined its chemical composition through time, with incubation times ranging from as little as 5 d to as much as 56 d. Within a single medium type, fungal N:C generally increased with time from the first sampling to the last. In Figure 1.7, we have plotted all the data from all sampling dates (which accounts for some of the variation observed at particular N:C values). This interesting example shows that the linear and logarithmic plots can give very different impressions. On the linear axes (Fig. 1.7A), N:C in the fungus seems to show weak or no homeostasis at low N:C in the medium, but fungal N:C is not as variable with substrate N:C at higher values in the medium, where fungal N:C seems to approach a strict homeostasis. In contrast, the logarithmic axes (Fig. 1.7B) suggest a constant homeostatic adjustment over the entire range of the data (the data are reasonably fitted by a straight line). Given the inherent "bendiness" of homeostatic relationships on linear axes (see Fig. 1.4B), we suggest that logarithmic axes are superior for diagnosing the existence and

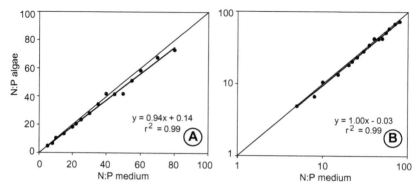

Fig. 1.5. Absence of homeostasis as seen in the N:P of *Scenedesmus* algae at growth rate equilibrium as a function of the N:P in the medium supplied. Cellular nutrient ratios are almost identical to the ratio in the surrounding environment (slope is near 1, intercept is near zero), which is apparent both in the plot with linear axes (A) and in the plot with logarithmic axes (B). Both the regression line and the 1:1 line are plotted. The regulatory coefficient H from Equation (1.3) is the inverse of the slope fitted to the log-log plot, which in this case is 1.00, indicating an absence of homeostasis. Based on Rhee (1978)

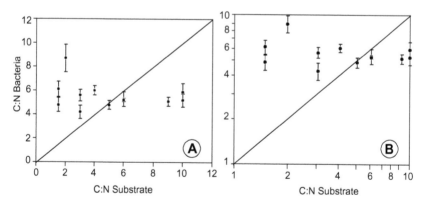

Fig. 1.6. A strict homeostasis. Bacterial C:N (mean ± standard deviation) does not vary systematically over a wide range of substrate C:N. The trend is not statistically significant in either linear (A) or logarithmic (B) plots. The regulatory coefficient coefficient H can be taken to be essentially infinity, given the apparent zero slope. For a description of homeostasis of bacterial N:P ratios, see Fig. 5.8. Based on Goldman et al. (1987b).

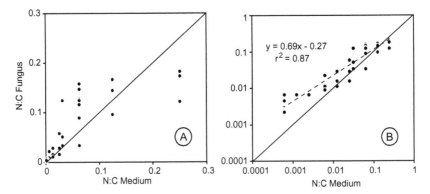

Fig. 1.7. Homeostatic regulation of fungal N content. On the linear axes (A), it appears that there are different degrees of regulation in different ranges of N:C in medium. When medium N:C < 0.05, fungal N:C varied strongly with the stoichiometry of the growth medium ($y = 1.65x + 0.004$, $r^2 = 0.56$). However, an upper bound seems to be approached, and when medium N:C > 0.05, fungal N:C seems to be less sensitive to medium N:C ($y = 0.52x + 0.05$, $r^2 = 0.46$). In contrast, on logarithmic axes (B), there is less of a convincing case for changes in homeostatic regulation in the different ranges. Nevertheless, the slope is clearly less than 1, indicating homeostasis, although not strict. The regulatory coefficient H is 1.4 (1/0.69), indicating a weak homeostatic regulation. Based on Levi and Cowling (1969) (assumes C is 0.45 of mass).

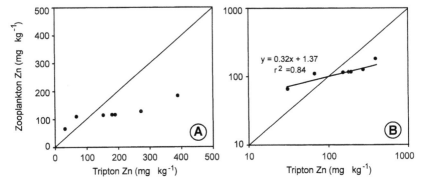

Fig. 1.8. Zinc content in zooplankton and tripton in lakes on linear (A) and logarithmic (B) axes. The regulatory coefficient H is 3.1, a relatively strong homeostasis. Based on Zauke et al. (1998).

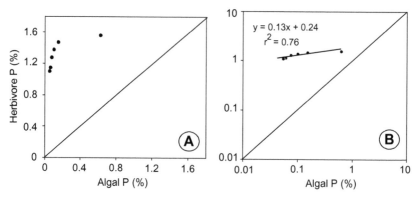

Fig. 1.9. Phosphorus content in *Daphnia* and algal food on linear (A) and logarithmic (B) axes. Although the linear plot makes it appear that there is a breakdown in homeostatic regulation at low food P content, a close to strict homeostasis is evident on the logarithmic axes. The regulatory coefficient H is 7.7, a strong but not strict homeostasis. Based on DeMott et al. (1998).

degree of homeostasis. A constant regulatory parameter H does different things on the linear axes in different portions of x,y space.

Two studies on zooplankton complete our examples. Zauke et al. (1998) measured the concentrations of zinc in freshwater zooplankton and tripton (settling particles) in several lakes in north central Poland. We will take the values in tripton to be a measure of the stoichiometry of the food available to the zooplankton. Zinc content was much less variable in the zooplankton than in tripton, and Zn content in zooplankton was strikingly consistent at intermediate Zn levels in tripton (Fig. 1.8). Both linear (Fig. 1.8A) and logarithmic (Fig. 1.8B) axes give a very similar impression. The zooplankton regulate at or very close to a strict homeostasis in the intermediate range of resource stoichiometry. At very low or very high Zn in the resources, homeostatic regulation breaks down. This example is unique in our set in suggesting a breakdown in consumer stoichiometric homeostasis at high levels of the element (Zn) in the resource. This may be the rule under conditions of toxicity. In a different study on zooplankton stoichiometry, DeMott et al. (1998) raised the cladoceran herbivore *Daphnia* (you will be treated to a great deal of information on this taxon in this book) in the laboratory on algal foods of different P content. Their results plotted on linear axes (Fig. 1.9A) seemingly demonstrate a great sensitivity of consumer P content to food P content at low food P content. But beware— and recall the bendiness of Figure 1.4B. When replotted on logarithmic axes, the strongly homeostatic nature of P in *Daphnia* is evident.

These examples illustrate some of the patterns of homeostatic element regulation in different organisms. Further examples will be encountered later in the book (e.g., Figs. 5.8 and 5.9). In Chapters 4–6 we will continue to discuss patterns of nutrient content in Metazoa and the ecological consequences of their physiological regulation. The concepts of homeostasis and stoichiometry in organisms are at first sight complex but are reasonably well captured by some simple theory. The examples we saw serve to illustrate the basic set of generalized patterns of variation in nutrient content of an organism as a function of nutrients in its environment. The degree of homeostasis is a function of several major things: the consumer, its life stage, the chemical, and the range of the data perhaps are the most important of these. The main point of considering these examples is to realize that creatures are, generally, *not* what they eat. Life requires a complete set of sustaining functions: mechanical support, metabolism, gene replication, production, reproduction, etc. Biochemical means to each of these ends are limited (Chapter 2). Hence, the need for life-sustaining processes creates limitations in the possible chemical formulas for living things. If organisms were what they ate, this would be a short book. But they aren't, and it isn't. Instead, organisms are what they first eat but then do not egest, defecate, excrete, lactate, exhale, or otherwise release back to the external world. In general, these are actively regulated processes, often under close control by the organism's physiology and thus subject to evolution by natural selection. Homeostatic regulation of stoichiometry can occur in multiple ways. It can involve food choice, habitat selection, assimilation, or excretion. Examples of each of these will be seen at different points in the book.

Homeostasis makes ecological stoichiometry interesting. If every living thing were what it ate, most of the phenomena discussed in this book either would not occur or would be considerably trivialized. Our message is that this active regulation of matter processing within organisms underlies myriads of ecological phenomena. So what general patterns in homeostasis might there be? How can we make some sense out of all the possible differences in stoichiometry in organisms? At this point, we can offer a few generalizations and propose some hypotheses. Hopefully, in the future as more data are collected, additional patterns will be revealed. We need much more information on the patterns of homeostatic regulation of different chemical resources in different organisms. These are fundamental data! A few patterns seem well enough defined to consider them as generally valid.

As we have just seen, some elements are more stoichiometric than others. Macroelements such as C and N are more stoichiometric than microelements (e.g., Goldman 1984). Carbon, for instance, is roughly 40–50% of the dry biomass of most living things (when we neglect massive noncarbonaceous support structures). Reasons for the relative constancy of C and

N content are discussed in Chapter 2. Trace elements have some particularly interesting patterns of concentration. Baines and Fisher (2001) showed that Se concentrations in algae (per unit volume) varied across algal species by almost four orders of magnitude when algae were exposed to common concentrations of Se in the water. They also found that cellular Se concentration varied only by two- or threefold when exposed to selenite concentrations that varied by 30-fold. The stoichiometry of trace elements provides enough variability that it has been used as a tool to trace the dispersal of mobile life stages of certain organisms (DiBacco and Levin 2000). Another major general contrast occurs across nutritional strategies. Recall that autotrophs are organisms that use an energy source such as light to fix organic carbon from inorganic carbon. In contrast heterotrophs are organisms that rely on already fixed carbon sources both for structure and for energy. As we will amplify in the coming pages, autotrophs (Chapter 3) generally are less homeostatic than heterotrophs, particularly in the case of multicellular metazoans (Chapter 4). We are still lacking a good concept of how poorly studied trophic strategies, such as bacterivory, detritivory, and the like, fit into this general pattern. As we will see (Chapters 5 and following), the consequences of these facts of life for ecological dynamics are considerable.

It is also important to emphasize that the degree of homeostasis depends on both the element under consideration and the organism involved. In other words, different degrees of variation are associated with both the identity of elements and the identity of organisms. There are a number of ways in which we could think about a nonstrict homeostasis. For example, in Chapter 3, we will look at models that assume strict homeostasis within certain physiological pools (e.g., structure vs. energy reserves) but allow the relative proportion of those pools to vary with time (other examples are given by Kooijman 1993, 1995).

This book will review and develop considerable theory related to how homeostasis and other stoichiometric constraints impinge on ecological systems. We propose the term "stoichiometrically explicit" to refer to such theory, because these models formally incorporate the limitations imposed by the mass balance of multiple elements during ecological interactions. We chose this term to be analogous to "spatially explicit" ecological models that recognize how space and rates of movement through space constrain ecological dynamics. Stoichiometrically explicit models have the advantage of preserving mass balance at all points in the model; alternatives, such as allometric expressions of nutrient release, lack that reality check. As Andersen (1997) described it, if zooplankton released nutrients at a fixed rate relative to their body size regardless of food consumption or actual growth (as they do in some food-web models in the literature), under various food conditions animals may be represented as synthesizing new nutrient atoms in their bodies! This would be an unacceptable model property. Instead we need models that obey conservation laws. Such approaches should acknowledge that animal biomass is constructed of multiple elements in relatively fixed proportions (Chapter 4) and that animals can independently adjust the efficiencies with which they retain various elements (Chapter 5). Thus, the rates and ratios of elements released by the animal (Chapter 6) will be a function of the elemental composition of the food being ingested (Chapter 3), the elemental composition of the animal biomass being formed, and the efficiency of retention by the animal of nutrient elements when they are limiting.

In spite of the importance of homeostasis as an ecological and evolutionary force, few have considered the fundamental, ultimate reasons that some organisms are more variable in their chemical content than others. We offer a variety of hypotheses later in this book. Others who have considered the question include Williams and Fraústo da Silva (1996), who suggested that there was a broad evolutionary trajectory of increased homeostasis in the progression from early procaryotes to later procaryotes, then to unicellular eucaryotes, and finally to multicellular eucaryotes. They related this hypothesized trend to such features as the evolutionary development of ion pumps in membranes, increased cellular compartmentalization, and increased biochemical complexity. Basic thermodynamics suggests that for organisms to maintain a chemical content different from what they have available, energy must be expended. Assuming that energy expenditures impact individual fitness, homeostasis needs to be considered from the standpoint of adaptive reasoning. Taghon (1981) and Calow (1982) have written about the apparent evolutionary paradox of the evolution of constrained chemical content. An unanswered question is, "What are the advantages that outweigh the known costs of homeostasis?" Why does it evolve? Perhaps explicitly acknowledging the chemical nature of living things will help us better understand the forces driving and constraining the evolution of biological complexity. There are big unanswered questions about the nature and evolutionary history of the variation of chemical content in organisms. Our hope is that this book will suggest some possible answers and, by demonstrating how important these patterns actually are, stimulate others to delve into the subject in the future.

YIELD

One of the key stoichiometric concepts we will encounter in many places in this book is **yield**. Yield has a common biological or agricultural meaning: it is the amount of biomass, e.g., of a crop, obtained from a given unit of investment, such as resource or effort. Examples are bushels of corn per kg N fertilizer applied. In the chemical literature, "stoichiometric yield" also has a precise definition. Stoichiometric yield is the amount of a product one obtains from a unit of reactant. For example, in photosynthesis, the maximum yield of glucose is one molecule for every six molecules of carbon dioxide. In a simple chemical reaction involving several reactants and several products, the reactant that is present in the smallest equivalent stoichiometric amount limits how much product will be produced; it is called the "limiting reagent."

Stoichiometric yield can be helpful in understanding more complex systems as well. Consider the entire suite of biochemical reactions involved during the growth of a cell. Zeng et al. (1998) studied a number of aspects of stoichiometry and yield in mammalian cell cultures. They compared the rate of lactate production to the rate of glucose uptake using an empirical analysis of data obtained from several different studies. Plotting these two variables yielded an apparent straight line with a slope of 1.6–1.7 (Fig. 1.10A). This plot shows one way of interpreting a stoichiometric coefficient: it is the slope of a plot of reactants versus products, such as in Figure 1.10A, and it is also a yield. Initially, inspection of Figure 1.10A suggests that there is a constant yield of lactate from glucose. However, foreshadowing other situations we will see later, when the data were studied further, it was revealed that the stoichiometric yield was not constant. There were differences in lactate yield at low glucose levels, which can be seen by plotting the ratio lactate formation:glucose uptake (which now has dimensions of a molar ratio) against residual glucose concentration in the cultures (Fig. 1.10 B). Closer inspection of Figure 1.10A reveals a tight cluster of points above the trend line at very low glucose. These points are made more obvious by plotting the ratio, as in Figure 1.10 B. By examining several relationships such as this, the authors concluded that key stoichiometric yields in mammalian cells are not constant at low resource concentration, even though previous research had assumed a constant yield under all conditions.

The concept of yield is very much a part of ecological stoichiometry. Consider again the patterns of nitrogen content in a fungus as a function of the growth medium (Fig. 1.7). At low medium N, fungal N content was low. Inverting, we could say that the yield of fungal biomass per unit N was high in the low-N medium. In contrast, yield was low when the medium was N rich. Yield is a useful concept at the organismal level, such as when we talk about evolution working on growth production (Chapter 4). Or, for another example, the yield of animal biomass from a unit of nutrient input to a system may be important if animals are being harvested, or if they are important keystone predators in the system. It is also useful at the ecosystem level when we talk about nutrient use efficiency of whole ecosystems or constraints on C sequestration in the biosphere (Chapter 8). For exam-

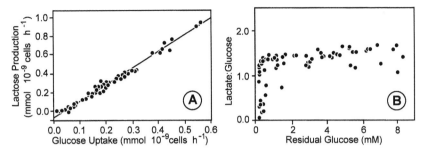

Fig. 1.10. Stoichiometric yield of an end product (lactose) from a reactant (glucose) in mammalian cell culture. A. Plotting the rate of formation of product vs. the rate of consumption of reactant generates a line with an apparent constant slope. B. Departures from constant stoichiometry are seen when one plots the ratio of lactate:glucose vs. the residual glucose concentration in the cell culture medium. At low glucose concentration, lactate yield per glucose drops. Based on Zeng et al. (1998).

ple, in pollution management and control, the yield of algal biomass obtained per unit of nitrogen or phosphorus incorporated into biomass can have great consequences; this yield may vary substantially across ecosystems and across different rates of N and P supply. Biological systems from single cells (Chapter 3) to multicellular organisms (Figs. 1.7 – 1.9) to communities (Chapter 7) to whole ecosystems (Chapter 8) often follow a commonsense pattern: the biomass yield per nutrient consumed is greater when nutrients are scarce. Fundamental patterns of homeostatic regulation of nutrient content thus have many implications for diverse ecological questions.

THE REDFIELD RATIO

Although we should be able to apply stoichiometric principles to all stable elements in the periodic table, the three elements C, N, and P are a special focus of this book. We will see some reasons why in an upcoming chapter. Perhaps the most famous reference point in ecological stoichiometry deals with C, N, and P in oceanic ecosystems. This is the **Redfield ratio**, named after Alfred C. Redfield (1890–1983), an oceanographer from Harvard and the Woods Hole Oceanographic Institute. Redfield discovered an unexpected congruence in C:N:P in numerous regions in the world's oceans and then used that congruence to infer something about the large-scale operation of the global biogeochemical system (Redfield 1934,

1942, 1958; Redfield et al. 1963). The Redfield ratio is such a cornerstone that we will cover it in some detail here in the beginning of the book.

Redfield found that the atomic ratios of carbon, nitrogen, and phosphorus in marine particulate matter (seston) were the same as the ratios of the differences of dissolved nutrients in those waters. In other words, when he plotted values of one of dissolved C, N, or P against one of the others, he obtained straight lines with slopes equal to the corresponding ratios in particulate matter (Fig. 1.11B). Redfield found the elements to be related by the ratio C_{106}: N_{16}:P_1 (Fig. 1.11A). The constancy of the C:N:P ratio of plankton was soon "embraced by many biological oceanographers and geochemists as canonical values, comparable to such physical constants as Avogadro's number or the speed of light in a vacuum" (Falkowski 2000).

Redfield's congruence in nutrient ratios between plankton and their aquatic medium indicated a balanced flow of C, N, and P in and out of the biota. The "Redfield ocean" is a biological circulatory system with constant C:N:P stoichiometry moving vast quantities of constant proportions of these three elements vertically over thousands of meters. A second congruence was that the line describing the N and P data had a zero intercept, indicating that these two elements would be depleted from ocean waters simultaneously (Fig. 1.11B). The same was not true for carbon: there was a surplus of carbonate when N and P were depleted.

Simultaneous depletion of N and P was surprising. There is no *a priori* reason to expect ocean water to contain N and P in proportions identical to biological demand. Why then should this measure of the chemistry of the ocean—such a vast proportion of the Earth's surface and subjected to major influences from geology, meteorology, and others—have an N:P ratio that matches biological demand? Redfield's (1958) answer was that the biota itself determined the relative concentrations of N and P in the deep sea. He suggested that it was P that ultimately determined the biological productivity of the world's oceans, and that biological feedbacks adjusted the level of N so that its availability matched the availability of P (Falkowski et al. 1998). Similar arguments were later applied to soils (Walker and Adams 1958, 1959). Redfield's findings were important in a very broad context: his work was instrumental in fostering a view that the ocean's biota has a major influence on the chemistry of even this vast volume of water.

Today's larger data sets and more precise methods have been used variously to support or modify Redfield's proposed ratio of C_{106}:N_{16}:P_1 in the ocean (references cited in Hoppema and Goeyens 1999), although we hasten to add that even the proposed deviations are modest in comparison to the big differences often observed in C:N:P in freshwater systems (Chapter 3). The concept of a constant Redfield stoichiometry in the offshore ocean has generally held up to modern data (Copin-Montegut and Copin-

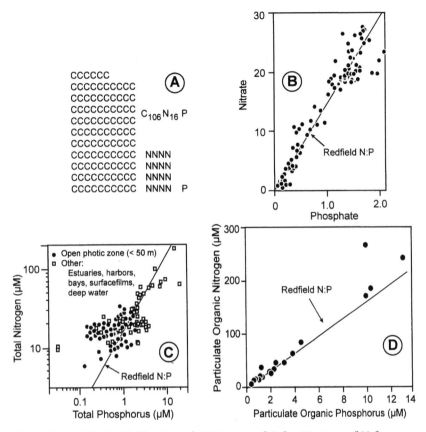

Fig. 1.11. A. The Redfield ratio with 106 atoms of C for 16 atoms of N for every one atom of P is a very famous ratio in ecological stoichiometry. This figure shows several examples of Redfield stoichiometry for N and P in different chemical fractions in marine waters. B. One of Redfield's original observations was that nitrate nitrogen and phosphate phosphorus (both dissolved) in waters of the western Atlantic have a proportionality of approximately 16:1. C. Total nitrogen and total phosphorus cluster roughly around Redfield proportions as well. Suggested systematic departures in open photic zone samples compared to others will be discussed further in Chapter 8. D. Marine particulate matter also generally shows Redfield N:P proportions. Panel B is based on Redfield et al. (1963), panel C is based on Downing (1997), and panel D is based on Copin-Montegut and Copin-Montegut (1983).

Montegut 1983; Karl et al. 1993; Hoppema and Goeyens 1999), although some interesting details have been layered on the classic picture of the Redfield ocean. For example, Downing (1997) proposed that there are systematic deviations in TN:TP ratios (TN and TP stand for total nitrogen and total phosphorus and they equal the sum of all nutrients, particulate and dissolved, in a water sample) between coastal and open ocean sites (Fig. 1.11C), suggesting that near-shore environments should become depleted first of N and off-shore sites should become depleted of P. Temporal trends on the scales of years to decades in deep-water N:P ratios (Pahlow and Riebesell 2000) and in riverine supplies to coastal zones (Justic et al. 1995) have been hypothesized and attributed to human influences on global biogeochemistry. Climatic couplings that have some influence on the composition of marine particulate matter have also been suggested (Karl et al. 1995); these will be discussed further in Chapter 8. Circumstances under which phytoplankton attain the Redfield ratio have been intensively studied (Chapter 3). And finally, although Redfield and colleagues considered both phytoplankton and zooplankton to have essentially the same C:N:P composition, systematic differences between these two ecosystem components and across the freshwater-marine contrast have been observed (Elser and Hassett 1994).

Although algae need not necessarily grow with C, N, and P in Redfield proportions, in the bulk of the ocean this seems to happen, and the Redfield ratio of $C_{106}:N_{16}:P_1$ has also been expanded to include comparable quantities of O and H. From this, a general equation for phytoplankton growth has been developed and related to biochemical fractions (carbohydrates, lipids, etc.) (Vollenweider 1985). Stumm and Morgan (1981) wrote this equation as,

$$106CO_2 + 16NO_3^- + HPO_4^{2-} + 122H_2O + 18H^+$$
$$(+ \text{ trace elements and energy})$$
$$P$$
$$\leftrightarrow$$
$$R$$
$$\{C_{106}H_{263}O_{110}N_{16}P_1\} + 138O_2 \tag{1.5}$$

where the curly brackets indicate algal biomass, and where movement to the right indicates photosynthesis (P) and movement to the left indicates respiration (R). Rearranging, algal biomass can also be expressed as

$$\{(CH_2O)_{106}(NH_3)_{16}(H_3PO_4)\}.$$

This expanded view with more elements allows for a deeper understanding of how cycles of C, N, and P relate to redox processes, a critical biogeochemical topic discussed briefly in Chapter 8. In Chapter 3, we will once again return to questions of balanced growth, C:N:P, and their links to biochemistry associated with photosynthetic production.

CONVENTIONS AND CONCERNS ABOUT ELEMENT RATIOS

We will make extensive use of element ratios. Many of the raw data in this book in fact are composed of different sorts of ratios (e.g., P as a percentage of dry mass, C:N, etc.). Why? For one thing, stoichiometry is about proportions, plain and simple. We couldn't escape looking at ratios even if we wanted to. For another, ratios have been important in one of the most successful theories predicting ecological outcomes from mechanistic details. Resource ratios have a long history in ecology (Rodhe 1948). Resource competition theory (Tilman 1982; Tilman et al. 1982; Grover 1997) has made extensive use of ratios of different kinds—fluxes, concentrations—to predict patterns of species dominance, coexistence, and diversity in communities (e.g., Smith and Bennett 1999; Interlandi and Kilham 2001). So, although the technicalities of using ratios can be tricky, there are good ecological reasons to use them. In basing so much of what we do on ratios though, some questions will naturally arise (Lampert 1999). Ratios are proportions. A molar C:P of 200 means that there are 200 atoms of C for every atom of P, but it does not say anything about absolute amounts of either C or P. One must realize what ratios do, and not expect them to do inappropriate things. The C:P of food, for example, is basically irrelevant to animal growth if there is next to nothing to eat.

To equip the reader to think about ratios throughout the book, we will touch on some technical issues here. This section is an "aside" and skipping it will do no particular harm to the casual reader looking mainly for concepts. A ratio is a complete description of relative proportions between two substances, but there are some hidden traps. Ratios are sometimes presented on a mass:mass basis ("mass ratios") and sometimes they are presented on a mole:mole basis. The latter are referred to as "atomic" or "molar" ratios. Either ratio is a complete description, but they are not numerically identical. Which is preferable? Mass units are appealing because we are accustomed to thinking about organism body mass or estimating total biomass of particular species or components in ecosystems. "Percent phosphorus," for instance, is derived from a ratio of the mass of phosphorus to the total mass. However, chemical reactions proceed by atom-to-atom interactions. We prefer molar ratios. Our main reason is that they minimize confusion when thinking about ions or compounds. One mole of NO_3^- is an unambiguous statement of quantity, but in discussion of the nitrogen cycle, that same mole of NO_3^- would represent 14 g of N or 42 g of nitrate. In this book, any ratios not specifically labeled should be taken to be molar ratios. Another source of confusion is that different studies may present the proportion of the "minor" element (e.g., N or P) relative to C (P:C or N:C) while others present the proportion of carbon relative to the minor element (C:N, C:P). Following the precedent of Redfield, we tend often to present data in the order of C:N:P (including C:N, N:P, and C:P). However, in keeping with the original studies sometimes we invert these. We apologize in advance for any mental knots this creates.

We also need to clarify the idea of "nutrient content" of biota. It is commonplace in the literature to express the elemental composition of an organism in terms of its "nutrient content." There are two ways in which this is generally done. In the first way, the whole-cell or whole-organism amount of an element is reported. For example, in phytoplankton ecology this often involves the expression of nutrient content as a "cell quota" in terms of mass or number of atoms per cell (generalizing, we will refer to the quantity of any particular substance within any individual as a **quota** and use the symbol Q; see the Appendix). We will define and work with the concept of the cell quota more formally in Chapter 3. While useful from the perspective of population dynamics, this approach has limited utility when diverse organisms need to be compared or when phenomena at higher levels such as the community or ecosystem are considered. In general, we will convert cell-quota data to elemental ratios using information about cell size and size-to-carbon conversions. Second, nutrient content is very frequently presented as a percentage or other proportion of dry weight. We will use the symbol Γ_x to represent the proportion of the total mass made up by a single substance. Such data are readily comparable across biota but are not directly stoichiometric. Indeed, in some cases they can mask significant stoichiometric variation, as in the case of organisms such as diatoms whose biomass comprises significant amounts of elements other than C. In general, we will convert data presented on a per dry weight basis to C:nutrient ratios using pertinent C:dry weight conversion factors. As mentioned above and discussed more fully later in the book, the range of variation in C content (e.g., % C by weight) is relatively narrow and thus such conversions are generally reliable within a circumscribed group of organisms.

Ratios present other mathematical issues. We often want to compare the variability of ratios, either directly when we talk about how different ecosystem components vary in their stoichiometry, or implicitly in statistical tests such as ANOVA (analysis of variance). To understand these things, one must know about propagation of errors. Error propagation (Bevington 1969) is a means of calculating the uncertainty of a composite calculation from the uncertainty of the component parts. Consider the composite value z calculated as a ratio of u and w,

$$z = \pm \frac{au}{w}, \tag{1.6}$$

where a is a constant and u and w are variables measured with error. Error propagation follows the following formula:

$$\frac{\sigma_z^2}{z^2} = \frac{\sigma_u^2}{u^2} + \frac{\sigma_w^2}{w^2} - 2\frac{\sigma_{uw}^2}{uw}, \tag{1.7}$$

where the σ indicate standard deviations and the last term refers to covariance between the two variables u and w. If one can assume that measurements of u and w are independent, the third term drops out. As an example, consider the following independent measurements and their associated standard deviations (errors):

$$C = 100 \pm 1,$$
$$P = 1 \pm 0.1.$$

The C:P equals 100, but what is the variability associated with this ratio? To find the variability of the composite, perform the calculation

$$\sigma_{\text{C:P}} = 100 \sqrt{\left(\frac{1}{100}\right)^2 + \left(\frac{0.1}{1}\right)^2}. \tag{1.8}$$

If you do the math, you'll find this works out to 10.05. The relative errors of the individual measurements were 1% (for C) and 10% (for P). The fractional error of the composite (10.05/100) converted to a percentage (10.05%) is slightly larger than the larger of the two individual errors but smaller than their sum. This example points out one disadvantage to ratios; the relative error will never be less than the error associated with the variable with the greatest error. In this example, if you were trying to learn something specifically about C, working with C:P would be a bad approach. Another property of the variance of ratios is that the relative error of a given ratio will be the same as the relative error of the inverse of the ratio. In other words, the coefficient of variation (standard deviation divided by the mean) of P:C will equal the coefficient of variation of the C:P ratio. If one is about 10%, so will its inverse be.

One unfortunate aspect to working with ratios is that they raise a whole set of important statistical problems. The literature is full of warnings and prescriptions (Atchley et al. 1976; Atchley and Anderson 1978; Pendleton et al. 1983; Tonkyn and Cole 1986; Buonaccorsi and Leibhold 1988; Prairie and Bird 1989; Jackson et al. 1990; Raubenheimer and Simpson 1994; Berges 1997; Knops et al. 1997). Certainly, it is easy to make bad blunders. We urge caution and care. However, some of the arguments against using ratios in statistical tests seem to us to sacrifice scientific insight on the altar of statistical purity. For example, we have often read that if only one of the two variables in a ratio is strongly associated with some predictor variable, then it is incorrect to conclude anything about the ratio. Say z is related to u, but not to w. Some argue that examining $u{:}w$ versus z is wrong. We disagree with this point of view. We view such a situation as evidence for a changing balance between u and w. If one variable changes, but the other does not, the balance between the two changes. Although it may also be correct, and it is probably statistically more transparent, to toss out the ratio and work only with the variable that changes, this is unacceptable if you are in fact interested in what the ratio does.

For example, imagine that we grew algae in chemostats at varying light intensity and measured algal C and P concentration in response to light. Imagine next that algal P was constant with increasing light intensity but algal C increased (i.e., algal C:P went up). Now imagine that we wonder how zooplankton growth would relate to light intensity if these algae were fed to zooplankton. Should our focus remain only on the algal variable that responded to light (algal C)? But what if the animals were sensitive to the P content of their food? We will see just such an experiment in Chapter 7 where we need to know about changing C and P balance in light gradients. This example calls attention to one reason why working with ratios is important: they capture the important aspect of dilution of potentially key nutrients in food biomass and thus become a means for directly incorporating food quality into food-web theory. Thus, we believe that a strong focus on elemental ratios has major heuristic, conceptual, and theoretical benefits. By saying all this, however, we are not trying to give license to incorrect analysis or artificial inflation of r^2 values, both of which are easy to do! This book does not attempt to prescribe statistical solutions to these problems. The interested reader can refer to the papers cited above and should become familiar with this literature.

SOME CONVENTIONS ABOUT GROWTH RATE

A major theme of this book is that organism growth rate and elemental composition are closely linked (Chapters 3 and 4). We are also very much interested in the phenomenon of nutrient limitation of growth rate, both in autotrophs and in consumers. Unfortunately, there are many contradictory ways that organism growth rate and nutrient limitation are expressed in the literature. In particular, expressions differ systematically between terrestrial and aquatic literature. To cover both subjects in one place, we need to set some conventions. Let us define two terms. The first is **specific growth rate** (μ; units of time^{-1}), which refers to the exponential rate of change of biomass of the organism, normalized to its biomass. In continuous time,

$\mu = dM/Mdt$ where M is the biomass. In practice, one measures the specific growth rate by measuring the biomass at two times and applying the formula for discrete time. In discrete time, μ is given by $\ln(M_t/M_0)/t$ where M_t is the biomass at time t and M_0 is the biomass at the start of the interval. Throughout the book when we use the term "growth rate" we mean "specific growth rate." This terminology is conventional in the aquatic literature. However, in the terrestrial plant literature this parameter is commonly referred to with the term "relative growth rate" (e.g., Grime and Hunt 1975). Here, we use the definition of the term **relative growth rate** (RGR) as the proportion of an organism's **maximum growth potential** (μ_m) that is actually achieved in any particular situation. In this book, RGR (μ/μ_m) is a measure of the severity of some limitation of growth rate and varies from 0 (severe growth limitation) to 1. We beg the pardon of our colleagues most familiar with the terrestrial plant literature and hope the change in lingo will not be too stressful.

A LOGICAL FRAMEWORK

Some of ecology's major operative concepts have been criticized as being "tautologies," "semantical," or with other dismissive terms. Perhaps this is because we ecologists do not pause often enough to try to organize our thoughts into a formal logical structure. Perhaps doing so too frequently would be a waste of time and ink. However, logical systems can clarify our thinking and tell us how it is we know what we think we know. Let us define some epistemological terms in simple language. An axiom is a statement taken to be true—in other words, a proposition taken to be self-evident and not needing proof (e.g., the whole is greater than a part). A theorem is a statement that is not self-evident but that can be proved using axioms and rules of logic. Reasoning from a premise to a specific conclusion is called "deduction," as is reasoning from the general to an unknown specific case. "Induction," on the other hand, is a form of reasoning where a general statement is constructed from a number of specific facts or individual cases.

Reiners (1986) presented a logical structure for ecological stoichiometry. He did not use the term "ecological stoichiometry." He said he was offering a "complementary" view of ecosystems (based on matter) to go along with a classical view built on energetics. Reiners' complementary model contained many of the core concepts we will discuss, so considering his logical reasoning does double duty for us here in the beginning of the book. He assembled a series of statements ranging from the cellular level to the global level into a logical flow chart using axioms to derive theorems (Fig. 1.12). This flow chart shows the way that lower-level stoichiometric processes (occurring at the cellular, population, and community levels) create higher-level specific pro-

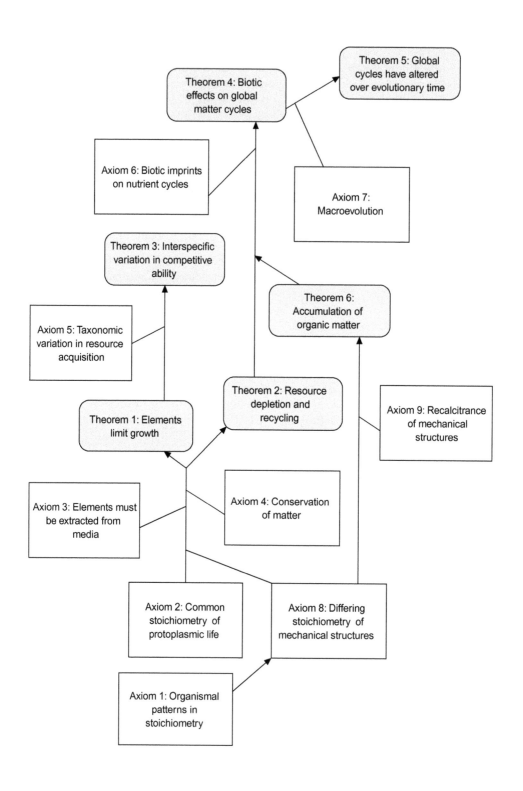

cesses and patterns (at the ecosystem and global levels). We begin with the first four axioms (Reiners' exact wording is used in the text below—we have paraphrased these axioms and theorems in the accompanying figure).

Axiom 1: *Groups of organisms have regular chemical stoichiometries.*

Axiom 2: *Protoplasmic life has a common chemical stoichiometry.*

Axiom 3: *Organisms have evolved mechanisms for extracting elements from media to synthesize biomass.*

Axiom 4: *Law of conservation of matter.*

These four statements relate to material we will cover primarily in Chapters 2–4. We can take these statements about "protoplasmic life" to refer to mass flow in microbial production. At the time he wrote his paper, Reiners lacked the information we now have on stoichiometric patterns in many organisms. He mainly had the observations of Redfield (discussed above) and a compilation of element data by Bowen (1979) with only a few data for living creatures (Bowen's compilation remains the single most comprehensive). Thus it is understandable that Axioms 1–4 describe the stoichiometrically balanced ocean with a single biotic C:N:P. Reiners had an overly simplistic outlook on the stoichiometry of organisms lacking major structural components, those that he called "protoplasmic life." As we have just seen (e.g., Fig. 1.5) and will continue to see throughout the book, microbes do not in fact have a "common chemical stoichiometry." Neither are they strictly homeostatic. Instead, they have differing adaptations for nutrient storage and have different C:nutrient balances depending upon such factors as the relative availability of light and nutrients (Chapter 3). We also understand much better today how differential biochemical investments in storage, proteins, and RNA result in differential cellular stoichiometry (Chapters 3 and 4). Finally, we are beginning to understand differences in the stoichiometry of autotrophs and heterotrophs, even at the microbial level. As we will reinforce in chapters to come, the biosphere in general is not like the Redfield ocean.

It may be disconcerting to begin with statements labeled as axioms, meaning that they should be self-evident without proof, but then immediately suggest additions and modifications. Are these statements axioms after all? What if we take them to be approximations? Table 1.1 and numerous other examples in this book make the point that the stoichiometry of the inorganic world can certainly be extremely variable in comparison to "protoplasmic

---

Fig. 1.12. A logical structure for stoichiometric theory. Clear boxes are axioms and shaded boxes with rounded corners are theorems. A small number of logical steps takes us from axioms about biochemistry to theorems about planetary biogeochemical cycling. Based on Reiners (1986).

life." From this very broad perspective, the members of the world's biota are in fact regular in composition. Hence, we may be able to predict something about mass flow and resource limitation from knowledge of inorganic nutrient supply alone, independent of these biological details. Few if any interesting statements in ecology can be taken to be axiomatic in every possible context.

Reiners' logical scheme recognizes differential organism stoichiometry primarily due to differing mechanical structures.

Axiom 8: *Mechanical structures of variable stoichiometries different from those of protoplasmic life have evolved over geological time.*

With this, Reiners recognized that the elemental composition of a tree differs from that of a mammal and those differ from that of an insect. From Axioms 1–4 and 8 one can deduce two theorems that relate organism growth to the mass balance of nutrient uptake.

Theorem 1: *Organic synthesis and metabolic rate are limited by the supply rate of essential elements.*

Theorem 2: *Net change in biomass will alter the amount of essential elements in surrounding media.*

Theorem 2 can be deduced logically from the preceding axioms. Theorem 1, on the other hand, has an unstated premise that matter matters. In other words, its premise is that organisms will be sufficiently numerous and active to deplete element resources to limiting levels. An ecological population held in low abundance by a predator or by stressful conditions could be considered a counterexample to Theorem 1. In addition, Reiners' "complementary" view is not about energetics. Theorem 1 might be refuted if energy, not elements, limits organic synthesis and metabolic rates. With these caveats and limitations, Theorem 1 follows by logic. These theorems summarize the stoichiometric constraints on organism function that are an inevitable result of cellular allocation, the finite resources of the environment, and mass balance. We will cover these topics in detail later in the book, and for now we shall press on to add another of Reiners' axioms.

Axiom 5: *There are differences among species in the means and rates at which they can sequester limiting elements.*

This axiom refers to the existence of many different adaptations for resource acquisition, including differential nutrient uptake, differential foraging efficiency, etc. We will discuss those adaptations further in Chapters 3, 4, and 7. From all this, you get another theorem.

Theorem 3: *There are differences among species in ability to live in particular environments or to compete for chemical resources.*

This theorem is the core of resource competition theory (Tilman 1982; Grover 1997). It relates to patterns we will discuss in Chapter 5.

We now progress to the ecosystem level by adding two more axioms.

Axiom 6: *Biological effects on the availability and chemical form of elements are unique in kind and magnitude.*

Axiom 7: *The world biota has changed quantitatively and qualitatively through geological time.*

Whether the statement that biological effects on elements are unique in kind and magnitude is an axiom or should be thought of as a theorem derived from the preceding axioms seems debatable. Nevertheless, the logic flow chart shows us how consideration of mass balance in organism growth (Axioms 1–6) ties together the concepts of resource limitation (Theorems 1 and 2 and Chapters 3 and 5) and nutrient cycling (Axiom 6 and Chapter 6).

From all the preceding, we can obtain two theorems about global biogeochemical cycles.

Theorem 4: *The world biota drives and regulates the global biogeochemical cycles.*

Theorem 5: *Global biogeochemical cycles have been altered by life over time.*

A stunning aspect of this logical scheme chart is how few steps it takes to get from statements about cellular allocation (Axiom 1) to statements about the largest spatial and longest temporal scales relevant to Earth's biota (Theorem 5). This amazing sweep is part of ecological stoichiometry's allure. We will come back to global processes in Chapter 8. Another ecosystem-level statement is this next axiom.

Axiom 9: *Mechanical structures are resistant to decay.*

Rates of mineralization are very dependent on element balance, as well as on other chemical aspects. We cover this material in Chapters 6 and 7.

Theorem 6: *Accumulation of resistant mechanical structures alters soils and sediments.*

This theorem refers to the buildup of organic matter over long time scales. The subject of carbon storage will be described in Chapter 8.

Reiners' flow chart illuminates an intellectual pathway from atoms to ecosystems, with ecology and evolution both occupying prominent places. We are unaware of any similar logical structures that have been spelled out in ecology. Where does this consideration of logic take us? For one thing, it lays bare the connections among different components of an overall theory of how biological systems are controlled by and, in turn, control mass flow. It is interesting that an epistemological structure for stoichiometry does not consist of a series of hypotheses; rather, it uses the language of mathematics: axiom and theorem. Mass balance, after all, is nothing less than the law of mass conservation and hardly needs to be tested. Because ecological stoichiometry in its broadest sense is a version of thermodynamics, it is not a hypothesis or series of hypotheses. Rather, we view ecological stoichiometry as a lens, a means of organizing thoughts, a hypothesis-generation machine, and a window to interesting connections in the biotic and abiotic worlds. It is a tool to use in appropriate places. Let's get to it.

THE STRUCTURE OF THIS BOOK

The organization of this book follows more or less the movement of materials through ecological systems and the scale considered increases as the book progresses. We start in Chapter 2 with an overview of the biological chemistry of the major elements, and how they are coupled in the structures of the major molecules used by organisms. Next, in Chapter 3 we consider the incorporation of C, N, and P into photoautotrophic organisms (both algae and higher plants) and how various environmental and interspecific factors generate variation in C:N:P stoichiometry at the base of food webs. In Chapter 4 we examine the stoichiometry of metazoans and how and why the C:N:P composition of higher animals differs among and within species of animals. The emphasis is primarily on invertebrates but we also consider vertebrates, with their special requirements for minerals in the formation of bone. Interspecific interactions enter the picture in Chapter 5, where we analyze autotroph-grazer interactions from a stoichiometric perspective. Here we begin to see the profound consequences for secondary production when variable and nutrient-deficient autotroph biomass is consumed by homeostatic herbivores. In Chapter 6 we will consider an end product of this interaction: nutrient recycling by consumers. We will see that the relative imbalance in elemental composition between consumed food and the requirements of the consumer has direct effects on the rates and ratios of nutrients that they recycle. These consequences in turn can affect nutrient availability in the system and a complex set of reciprocal interactions is established. In Chapter 7 we further consider the consequences and feedbacks established by stoichiometric balance, discussing food-web structure, trophic dynamics, interaction strength, and a variety of other characteristics of communities. Ecosystems are the subject of Chapter 8, where we examine the broadest-scale patterns in ecological stoichiometry. We also see how the small-scale coupling of elements in biological molecules has effects on a grand scale, impinging even on global biogeochemical cycles. This stoichiometric coupling has ramifications for global climate and the potential outcome of humankind's unplanned experiments with our global habitat, in which the cycles of C, N, and P are disrupted simultaneously but to differing degrees and in different places.

A collection of symbols and their definitions appears in the Appendix. Each chapter includes a "Summary and Synthesis" section that repeats major points and integrates across concepts within the chapters. In addition, Chapters 3–8 include a list of key unanswered questions. We hope that by calling attention to those things that we feel are unanswered or require more information or deeper study we will accelerate future research. Hence, we have called these "Catalysts for Ecological Stoichiometry." The book closes (Chapter 9) with a comprehensive integration of major themes presented in various chapters and with a discussion of the general utility and potential future for stoichiometric theory in ecology and biology.

SUMMARY AND SYNTHESIS

This chapter laid the foundation for what is to come. It explored the nutrient contents of organisms as compared to the rest of the material world. Stoichiometry represents organisms as single abstract molecules where constituent elements exhibit mass conservation.

The rules determining the constraints on organismal chemical content are complex and vary with the organism, its feeding mode, and the chemicals involved. We derived a new, precise definition of stoichiometric homeostasis and showed how it can be used to diagnose regulation of nutrient content in biota. In our empirical examples, absence of homeostasis was seen in the N:P of autotrophs. A strict homeostasis was seen in the C:N of heterotrophic bacteria. Other taxa and chemical elements exhibited intermediate degrees of homeostasis.

We defined the key terms to be used throughout the book. Stoichiometry relies on mass conservation laws. Substances that are subject to stoichiometric balancing are conservative, like elements. They are neither created nor destroyed (obviously, we are dealing only with normal chemical reactions here).

A key reference point is the Redfield ratio (C:N:P = 106:16:1). The offshore ocean is a stoichiometrically balanced system with those three elements moving between the biotic and abiotic world in balanced proportions.

Ecological stoichiometry can be placed in an epistemological structure that uses the language of mathematics (axiom and theorem). It is about constraints of matter and energy at scales ranging from the atom to the biosphere.

KEY DEFINITIONS

These are "our" definitions and we have taken care to use them consistently throughout the book. The reader should be advised when reading the primary literature that not all studies use the same definitions. For example, in some papers, the term "Redfield ratio" is used to refer to any C:N:P ratio.

Balanced growth—Equal specific rates of change of elements during growth (dX/Xdt where X is any chemical substance). Under balanced growth, nutrient content does not change. Note also that chemical composition can be regulated actively (homeostatically) at a fixed level even in an organism that is not growing; thus, "strict homeostasis" is not interchangeable with "balanced growth."

Ecological stoichiometry—The balance of multiple chemical substances in ecological interactions and processes, or the study of this balance. Also sometimes refers to the balance of energy and materials.

Elemental imbalance—Dissimilarity in nutrient content between two things, such as consumer and food resources or between an autotroph and the inorganic medium. If consumer and resource have identical stoichiometry, they are perfectly balanced. The greater they differ, the more their imbalance.

Homeostasis—Physiological regulation of an organism's internal environment reducing changes within the organism. In stoichiometry, homeostasis results in a narrowing of variation in chemical content in an organism compared to the resources it consumes.

Homeostatic—Regulated by a negative feedback so that the state of the organism (e.g., its nutrient content) is less variable than predicted based on external variation.

Law of conservation of matter—In an ordinary chemical reaction, matter and component elements are neither created nor destroyed.

Law of definite proportions—Generally attributed to Proust, who in 1797 wrote that iron and oxygen combined in a fixed ratio of FeO_2. This law states that the relative amount of each element in a particular compound is always the same, regardless of preparation or source.

Maximum growth potential (μ_m)—The specific growth rate (dM/Mdt where M is biomass) achieved under conditions of resource surplus such that an organism is growing at its full, genetically determined, capacity for those physical conditions.

Nonhomeostatic—Lacking homeostatic regulation. In ecological stoichiometry, nutrients in nonhomeostatic organisms track their availability in the external surroundings.

Nutrient content (or composition)—The amount of nutrients in an organism. Nutrient content is generally expressed in units such as g P per dry weight, (or as a percent). We represent nutrient content using the symbol Γ_x where x is an element. Since biomass C content is relatively constant as a function of dry weight for most living things, nutrient content is generally proportional to the nutrient:C ratio.

Quota (Q)—The amount of any chemical substance in a single individual organism.

Redfield ratio—Proportions of elements equal to 106 atoms of C per 16 atoms of N per 1 atom of P.

Relative growth rate (RGR)—The proportion of an organism's maximum growth potential that is actually achieved, μ/μ_m.

Specific growth rate (μ)—The exponential rate of change of biomass of the organism normalized to its biomass ($\mu = dM/Mdt$ where M is biomass). In discrete time, μ is given by $[\ln(M_t/M_0)]/t$ where M_t is biomass at time t.

Stoichiometric equilibrium—see "balanced growth."

Stoichiometry—1. A branch of chemistry that deals with the application of the laws of definite proportions and the conservation of mass and energy. 2. The quantitative relationship between constituents in a chemical substance.

Strict homeostasis—In living and nonliving things, having elements occurring in fixed, definite proportions.

Yield—The amount of something produced relative to an investment. Yield is often given relative to a resource or a reactant, e.g., biomass produced from a given quantity of nutrients. Yield of biomass per nutrient invested is the inverse of nutrient content.

2

Biological Chemistry:
Building Cells from Elements

Phosphorus has a special function among the substances assimilated by living things: carbon, hydrogen, oxygen and nitrogen are the elements of biological frameworks, phosphorus is the instrument of their manufacture.—Pautard (1978)

The fact that the chemical composition of organisms differs in many ways from that of the nonliving world (Table 1.1) implies a "natural selection of the elements" (Williams and Fraústo da Silva 1996). In other words, the elements used extensively by biological systems are not an unbiased sample from the periodic table or from the abiotic world. Indeed, a distinct stoichiometric signature of living systems is one tool that has been used to separate life from nonlife in ancient fossilized material (Watanabe et al. 2000). The physical chemistry of an element directly determines the types of interactions it has with other elements and thus the sorts of chemical and biochemical functions that element can perform for the organism (the self-maintaining and self-regenerating suite of chemical compounds and reactions with which it is associated). Early chemical evolution and subsequent biological evolution in response to major changes in environmental conditions (e.g., changes in atmospheric oxygen levels or aqueous redox conditions) involved selection of elements from the array presented by the external geochemical environment. Both in the universe as a whole and in earthly biomass, lighter elements predominate (Table 1.1). Also, atoms with even atomic numbers are generally more numerous than those with odd numbers (a consequence of the creation of heavy elements via fusion of He nuclei). The interactions of molecules constructed from these elements were among the first steps of chemical evolution that set the stage for the subsequent history of life, with its myriad variations in structure and function made possible by the mechanical and catalytic talents of biological materials (Morowitz 1992).

Examination of this natural selection of elements helps shine light on the often obscure interface between evolutionary biology and ecosystem science. The physical chemistry of the elements is an ultimate constraint on evolutionary outcomes: no biological process can arise outside of the bounds imposed by the physical-chemical talents possessed by these variants of matter. Further, the abundances of the elements impose an envelope within which evolution must operate. Evolution should not favor using a vanishingly rare chemical element for a significant component of biomass, even if it is well-suited for some function. Extreme resource scarcity would lower the fitness of any organism that attempted to do that. As in the rest of evolution, the natural selection of chemical elements should favor locally optimal solutions to adaptive problems, given available variation and the constraints of history and protogenetic systems.

Chapter 1 introduced us to the ideas that organism chemical composition differs among species and is under physiological control. Here we begin to explore some reasons for these patterns of variability, constraint and control. We first present a brief chemical taxonomy of the elements, emphasizing C, N, and P. We then examine the importance of these elements in major classes of biological molecules. Because we find that different classes of biological molecules have distinctive elemental composition, we can show relationships between biological function and organismal C:N:P stoichiometry. We will end the chapter by assembling cells from component elements and molecules, thereby taking the first big step in establishing the underlying basis of variation in elemental composition in organisms.

THE BASIS FOR SELECTION OF CARBON, NITROGEN, AND PHOSPHORUS IN BIOCHEMICAL EVOLUTION

The periodic table is a big, but well organized, place. The large number of biologically important elements (Fig. 1.1) necessitates the development of schemes addressing chemical diversity in order to organize thinking about the biological roles of different elements. A variety of such schemes have been proposed. Since life requires water, most of this analysis has focused on the properties and abundance of chemical elements in aqueous solution. Aquatic geochemists have distinguished dissolved chemicals with categories primarily based on their environmental concentration and the extent to which they are used by organisms. For example, Moss (1998) discusses the "major ions" (elemental cations such as Ca^{2-}, K^+, Na^+, and elemental and oxyanions such as Cl^-, SO_4^-, and HCO_3^-), "nutrient elements" (elements often limiting to organisms, such as C, N, P, Si, and perhaps Fe and Mg), and "trace elements." Trace elements are present at very low concentrations in the environment (below 1 part per billion), and include some elements used at low levels by organisms (Cu, V, Zn, Br, Co, Mo) as well as elements that are potentially toxic (Hg, Sn, Cd).

TABLE 2.1

Primary functions and the chemical elements (or associated ions) involved in performing them for living things (modified from Table 6.2 of Fraústo da Silva and Williams 1991). Elements with a relatively minor role associated with a particular function are indicated in parentheses

Function	Elements	Chemical Form	Examples
Structural (biological polymers and support materials)	H, O, C, N, P, S, Si, B, F, Ca, (Mg), (Zn)	Involved in chemical compounds or sparingly soluble inorganic compounds	Biological molecules (Table 2.2); tissues; skeletons; shells; teeth; etc.
Electrochemical	H, Na, K, Cl, HPO_4^{2-}, (Mg), (Ca)	Free ions	Message transmission in nerves; cellular signaling; energy metabolism
Mechanical	Ca, HPO_4^{2-}, (Mg)	Free ions exchanging with bound ions	Muscle contraction
Catalytic (acid-base)	Zn, (Ni), (Fe), (Mn)	Complexed with enzymes	Digestion (Zn); hydrolysis of urea (Ni); PO_4 removal in acid media (Fe, Mn)
Catalytic (redox)	Fe, Cu, Mn, Mo, Se, (Co), (Ni), (V)	Complexed with enzymes	Reactions with O_2 (Fe, Cu); nitrogen fixation (Mo); reduction of nucleotides (Co)

Fraústo da Silva and Williams (1991) created a taxonomy of the elements more explicitly aimed at critical biochemical function (Table 2.1). They distinguished among elements required for structure, electrochemistry, mechanics, and catalysis. The macroelements H, C, N, O, S, and P are the dominant components of structural macromolecules while various metals (Fe, Cu, Mn, Mo, Se) are key catalysts of biological redox reactions. It is interesting to note the specificity of biological functions for various elements. For example, the cation Ca^{2+} is primarily involved in mechanical responses (e.g., muscle contraction) but it is not involved with electro-chemical processes (e.g., nerve impulses) where other cations (Na^+) are involved. Also note that the macroelements (H, C, N, O, S, and P) are involved primarily as organic molecules contributing to organism structure. However, P is an exception to this rule because HPO_4^- is also involved in electrochemical (ATP metabolism) and mechanical functions.

The biological deployment of a variety of elements is frequently far out of proportion to their abundance. For example, consider the critical job of Co^+, which catalyzes reduction of nucleotides during nucleic acid syn-thesis and is also the scarcest element in a human (Fig. 1.2). However, our emphasis in this book is on processes associated with elements that are both important in living systems and whose environmental cycling is influ-enced greatly by biology. For our purpose the most important are elements involved in "structural" functions (Table 2.1), which generally are the ele-ments present in macromolecular polymers because these structural com-ponents dominate the biomass of organisms. Out of these, the macroele-ments (C, N, P) whose cycling in the environment is strongly regulated by biological processes are particularly important. So we will emphasize these three elements, but we realize that other elements are critical to biological function and that biological processing often regulates the dynamics of those microelements as well. Eventually, ecological stoichiometry may be just as useful in understanding the cycling of many other elements in the environment and their role in ecological interactions, but unfortunately there is far too little information available at the moment to go very far. We also point out that the taxonomy of Fraústo da Silva and Williams (1991) (Table 2.1) generates only a very broad set of categories (periodic phyla?), obscuring significant diversity in function within each category. The biological roles of elements associated with the structural function category differ profoundly: nitrogenous enzymes are involved in energy capture and release; phosphorus-rich nucleotides are involved in the pro-cessing of genetic information; carbon-rich molecules are involved in en-ergy storage; and phosphorus-rich components are involved in mechanical support in vertebrates (bone).

Carbon, nitrogen, and phosphorus are three of the main constituents in biological structural molecules. However, C, N, and P are not particularly abundant on Earth or in the universe as a whole and thus it seems that living things made a very discriminating selection of elements from the environment (Fig. 2.1). In terms of C, N, and P content (again, by dry mass), the human body is 57% C, 6.4% N, and 2.8% P (Bowen 1979) (note that Lotka's data in Fig. 2.1 as well as the recipe for the "human molecule" in Chapter 1 are for wet mass). The Earth's crust contains only 0.18% C, 0.03% N, and 0.11% P (Bowen 1979) although of course Earth's atmo-sphere has significant concentrations of C and especially N that are so rare

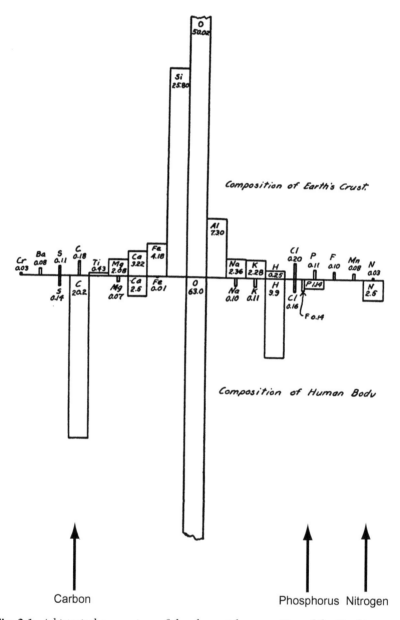

Fig. 2.1. A historical comparison of the elemental composition of the Earth's crust (upper bars) and of the human body (lower bars). The disproportionate abundance of C, N, and P in the human body is apparent (disregard oxygen and hydrogen, which dominate due to their preponderance in mineral oxides and in water). Our book considers the evolutionary and ecological ramifications of the fact that natural selection has made this highly nonrandom selection from the periodic table and from the available elements on the Earth. Based on Lotka (1925).

in the crust. The bulk elemental composition of life is clearly not a random sample of available elements. Why? What features of C, N, and P are so important for biological function that these elements have been so intensely distilled from the nonliving world?

Many of these issues have been covered extensively elsewhere; our overview comes mainly from Fraùsto da Silva and Williams (1991), Williams and Fraùsto da Silva (1996), and Cox (1995). First, let us look at the familiar example of carbon. The electron orbital structure of the C atom permits it to form four stable, high-energy, covalent bonds with a number of chemical elements, including itself as well as N, O, H, and S. From this column of the periodic table (group 14), both C and Si are almost equally abundant in the solar system, and Si is even more abundant than C in the Earth's crust (Cox 1995). But, of all the solid nontransition elements in the periodic table, C has the highest binding energy (expressed as kcal mol^{-1}; Williams and Fraùsto da Silva 1996), suggesting that one feature that favored the selection of C for use in biology was its ability to store energy. The high degree of bonding flexibility of C also includes the possible formation of double and triple bonds of C with itself, leading to a variety of bond angles that generate considerable architectural flexibility. On the other hand, C cannot easily bond directly with P and thus the involvement of P in biochemistry is almost always mediated via O in the form of phosphate groups. An interesting exception to this rule is the phosphonates, a class of organic molecules with P bonded directly to C. Certain classes of phosphonates are involved in subunits in certain phospholipids and in some organisms phosphonates can be a major form of P in developing embryos (for example, in the planaborid snail *Helisoma*; Miceli et al. 1980). Interestingly, phosphonates appear to be a major component of relatively refractory dissolved organic P in the oceans (Kolowith et al. 2000). Nevertheless, the biological significance of phosphonates is not yet clear, and in this book our focus will be on the major molecules in which P is bonded to C via PO_4 in a monoester linkage. In any case, C seems a natural choice from the periodic table for forming a sufficiently flexible, diverse, energy-rich, and stable chemical backbone in evolving biomolecules.

Next consider nitrogen. The fact that N can assume a diversity of chemically stable redox states (including NH_4^+ and NH_3, N_2, N_2O, NO, NO_2^-, NO_3^-) has great biogeochemical significance. However, its most ubiquitous form, gaseous N_2, is difficult for cells to incorporate directly, as doing so requires a six-electron reduction and considerable expenditure of energy (see also Chapter 8). Catalytic N_2 fixation is confined to a very limited set of organisms. Most organisms incorporate N into biochemical metabolism in the reduced state of NH_3 (some after reduction of oxides and oxyanions of nitrogen, e.g., NO_3^-), which is then used in synthesis of monomers (e.g., glutamine), precursors in protein, nucleotide, and nucleic acid production. In turn, the nature and sequence of the repeating mono-mers afford these polymers different degrees of stability, catalytic activ-ities, and informational content, contributing to the polymer's retention over evolutionary time. An additional key feature of the biological chemis-try of N is the fact that the amine group ($-NH_2$) that forms the core of amino acids is basic. At neutral pH, this group is generally protonated ($-RNH_3^+$). Thus, amino acids are one of the few monomers in biological chemistry capable of retaining a positive charge under normal cellular condi-tions, a key factor affecting intra- and intermolecular interactions in proteins. Also contributing to the folding and binding properties of proteins and nucleic acids is the fact that the N—H bond is polar and thus can form hydrogen bonds. Finally, N is also a good coordinating atom for metals (hence its association with Fe in hemoglobin and Mg in chlorophyll, helping to escort electrons during respiration and photosynthesis, respectively).

Finally, phosphorus. Phosphorus is ubiquitous in cellular metabolism and biological structure. In the form of phosphate esters and anhydrides, P is present in genetic polymers (RNA, DNA), in coenzymes (e.g., thiamine pyrophosphate), in intermediary metabolites (glucose-6-phosphate), and in the principal vehicle of biochemical energy (ATP). In contrast to this bio-logical ubiquity, P is relatively rare in the geosphere, particularly in the oxic aqueous solutions upon which most of life depends. Abiotic synthesis of high-energy phosphates or phosphate esters also is thought to occur at very low rates (Keefe and Miller 1995). What special features does P pos-sess that led to such a high degree of cellular reliance in modern biological systems? The chemical basis of the central role of P in biological chemistry has been reviewed by Westheimer (1987, 1992), who noted two key fea-tures that likely resulted in its selection during early protobiological evolu-tion. First, chemical intermediates in metabolic cycles need to possess a charge to be efficiently retained within lipid membranes (lest they diffuse out and be lost from the cell). Phosphate, even when joined via phospho-monoester bonds to molecules, retains a negative charge. Second, the free charge in replicating polymers needs to be negative in order to repel nu-cleophilic chemicals (such as hydroxide ions) that would rapidly hydrolyze the molecule. Thus, the strong negative charges associated with the phos-phate group provide a critical level of stability for polymeric molecules, especially nucleotides, necessary for chemical continuity (prebiotic ge-netics) and further accumulation of chemical functions during evolution. It is interesting to note that sulfate can perform a similar bridging function in polymers but the chemical stability of sulfate linkages is very low com-pared to phosphate (Williams and Fraústo da Silva 1996). For these rea-sons, it seems, nature "chose" phosphate and chose well. One consequence of this choice is easily discerned by noting the ubiquity of free sulfate and the rarity of free phosphate in the modern environment.

Having now seen how the physical properties of elements selectively lend them to differing biological functions, let us undertake a concerted survey of the major biological molecules, with an eye toward their elemental content.

THE ELEMENTAL COMPOSITION
OF MAJOR BIOCHEMICALS

If we are to understand the reciprocal interactions between the chemistry of life and the chemistry of the nonliving world, we need to consider the elemental composition of the biochemical tools life has evolved. One basic question for us in this context is, "Do the biochemicals associated with different life functions differ significantly in their C:N:P stoichiometry?" If they do, we might be able to connect biological function and elemental composition, and thus couple important evolutionary and ecosystem processes (Reiners 1986). Our approach will be to consider major classes of biochemical compounds containing C, N, and P; we will also focus on some of the ways organisms process and store chemical elements in inorganic form, including the formation of external and internal structures.

Consider several features of the major classes of biological materials containing C, N, and P (Table 2.2). Note that classes of molecules with distinct biological functions differ quantitatively and qualitatively in their mix of elements. Prominent among these is the fact that P plays a role in the biology of nucleic acids but is for the most part absent from proteins (activated phosphoproteins are not a large component of biomass). Second, inorganic deposits lack N but may contain significant amounts of P and C. In the following, we consider these biological materials in more detail to determine the extent to which each is important in affecting organismal elemental composition and thus how contrasting biological functions have differing elemental demands.

Lipids

Dictionary definitions of lipids revolve around the fact that they are water insoluble but soluble in nonpolar organic solvents. This operational definition offers little insight into their chemical composition and so a more useful characterization is that lipids are fatty acids and their derivatives, and substances biosynthetically derived from these compounds. Five categories of lipids are usually distinguished (Lehninger et al. 1993): triacylglycerols (an important energy storage lipid group), waxes, phosphoglycerides (the primary type of lipid involved in cell membranes), sphingolipids (involved primarily in nerve cells), and sterols (cholesterol, for example).

TABLE 2.2
The dominant classes of ubiquitous biological materials that contain C, N, or P. Parentheses indicate that the element indicated is found only in certain forms of the biological material considered

Biological Material	Examples	Functions	Comments
Protein: C, N, (P)	Collagen, RUBISCO	Structure, regulation, communication, metabolism	Average N content of the 20 amino acids in proteins is 17%
Nucleic acids: C, N, P	DNA, mRNA, tRNA, rRNA	Storage, transmission, and expression of genetic information	DNA content is conservative; RNA:DNA is typically greater than 5:1 by mass; rRNA dominates total RNA
Lipids: C, (N), (P)	Phospholipids, neutral lipids	Cell membranes, energy storage	Carbon-rich molecules; phospholipids are relatively minor components of cells (<5% of total mass)
Energetic nucleotides: C, N, P	ATP, phosphocreatine	High-energy carrier or storage molecules	ATP is only approximately 0.05% of invertebrate body mass
Carbohydrates: C, (N)	Glucose, starch, glycogen, cellulose, lignin, chitin, peptidoglycan	Energy storage, structure	Generally lacking in N and P (peptidoglycan and chitin are exceptions but still have relatively low % N)
Pigments: C, N	Chlorophyll	Light absorption	Chlorophyll is 6.5% N
Inorganic materials: C, P	polyphosphate, hydroxyapatite, $CaCO_3$	Nutrient and energy storage, mechanical support and protection	In some situations, polyphosphate can be a major storage pool in bacteria, algae, fungi, and some higher plants, but not in metazoans

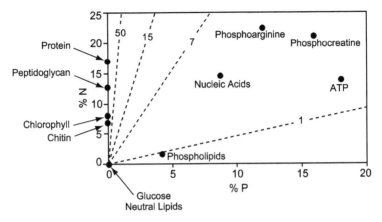

Fig. 2.2. Biochemical stoichiometry. This diagram illustrates the % N and % P of important biomolecules. In most cases, estimates were made from the biochemical structure for each molecule. However, for proteins and phospholipids estimates were made by determining the average composition of the monomers (amino acids in the case of proteins) or of various types of the molecule (e.g., different kinds of phospholipid).

In the following we focus on triacylglycerols, waxes, and phosphoglycerides ("phospholipids"), because these are the lipid classes most likely to contribute significantly to organism biomass.

The generic structure of triacylglycerols involves three fatty acids linked to a glycerol molecule via ester bonds. These molecules function primarily in energy storage and are commonly found in the cytosol. In many taxa, large deposits of triacylglycerol can be found in specialized fat cells (adipocytes). Triacylglycerols are much better suited for energy storage than are carbohydrates because they yield over twice the energy, gram for gram. Due to the differences in the size of the pools and their specific energy content, the human body can store only about a day's worth of energy in the form of glycogen (a carbohydrate), but lipids in an obese person can store sufficient energy for several months of basal metabolism.

Fatty acids and glycerol contain neither N nor P. Accordingly, triacylglycerols, waxes, as well as sterols are 0% N and 0% P (Fig. 2.2). The C content of fatty acids is high compared to other major biomolecules. For example, the C content of a typical triacylglycerol (tripalmitin) is 75%; therefore, increased triglyceride content in a cell or organism will tend to raise the C content. So changes in the contribution of triacylglycerols to organism biomass will affect % C but will alter % N and % P proportionately with no effect on organismal N:P.

The second major class of lipids is the phospholipids, the primary component of biological membranes. A prominent feature of these lipids is the presence of a phosphoester linkage between the fatty acid subunits and a variable terminal functional group. In some lipids, the phospholipid group involves a phosphonate linkage (direct C-P bond), rather than a phospho-ester bond; this may afford the lipid a greater degree of stability given the inherent stability of phosphonate. The functional groups in phospholipids frequently contain N atoms. Thus, phospholipids are a potential pool of C, N, and P. To quantify the elemental composition of these contributions we calculated the C, N, and P content of five important phospholipids (phos-phatidylethanolamine, phosphatidylcholine, phosphatidylglycerol, phospha-tidyldiglycerol, sphingomyelin). The average C, N, and P content for these five phospholipids is, respectively, 65, 1.6, and 4.2% (39:0.8:1 C:N:P, Fig. 2.2). Phospholipids have low N but relatively high P content. Increases in relative contributions of phospholipid to cells will thus tend to raise organismal C content but lower its N:P.

The third major domain of life, the Archaea, is distinguished, in part, by unique membrane lipids (Langworthy 1985). As we have described, lipids in the other two major domains, Bacteria and Eukarya, are characterized by straight-chain fatty acids and fatty acid ester–linked glycerols. In contrast, lipids in Archaea are characterized by isoprenoid hydrocarbon and isopranyl glycerol ether–linked lipids. Particularly unusual are the diether and tetraether lipids, which have no analogues in other organism groups. Most of these peculiar lipids are 40 carbons long and have polar functional groups at each end. It is thought that this allows the lipids to span the entire membrane, providing considerable membrane rigidity, which may contribute to the astounding abilities of Archaea to thrive in extreme environments (De Rosa et al. 1991). Despite these fundamental biochemical differences associated with the lipids of the Archaea, the C:N:P stoichiom-etry story is little different from what we have already discussed. Archaeal lipids have generally high C content, low N content (actually, zero N content in Archaea), and moderately high P content. Thorough characterizations of the structures and relative abundances of various lipids in diverse Archaea are difficult to find and so we cannot attempt a calculation of the C:N:P stoichiometry of the average Archaeal lipid. However, the discovery that Archaea are widespread and abundant in the world's oceans (Olsen 1994; Karner et al. 2001), in addition to their known dominance in extreme environments, indicates that achieving a better understanding of the biochemical and C:N:P stoichiometry of Archaea should be an important priority in coming years.

Given these differences in C, N, and P in different lipids, we must consider whether lipids are an important enough constituent of biomass to appreciably effect an organism's elemental composition. We also need to know if the contribution varies significantly among organisms. First consider phospholipids. The primary cellular location of phospholipids is in membranes, including the plasmalemma, the nuclear membrane, the endoplasmic reticulum (ER), and in the external membrane and internal lamellae of mitochondria and chloroplasts. There is good reason to believe that the percent contribution to biomass of membrane structures will be relatively invariant in different cells. For example, while increasing cell size would lower the contribution of plasma membrane to total cell biomass (the cell surface area:volume ratio decreases with increased cell radius), the increasing cell volume would need to be filled with a more elaborate network of membranous vesicles of the ER, compensating for decreased surface membrane lipid. The literature correspondingly indicates that phospholipid content does differ little among organisms, as in studies of marine vertebrate and invertebrate animals indicating that most phospholipid can be considered "structural" and associated with membranes (Reinhardt and Van Vleet 1986; Sargent and Henderson 1986). However, it has also been shown in the bacterium *Bacillus subtilis* (Lahooti and Harwood 1999) and in beans (Gniazdowska et al. 1999) that, under P-limited growth conditions, the relative contribution of phospholipids to membrane lipid can decline, as P-containing lipids are substituted with molecules lacking in P. Such a response can be seen as a means of even further reducing the minimal requirements of the organism for P (reducing minimal cell quota; see Chapter 3) but ecological ramifications of such genetic and biochemical responses remain largely unstudied. In addition, levels of phospholipid are relatively low in most organisms. In crustacean zooplankton, phospholipids comprise less than 6% of total biomass (Baudouin and Ravera 1972; Sargent and Falk-Petersen 1988) and thus phospholipid-P amounts to at most ~0.2% of body mass, a low value, considering the average P content of zooplankton (~1%, range 0.3–3%; Chapter 4). Thus, only in low-P organisms would phospholipid-P represent a relatively significant pool.

In contrast to phospholipids, storage lipids can contribute substantially to organism biomass, as when lipids accumulate in organisms during periods of high caloric intake. Gilbert (1967) reported lipid content for a wide variety of insect taxa. Lipid contributed, on average, ~25% of animal dry mass with generally lower values for adults relative to larvae and pupae. In addition, lipid content varied widely among taxa, with values as low as 5% (for larvae of the grasshopper *Schistocerca gregaria*) and as high as 85% (for larvae of the lepidopteran *Carposina niponensis*). Lipid content also varied with development, with higher levels in larvae than in adults. Ontogenetic changes can be dramatic. For example, in the lepidopteran *Malacosoma americanum* lipid percentage dropped from 82 to 25% as larvae developed into adults. Lipids can also be a significant component in crustacean zooplankton. Båmstedt (1986) compiled whole-body lipid content of marine calanoid copepods and reported that lipid content varied from less than 5 to 70% of body mass with a mean of ~25%. In this group of invertebrates, the majority of the body lipid is in the form of wax esters that serve primarily as metabolic energy reserves (Sargent and Henderson 1986). A pigmented copepod with high reserves of wax esters chemically resembles a swimming crayon! Cavaletto and Gardner (1999) summarized lipid data for five taxa of freshwater invertebrates (the amphipod *Diporeia*, tubificid and chironomid insects, the opossum shrimp *Mysis*, and the bivalve *Dreissena*). Total lipid content (as a percentage of dry mass) varied from 6 to 46% among these taxa, while the contribution of phospholipid to the total lipid varied from 9 to 63%.

Major differences in total lipid content may have nontrivial effects on organismal C content, given that other major macromolecules (proteins, nucleic acids) have lower C content (32–53%; see below) than lipids (~70% C). For example, an increase in lipid content from 10 to 75% of body mass would raise the carbon percentage from 50.6 to 64.6% (assuming that proteins and nucleic acids make up the remaining biomass and are present in a 3:1 proportion by mass). For an organism with a P content remaining constant at 0.5%, this change in C content would be sufficient to cause biomass C:P to increase from 250 to 325. This indicates that variable lipid content may, in some cases, need to be accounted for in refining stoichiometric models of organism growth (Chapter 4).

A group of chemicals closely related to lipids is alcohols; thus, this seems an appropriate place to address two functionally important molecules in plants: lignin and chlorophyll (and other pigments, such as carotenoids). The term "lignin" refers to a chemical family of polymers containing branched and/or cross-linked polymers based on coniferyl-alcohol-type monomers (McCarthy and Islam 2000). Lignin can comprise 20–40% of higher plant tissues, where it functions significantly in stiffening. Inspection of the chemical structure of coniferyl alcohol indicates that lignin is ~60% C while itself containing no N or P, which agrees well with direct measurements of % C in cork and in eucalyptus bark (59–62% C; Akim et al. 2000). Thus, increasing allocation to lignin in a material will tend to raise its C:nutrient while leaving N:P unaffected.

The chlorophyll molecule consists of a long phytol (alcohol) tail bonded to a nitrogenous tetrapyrrole ring structure with a single Mg atom at its core. The elemental composition of the chlorophyll molecule (Fig. 2.2) is relatively C rich and also somewhat high in N while lacking P (72% C, 7.9% N, and 0% P; C:N = 7.8). The chlorophyll content of plant tissue varies in response to environmental conditions (light, temperature, nutrient supply). Values for chlorophyll content in leaf tissue typically vary from 0.1 to 2.5%. Since chlorophyll is 7.9% N, chlorophyll-N thus comprises from less than 0.01 to 0.2%. Since leaf N is typically around 2% (Chapter 3), the N demands for chlorophyll production will generally represent only a modest fraction of the leaf's N allocation. The situation is likely the same in algae. Typical C:chlorophyll ratios in algae range from 40 to 200 by mass (Harris 1986). Assuming that algal biomass is ~45% C, a chlorophyll-rich cell with a C:chlorophyll ratio of 40 would be ~1.1% chlorophyll by mass, which translates to a contribution of chlorophyll-N to biomass of only ~0.09%. Since the N content of nutrient-sufficient, chlorophyll-rich, algal cells is typically around 5.7% (assuming algal C:N of 7; see Chapter 3), chlorophyll investment represents a relatively minor component of cellular N allocation in algae. Algae and higher plants can also invest heavily in accessory pigments for photoprotection (in high light) or for improved light harvesting (in low or chromatically narrow light). An important group of accessory pigments are the carotenoids, a family of molecules imparting red-orange coloration and having in common a 40-carbon polyene chain with various terminal groups (Lehninger et al. 1993). Thus, unlike chlorophyll, carotenoids do not contain N. Within the "algae," members of the Rhodophyta and Cryptophyta as well as the cyanobacteria can produce an additional class of photosynthetic pigments, the phycobilipigments. These compounds are actually not alcohols, but we will deal with them here anyway. Instead, they are linear tetrapyrroles that can bind via sulfide bridges to membrane proteins in the photosynthetic apparatus. The tetrapyrrole units contain N and examination of the chemical structures indicates that these molecules have a C content of ~65% and N content of ~10% with a C:N of 7.75. However, concentrations of such pigments in cells are generally low on an absolute basis, even when relative allocation is high. Thus, the contribution of phycobilipigments to organismal N demands is probably modest.

So high N demands for photosynthesis are more likely related to increased investment in photosynthetic enzyme proteins (see below), not to the N required for chlorophyll molecules or accessory pigments themselves.

Carbohydrates

Carbohydrates are "polyhydroxy aldehydes or ketones or substances that yield such compounds on hydrolysis" (Lehninger et al. 1993). Most of us know them as sugars and starches. Sugars are the central substrate of energy metabolism. The photosynthetic production of glucose and its respiratory degradation via the Krebs cycle are the core of the biochemical coverage of most introductory biology courses. However, in spite of the impression this gives, organisms are not sugar cubes! Carbohydrates themselves play a much wider role in biology than simply in energetic transformations or storage. Structural polysaccharides can comprise a considerable percentage of organismal biomass, particularly in plants. Cellulose in fact may be the most abundant biochemical on earth (Lehninger et al. 1993).

Let us look at the elemental compositions of polysaccharides important in biomass. Glucose and the more complex di- and polysaccharides such as starch and glycogen are based on a unit with the stoichiometric ratio of $C:H_2:O$. This makes them 37% C and 0% N and P. This C content is relatively low compared to lipids and reflects the higher O content of carbohydrates (in fatty acids the repeating subunit is —CH_2). Cellulose, which is an extremely long unbranched chain of more than 10,000 glucose monomers, is also ~37% C. The same is true of glycogen, amylase, pectins, and other similar large carbohydrates. More stoichiometrically interesting than cellulose and its relatives are several important structural carbohydrates containing N. Most bacterial cell walls are based on peptidoglycan, a complex polysaccharide containing nitrogenous tetrapeptide side chains. The elemental composition of the cell wall peptidoglycan of *Escherichia coli* is 32.3% C and 12.7% N (C:N = 2.2; Fig. 2.2). Some fungi as well as most invertebrates utilize chitin for structure. Chitin is a linear polymer of N-acetyl-D-glucosamine and thus also contains N. Chitin is constructed with the same building block as peptidoglycan but it lacks the peptide side chains. Chitin has a higher C content (41.4% C) and lower N content (6.9% N, C:N = 5.1; Fig. 2.2).

Structural polysaccharides contribute greatly to plant biomass and thus to their elemental composition. This is particularly true in large vascular plants where a majority of biomass is allocated to support structures built of cellulose and lignin. For example, in mature Scotch pine, 75% of tree biomass is in the trunk and branches with the remainder allocated to leaves and roots (Ovington 1957). Wood has a C:N ratio from 200 to more than 1000 (Levi and Cowling 1969). Fresh dead wood is less than 0.2% N averaged over an entire cross section and is maximal in the inner and outer bark, where it is ~0.5% N (Merrill and Cowling 1966). One effect of the dominance of polysaccharides in total organism biomass is to substantially reduce N and P content (and raise C:N and C:P ratios). For other types of organisms (e.g., bacteria and metazoan animals), polysaccharides are a less important contributor to biomass. In bacteria, polysaccharides comprise only ~10% of dry mass (data for *E. coli* and *Salmonella typhimurium*; Brock and Madigan 1991).

In invertebrates, structural polysaccharides (i.e., chitin) represent a variable but generally small pool. For example, for nine taxa of marine copepods, chitin contributed 2.1–9.3% of total animal dry mass (mean 4.6%; Båmstedt 1986). Similarly, chitin contributed ~3–6% of total body mass in late larvae and early pupae of southern army worms (*Prodenia ereidania*, a lepidopteran; Porter 1965). Since chitin is 7% N by mass, chitin-N represents only ~0.35% N, less than 1/20 of total animal N, which generally ranges from 8 to 12%. Thus, structural carbohydrates represent a somewhat variable but generally small element pool in most animals, especially when compared to plants. Variation in chitin content among animal taxa would have little impact on C content because the C content of chitin is intermediate relative to other major biomolecules, but it would raise organismal N:P due the presence of N and absence of P. At the level of tissues, chitin is primarily associated with invertebrate cuticles, which themselves contain considerable amounts of protein. Using a typical composition profile for insect cuticle (55% protein, 35% chitin, 10% lipid; Richards 1978), we can estimate an elemental composition for cuticle of around 55% C and 12.6% N (C:N = 5.1). Thus, insects with extensive cuticular investment would have low % P and elevated N:P.

Proteins

Proteins are polymers comprised of various sequences of around 20 nitrogenous amino acids. The identity, number, and specific sequence of the amino acids used in a given protein determine its biochemical properties. The 20 amino acids range in N content from 8.8% (tyrosine) to 35% (arginine). With an equal representation of the 20 amino acids in the major protein pools of an organism, the average protein would be ~53% C, 17% N, and 0% P (Fig. 2.2) with a C:N of 2.7. In the following sections we will use the value of 17% to estimate the N content of various biological structures containing protein. If the proteins involved in biological structures were composed of amino acids of particularly high or low N content, use of the 17% N value would be erroneous and more accurate estimates would have to incorporate the specific amino acid composition of the actual proteins involved in each structure. As a check, we quantified a small sample of the 20 ribosomal proteins for *E. coli* (the S4, S10, and S21 proteins) to directly estimate the average N content of the amino acids actually used in this major protein pool (Wittman and Wittman-Leibold 1974). Ribosomal proteins are known to be rich in basic amino acids but depleted in aromatic amino acids. We found that the sampled ribosomal proteins were 18.1% N on average, indicating that ribosomal proteins do not have an amino acid composition that is significantly skewed with respect to high- or low-nitrogen amino acids. The value of 17% N as representative of protein N content thus seems reasonable and corresponds well with the widespread rule of thumb for converting N to protein content by multiplying by 6.25 (1/6.25 = 0.16).

While the rule of thumb just described will likely prove satisfactory in most applications, recent data indicate that there are systematic differences in the elemental composition of proteins in response to particular nutrient limitations (Baudouin-Cornu et al. 2001). Using genomics databases for protein-coding regions for enzymes involved in carbon, nitrogen, and sulfur assimilation, these investigators showed that the amino acid profile of enzymes involved in a particular assimilatory process for an element were significantly biased in favor of amino acids with low content of that particular element. In other words, enzymes involved in N assimilation (connected with conversions of ammonia, urea, allontoate, and proline) have decreased numbers of N atoms in their residues in their side chains while there was no significant bias toward low-N residues in sulfur-assimilatory enzymes. Similarly, enzymes involved in sulfur metabolism were composed of amino acids relatively low in S while proteins connected with carbon assimilation were significantly depleted in C. This display of stoichiometric adjustment at the level of protein evolution points toward ecological influence on protein elemental composition. To put it in the words of the authors, "[The results suggest] that the elemental composition of biological polymers has been more generally subjected to ecological constraints than was previously thought, and that metabolic costs are among the variables optimized by natural selection."

Proteins have the highest N content of the biochemicals we have considered thus far (Fig. 2.2). Because they are a major component of the biomass of all organisms, proteins have a major role in total determining organismal elemental composition. Protein can comprise 35–75% of the body mass of crustacean zooplankton (Båmstedt 1986; Berberovic 1990), ~70% of the biomass of eucaryotic algae (Fogg 1975), ~50% of bacteria (Giese 1979), and ~30% in actively growing leaves of higher plants (Lawlor et al. 1989). The functions of proteins are diverse and extensively reviewed elsewhere (Lehninger et al. 1993). However, two classes of protein function are particularly important for our purposes: proteins that serve for structure and proteins that function as enzymes. In most cases structural proteins are the most important in ecological stoichiometry because they most commonly contribute significantly to the biomass of many organisms. An example is collagen, which is the most common protein in vertebrates and alone can be one-third of total vertebrate protein (Lehninger et al. 1993).

However, there is one very important case where enzymatic proteins contribute a significant fraction to tissue and even to whole organism elemental composition. Photosynthesis is a nitrogen-intensive process because of a heavy protein requirement. The single photosynthetic enzyme ribulose bisphosphate carboxylase (RUBP carboxylase or RUBISCO), which is the major carboxylating enzyme incorporating CO_2 into sugars in C_3 plants, can alone comprise as much as 60% of leaf protein (Groot and Spiertz 1991). According to Chapin (1991b), in a "sun leaf" in a plant with C_3 photosynthesis, 26% of N is involved in CO_2 fixation (the great majority of that is in RUBISCO), 25% is in nonchloroplast nitrogen, 19% is involved in light harvesting (chlorophyll and other pigments), 18% is involved in biosynthesis (RNA and ribosomes and other substances), 7% is put to structural and "other" uses, and 5% is involved in bioenergetics (including electron transport). Gross biochemical differences between C_3 metabolism, which uses large concentrations of RUBISCO (30–60% of leaf protein), and C_4 metabolism (5–10%) are well known and have stoichiometric implications as we will see in Chapter 3. If RUBISCO is 60% of leaf protein and assuming a typical leaf protein level of 30% of leaf mass, RUBISCO-N would itself be sufficient to produce a leaf N content of ~2.9%. This is a substantial value considering that leaf N content generally ranges from 2 to 5% N (Mattson 1980). These calculations, as well as direct determinations indicating that chloroplasts can contain 75% of leaf N (Stocking and Ongun 1962), illustrate the biochemical and cellular basis of the well-established relationship between leaf photosynthetic capacity and leaf N content (Chapter 8; Marshall and Porter 1991).

In Figure 2.2 we depict proteins as being free of phosphorus. This is not always true. There are a few stable proteins containing P, such as casein (milk protein), but these we consider special cases and do not analyze them further. In addition, many enzymes and gene regulatory proteins are temporarily phosphorylated. We consider such protein phosphorylation to be a phenomenon that is better characterized by consideration of ATP levels (ATP provides the phosphate molecule temporarily associated with the protein), which is where we turn now.

Energetic Nucleotides

If you were to abruptly demand of a biologist that she quickly name a biochemical containing phosphorus, the chances are very high that she would respond "ATP." This response may be natural, given the fact that the letter "P" comprises 33% of the typographical biomass of the acronym "ATP," which stands for adenosine 5'-triphosphate. This molecule has three components: a six-carbon sugar, an N-rich adenine, and a three-unit polyphosphate chain where exchangeable energy is stored. ATP is 28% C, 14% N, and 18% P (C:N:P = 3.9:1.7:1, Fig. 2.2) and is one of the most, if not the most, P-rich organic molecule in extensive use by most living things. There are other naturally occurring biomolecules with higher P content, for example, methylphosphonate (C:P = 1) or glycerol-P, but these are not as widely distributed as ATP. The related nucleotides phosphocreatine and phosphoarginine are also used for longer-term energy storage in many organisms; these too are extremely P rich (Fig. 2.2). However, a P-rich composition is not in itself sufficient to assure that a molecule will play a big role in determining organismal stoichiometry; it is also necessary that that molecule comprise a reasonably large percentage of total biomass. Although ATP is a critically important compound with extraordinarily rapid turnover rates, ATP levels in most organisms are insufficient to contribute greatly to whole-organism biomass. Båmstedt (1986) compiled values for marine copepods and reported that ATP levels ranged from 0.3% to 1.8% (mean 0.7%) while DeZwann and Thillart (1985) measured ATP levels in 22 species of terrestrial insects and found that ATP ranged from 0.02 to 2% of dry mass with a mean of 0.05%. These values translate into average ATP-P contributions of 0.12% for copepods and 0.01% for insects. Again, compared to typical % P values (~0.3–3%, Chapter 4), ATP-P at the very most might contribute one-third of whole-animal P content and in most cases will be far lower. ATP levels in plant tissues are generally low. For example, in the water fern *Spirodela*, foliar concentrations of ATP were ca. 170 nmole/g of fresh weight (Bieleski and Ferguson 1983). Assuming water content to be 15%, this equates to an ATP content of 0.006% of dry mass. Thus, as in animals, ATP likely does not contribute much to the P content of plants.

Nucleic Acids

Nucleic acids are macromolecules constructed from five different nitrogenous bases (adenine, guanine, and cytosine in DNA and RNA, thymine only in DNA, and uracil only in RNA). When bases are joined with a sugar and phosphate, these then comprise a nucleotide (Fig. 2.3A). Nucleotides are then joined in the nucleic acid chain via phosphodiester linkages between pentose sugar components (2′-deoxy-D-ribose in DNA, D-ribose in RNA) (Figs. 2.3B,C). The central dogma of molecular biology describes the translation of DNA genetic codes to mRNA, with tRNA and rRNA serving a catalytic function in converting that code to the amino acid sequence of proteins. Our interest is in the C:N:P stoichiometry of this process. Key stoichiometric quantities in this central dogma include the fundamental ratio of one phosphate per sugar per nitrogenous base, which is found in DNA and RNA. Codons, the information for a single amino acid, require six DNA nucleotides (three on each neighboring strand). In addition, a single tRNA which participates in the addition of a single amino acid to an elongating protein chain contains 75–95 phosphates. Finally, a single ribosome contains about 4,700 phosphates. We will not dwell further on the many details of the chemistry and biology of nucleic acids as many volumes have already done so. Instead we focus on their elemental composition and contribution to biomass and how these are associated with cellular growth and proliferation.

The N content of the nitrogenous bases varies from 21.5% N (for thymine, a pyrimidine) to 51.8% (for adenine, a purine). However, since in nucleic acids purines and pyrimidines are found in approximately 1:1 proportion (Lehninger et al. 1993) we can use the average elemental composition of the four bases (39.1% N, 43.1% C) in estimating the elemental

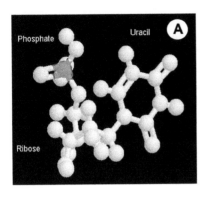

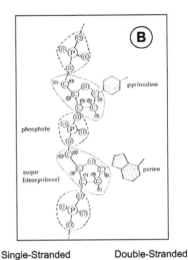

Stoichiometry of the Central Dogma
Nucleotides and Nucleic Acids
1 Phosphate : 1 Sugar : 1 Nitrogenous Base

Pyrimidines	Purines
Cytosine - N:P = 3	Adenine - N:P = 5
Uracil - N:P = 2	Guanine - N:P = 5
Thymine - N:P = 2	

DNA Codons
6 Phosphates : Amino Acid
(Double-Stranded)

mRNA Codons
3 Phosphates : Amino Acid
(Single-Stranded)

tRNA
75-95 Phosphates : tRNA

rRNA
~4700 Phosphates : Ribosome

Single-Stranded (RNA)	Double-Stranded (DNA)
A, G, C, U	A, G, C, T

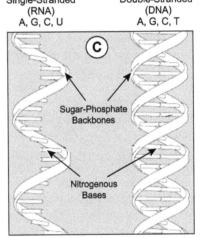

Fig. 2.3. Several representations of the stoichiometry of nucleotides and nucleic acids, emphasizing phosphorus. A. Ball and stick model of a nucleotide (nitrogenous base plus sugar plus phosphate). B. A portion of a single strand of DNA, showing the location of the sugar phosphate backbone relative to the nitrogenous bases. C. Comparison of single-stranded RNA and double-stranded DNA. The sugar:phosphate:base stoichiometry is the same for both. This figure also gives some of the key stoichiometric quantities involved in the DNA-RNA-protein central dogma of molecular biology and highlights the N- and especially P-intensive nature of storing and processing genetic information. Panel B is reprinted from Watson (1968).

composition of nucleic acids. Taking into account that a PO_4 group joins adjacent nucleotides via a sugar bridge, the calculation reveals that nucleic acids are 32.7% C, 14.5% N, and 8.7% P with a C:N:P of 9.5:3.7:1. Nucleic acids thus are low-N:P molecules (Fig. 2.2). These calculations reveal two key features of the elemental composition of functionally important macromolecules. First, proteins and nucleic acids differ only slightly in terms of N content (Fig. 2.2): 17 versus 14.5%, respectively. Thus, structures and organisms constructed of differing proportions of proteins and nucleic acids (holding other components constant) will vary within only a narrow range of % N and C:N. Second, nucleic acids are very P rich, with the second highest % P value of the major classes of compounds that contribute significantly to biomass. Proteins, as we have mentioned, are P poor. Thus, differing mixtures of proteins and nucleic acids, while exhibiting little variation in % C and % N values, will differ greatly in terms of P content.

Again, differences in elemental content in biomolecules will only matter at the whole-organism level if those biomolecules are significant contributors to total mass. To what extent, then, do nucleic acids contribute to the overall biomass of an organism? First consider DNA. Genome size (quantity of DNA per haploid genome) varies over a 10^5-fold range among major groups of organisms (Cavalier-Smith 1985; Alberts et al. 1994). However, genome size is not the same as nucleic acid content because it does not take total cellular mass into account. When we take that into consideration, we find that DNA content is remarkably constant as a function of cellular or organismal biomass across a wide array of organisms (Holm-Hansen 1969; Cavalier-Smith 1985). This constancy is well illustrated by the unicellular ciliates, the largest and most complex unicellular organisms known. Cell size and biomass vary by a factor of 10^5 from the smallest to the largest ciliate taxa. This huge variation would lead one to expect that, since each ciliate contains a single set of chromosomes and thus a relatively fixed quantity of DNA, DNA per cellular biomass should vary widely with cell size in ciliates. This is not the case. In fact, as cell size increases in ciliates the number of micronucleii increases (micronucleii contain copies of the nuclear genome and participate in normal cell function but do not initiate cell division). As a result, the quantity of DNA per cell biomass varies less than twofold across ciliate taxa (Cavalier-Smith 1985). The cellular underpinning of this constancy appears to be the necessity of the cell to maintain a relatively constant ratio of cell volume to nuclear material to maintain required rates of export of mRNA to the cytoplasm in directing cell function. In short, DNA content is a strongly conserved parameter in organisms.

Further, values from the literature indicate that DNA levels in many organisms are too low to appreciably affect whole-organism elemental content. For example, Maaløe and Kjeldgaard (1966) determined that DNA in *E. coli* contributes only 3–4% of cell biomass. DNA content of eucaryotes is even lower. In zooplankton, DNA contributes little more than 1.5% to animal biomass (Båmstedt 1986). From this, we calculate that DNA-P amounts to only ~0.13%, a small contribution. Thus, DNA can be treated as a relatively small and constant background contributor to organism C:N:P stoichiometry.

The same, however, is not true for RNA. Unlike conservative DNA, RNA levels vary considerably both among different species of organisms (Munro 1969) and within taxa as a function of development, ontogeny, and physiological condition (Dagg and Littlepage 1972; Båmstedt 1986). The same study mentioned above (Maaløe and Kjeldgaard 1966) found that RNA contributed 12–30% to *E. coli* biomass. RNA also represents up to 15% or more of biomass in some Metazoa (Sutcliffe 1970; Elser et al. 1996; Dobberfuhl 1999) and sometimes exceeds 40% of microbial biomass (Maaløe and Kjeldgaard 1966; Sutcliffe 1970).

Because of this high P content and large contributions to organism mass, the C:P and N:P ratios of biomass should reflect differences in naturally occurring concentrations of nucleic acids. A level of 15% RNA in biomass translates to ~1.3% P, a very significant pool compared to known P content values in animals (this calculation is examined in much greater detail in Chapter 4). Thus, it appears that RNA may be a promising place to seek answers when addressing variation in organismal C:N:P ratios, and especially variation in P content. For a generalized illustration, consider a mixture of 70% protein and 30% nucleic acid (structure A) versus another (structure B) composed of 30% protein and 70% nucleic acid. Applying the known N and P content of proteins and nucleic acids, structures A and B are estimated to have N:P of 13.7 and 5.5, respectively (a 2.5-fold difference). How much of that range reflects the slightly lower N content of nucleic acids relative to proteins? Very little. If we repeat the calculation but with the N content of proteins and nucleic acids fixed at 17%, the N:P values of these two structures are now 14.3 (structure A′) and 6.1 (structure B′), a 2.3-fold difference. These calculations provide us with a biochemical basis to expect that variation in an organism's nucleic acid content may have significant effects on its N:P and that this effect will be driven primarily by changes in P content.

Inorganic Materials

Up to now, we have emphasized the elemental composition of the "soft parts" of living systems. However, many organisms have hard structural parts that are critical components of their evolutionary strategy and these may appreciably alter organism stochiometry (Reiners 1986). These minerals can comprise a significant proportion of total organism biomass and may in fact dominate the whole-organism pool of some elements. Here we describe the elemental composition and biological role of some important biominerals in animals. The most familiar examples are the calcium carbonate–based shells of mollusks and other organisms, the siliceous frustules of diatoms, and the hydroxyapatite bones of vertebrates.

The molluscan shell includes an outer covering, the periostracum, which is composed of organic chemicals. Below this are several inner layers of calcium carbonate deposited as either calcite or aragonite (both with chemical formula $CaCO_3$) within an organic matrix. The unicellular Foraminifera also possess organic or $CaCO_3$ tests, as do the coccolithophorids (think chalk!). The biogenic activity linked to these biominerals is a key factor in the cycling of calcium and carbon, even at the global scale. For example, these two taxa are key players removing CO_2 from the atmosphere and depositing it in the deep ocean (Chapter 8; Legendre and Le Fèvre 1989). Stoichiometric studies incorporating both the hard and soft parts of key ecological interactors have appeared; for example, Arnott and Vanni (1996) examined the ecosystem role of the zebra mussel in this way.

The siliceous frustule of diatoms can be a large component of biomass (a typical Si:N in diatoms is ~1; Sommer 1986). The frustule is made up of opal, or amorphous hydrated silica ($SiO_2 \cdot nH_2O$). Although some other organisms have smaller Si requirements, this large Si requirement makes diatoms stoichiometrically unique and links them strongly with biogeochemical Si cycles.

Bone is another major mineralized biological material. On a dry mass basis, bone consists of roughly 60–70% of hydroxyapatite [$Ca_{10}(PO_4)_6$ $(OH)_2$] deposited in a collagen-based organic matrix (Simkiss and Wilbur 1989). Bone not only makes up the vertebrate skeleton, it is also involved in "bony materials" such as teeth and antlers. Bone has such a unique elemental signature, and its contribution to biomass can be so large, that consideration of the stoichiometry of bone will reenter our analyses several times in the book. We will explore the connections between bone investment and organismal stoichiometry when we discuss the relationships between metazoan body size and elemental composition in Chapter 4. We will relate those patterns to growth dynamics in Chapter 5 and we will touch on the role of vertebrates in biogeochemical cycling in Chapter 6.

CELL COMPONENTS: THE ELEMENTAL COMPOSITION OF CELLULAR STRUCTURES

Now that we have reviewed the functions and elemental composition of major biomolecules, the next step toward understanding total organism nutrient content will be to focus on subcellular structures. A major question here is, "Do cellular structures associated with contrasting biological functions have distinct elemental compositions?" If they do we may be able to connect aspects of cell biology with organismal C:N:P stoichiometry and thus continue to establish the stoichiometric requirements of major biological functions. To assess the elemental composition of subcellular structures we use two approaches. Our primary strategy is to calculate composition from the known atomic structure of the specific biochemicals making up various subcellular structures. However, in some cases (nucleus, mitochondria), we are able to obtain values derived from direct elemental analysis of structures obtained from living cells via cell fracture or other techniques.

Cell Walls

Bacteria and plants produce cell walls that may be an important proportion of organismal biomass; thus we must consider their elemental composition. For plants, this starts off as a simple calculation as the elemental composition of the primary constituents of plant cell walls (cellulose, hemicellulose, pectins) are of uniform C content and are free of both N and P. However, plant cell walls also contain varying amounts of lignins (as we discussed above, these are also molecules with no N or P but with ~60% C compared to 35% C for cellulose). Thus, we need to know the relative contributions of cellulose and its derivatives versus lignin in plant cell walls. Lignin seldom if ever exceeds 10% of cell wall mass (Goodwin and Mercer 1972) even if lignin content at the whole-tissue scale can exceed 10% (lignin is found elsewhere in plant tissues other than just in the cell wall). Thus, based only on variation in lignin, the C content of plant cell walls ranges only from 35% (0% lignin) to 38% C (10% lignin). This range of values agrees reasonably well with direct measurements of the C content of whole leaves (35–46% C, Rooney 1994), as other plant constituents (including lipids and proteins) would have higher % C than carbohydrates. Plant cell walls can also contain small quantities of protein (less than 5% of dry mass), which may be enzymes bound within the cell wall matrix or small quantities of structural proteins (Goodwin and Mercer 1972). A protein contribution of ~3% would be sufficient to raise the N content of plant cell wall material from 0% N to only 0.5% N (Fig. 2.4). Plant cell walls clearly are nutrient-poor structures and thus contribute toward high C:N and C:P in plant biomass.

Bacterial and fungal cell walls differ considerably from plant cell walls, both in the nature of the basic repeating carbohydrate component (peptidoglycan vs. cellulose) and in the rest of the matrix. The extracellular materials also differ among bacterial groups. In gram-positive bacteria, only a single extracellular layer, comprised mostly of peptidoglycan, is

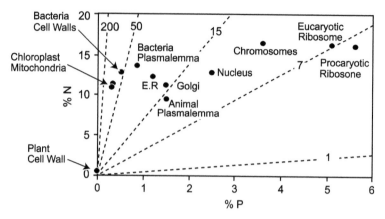

Fig. 2.4. Intracellular stoichiometry. This diagram illustrates % N and % P of major cellular structures. In most cases, elemental composition was calculated on the basis of reported values of biochemical composition coupled with estimates of the elemental composition of those biomolecules (Fig. 2.2). However, in the case of chloroplasts, mitochondria, and nuclei, values reflect direct measurements of organelles isolated via cell fracture techniques.

found, whereas in gram-negative bacteria a thick layer composed of lipid, polysaccharide, and protein overlays a thinner peptidoglycan layer. Estimation of the elemental composition of cell walls of gram-positive bacteria is thus relatively easy. We assume that peptidoglycan polymers are cross-linked with pentapeptide cross links (Lehninger et al. 1993), that this complex represents 90% of the cell wall in gram-positive bacteria (Brock and Madigan 1991), and that the remainder consists of teichoic acids (phosphorus-containing "polyols"; 33% C, 4.5% N, 5% P; Brock and Madigan 1991). Under these assumptions we estimate the elemental composition of cell wall material from gram-positive bacteria to be 38% C, 13.2% N, and 0.5% P (C:N:P = 193:58:1; Fig. 2.4).

Thus, in contrast to plants, bacterial cell walls are relatively nutrient rich. Their nutrient-rich composition may be a factor responsible for bacterial biomass being relatively nutrient intensive. We have already noted above that polysaccharides generally contribute less than 10% to *Salmonella* (a gram-positive bacterium) biomass. Since polysaccharides comprise ~50% of the peptide-fortified peptidoglycan layer of gram-positive bacteria, we might expect (assuming that all of the polysaccharide present in the cell is peptidoglycan) the cell wall to comprise as much as 20% of bacterial biomass. This is reasonably substantial and thus should be accounted for in establishing the core stoichiometry of bacterial biomass.

Cell Membranes

Various cellular organelles (mitochondria, chloroplasts, and the Golgi apparatus among them) have membranes as important components of their structures. In this section we consider the plasmalemma. In both eucaryotes and procaryotes these membranes serve as selectively permeable boundaries between the cell and the extracellular environment. The primary structure of cell membranes consists of a phospholipid bilayer and associated proteins (with a smaller carbohydrate component); the specific balance of these constituents is modified according to the nature and types of biochemical processing occurring across that membrane (Evans 1989). Consideration of a variety of biological membranes characterized in the literature indicates that the biochemical composition of cell membranes is somewhat variable: 25–56% lipids, 25–62% proteins, and less than 10% carbohydrates. Thus, membrane elemental composition should also be variable. Biochemical data for bacterial membrane indicates a high protein content that results in an elemental composition of 55% C, 13.7% N, and 0.86% P (C:N:P = 162:35:1; Fig. 2.4) while animal cell membrane appears to have a greater proportion of phospholipid relative to protein and thus a higher C and P content but a lower N content (59% C, 9.5% N, 1.5% P, C:N:P = 100:15:1; Fig. 2.4). This balance has links to biological function (e.g., the biophysical and transport functions of the membrane) and thus even within just membranes we see a link between elemental composition and biological function due to the elemental composition of major biomolecules.

Cytoplasm

The interior volume of the cell not occupied by organelles is referred to as the cytoplasm, which in animal cells can occupy as much as 50% of cell volume. The cytoplasm has two components: an aqueous component known as the cytosol, which includes solutes, and a proteinaceous internal framework known as the cytoskeleton. The latter, however, is not found in procaryotic cells. Besides the inorganic solutes involved in buffering and ionic regulation, the cytosol also contains a huge diversity of metabolic intermediates as well as the enzymes involved in intermediary metabolism. As a result, the protein content of the cytosol can exceed 20% (Lehninger et al. 1993). The cellular cytoskeleton is involved in moving and securing organelles within the cell and in determining the overall shape of the cell itself. Since it is likely that the importance of the cytoskeleton increases in larger, more differentiated cells compared to small, undifferentiated cells, it is possible that differentiated cells with higher cytoskeletal contributions have higher N content and higher N:P than undifferentiated cells lacking such structures.

Vacuoles

One of the characteristics distinguishing plant from animal cells is the prominence of vacuoles (this contrast will emerge as a major theme in the next chapter). In animal cells, vacuoles are small and function in sequestration and transport of waste materials or secretory products to the outside of the cell, while in plant cells the interior is often overwhelmingly dominated by a large central vacuole. In animal cells, lysosomes have a role in concentration of mineral elements (Berry 1996; Matsuzaki et al. 1997), but lysosomes do not occupy as prominent a fraction of cell volume as do plant vacuoles. This discussion will focus on plant central vacuoles.

Plant vacuoles have a twofold function. In terrestrial plants the vacuole is a major site for storage of water, providing cellular turgor pressure and permitting plant cells to grow to a large size without investment of potentially limiting energy or nutrients (Wiebe 1978). In such situations, the vacuole fills the great majority of the interior of the cell with only thin strands of cytoplasm crossing the cellular space and major organelles (chloroplasts, mitochondria) clinging to the periphery of the cell. Thus, in terrestrial plants an important function of the vacuole is to impart structural rigidity, which does not need to involve nutrient elements. The second function of the vacuole is storage (Alberts et al. 1994). A variety of materials is sequestered here, including simple sugars, amino acids, pigments, proteins (in seeds), and inorganic solutes (such as NO_3^-, PO_4^{3-}). Many of these would be harmful if present in the cytoplasm in high concentrations. The vacuolar concentrations of these substances are highly variable as a function of nutrient supply and growth conditions (Chapter 3). The sizable contribution of the vacuole to plant cellular volume coupled with its potentially great variation in biochemical and elemental composition provide a cellular basis to expect that plants will have highly variable elemental composition compared to animals.

Golgi and Endoplasmic Reticulum

The Golgi apparatus and endoplasmic reticulum (ER) are highly membranous and are likely to have elemental composition similar to the plasmalemma. The ER is a network of membranes that comprises ~15% of cell volume (Mieyal and Blumer 1981). The ER constitutes the intracellular cytocavitary network, provides sites for localization of ribosomes (rough ER), and is involved in important biochemical transformations such as detoxification and dehydroxylation reactions. The protein:lipid ratio of ER is high relative to that of other membranes (2.3:1; Becker 1986) and thus ER has high N content and N:P (56.6% C, 12.3% N, 1.2% P, C:N:P = 120:23:1; Fig. 2.4). The Golgi apparatus is intimately connected to the ER.

It mediates the flow of secretory proteins from the ER to the exterior of the cell. The Golgi apparatus has a protein:lipid ratio intermediate between those of the ER and of the plasmalemma (Becker 1986) and thus an elemental composition intermediate between these structures (57.4% C, 11.2% N, 1.5% P, C:N:P = 97:16:1; Fig. 2.4).

Mitochondria

In eucaryotes, mitochondria are the primary organelles of cellular energetics. They have a complex structure comprised of outer and inner membranes and a gel-like inner compartment (the "matrix"). The matrix is protein rich (~50% protein by mass) and also contains a small quantity of DNA (mDNA; ~15,000 base pairs in human mitochondria) as well as ribosomes (Becker 1986), remnants of the mitochondrion's endosymbiotic origin. The biochemical composition of the outer mitochondrial membrane is similar to that of the outer cell membrane but the inner membrane is extremely protein rich (80% protein, 20% lipid in liver mitochondria; Lehninger et al. 1993), a reflection of the large quantity of the enzymes of respiratory metabolism. The relative contribution of the inner membrane to the whole mitochondrion is a function of the degree of inner membrane folding, which itself is directly linked to the ATP generation capacity of the mitochondria and to the intensity of cellular respiration (Lehninger et al. 1993). Since the inner membrane is protein rich and thus has high N content and high N:P, we suggest that enhanced respiratory capacity is a biological feature that requires material investment of high N:P.

The structural and biochemical complexity of mitochondria complicates the direct calculation of mitochondrial elemental composition. Fortunately, direct measures are available based on cell fracture studies (Bowen 1979). These data support the idea that the respiratory process requires material investment of high N:P (Fig. 2.4): mammalian mitochondria were found to have elemental composition of 11% N and 0.31% P (no C data were reported) with N:P of 78.

Chloroplasts

Chloroplasts are complex structures responsible for photosynthetic energy capture and transformation in eucaryotic autotrophs. Chloroplast structure and function resemble those of mitochondria: namely, creation of proton gradients across inner membranes generates energy transformations during both photosynthetic (chloroplasts) and respiratory (mitochondria) metabolism. The chloroplast consists of a highly permeable outer membrane, a less permeable inner membrane in which specific transport proteins are imbedded, and an inner space called the stroma containing a third membrane system (thylakoid membrane) that is folded into thick stacks (grana) containing the photosynthetic pigments and other photosystem components.

Biochemical data for whole chloroplasts have been reported, thus permitting direct assessment of chloroplast C:N:P stoichiometry. Kirk and Tilney-Bassett (1978) summarized measurements of major biochemical pools (proteins, lipids, nucleic acids, and carbohydrates) for chloroplasts of spinach. We can use these values to estimate whole-chloroplast elemental composition. Doing so indicates that, consistent with the calculations earlier in this chapter regarding the N contribution of the ubiquitous enzyme RUBISCO, the elemental composition of the chloroplast is relatively rich in N and low in P (47.5% C, 11.3% N, 0.32% P, C:N:P = 377:80:1; Fig. 2.4). The similarity of chloroplast and mitochondria N:P (80 vs. 78, respectively) seems remarkable given the inherent differences in the biochemical pathways they house. Plant tissues with extensive allocation to chloroplasts compared to other cellular constituents must have a high N:P.

Chromosomes and Nucleus

In eucaryotes, genetic material consists of more than simply DNA and is organized into chromosomes via the complexation of DNA with histone and nonhistone proteins. This complex is known as "chromatin" and it has a protein:DNA ratio of 1:2 by mass (Lehninger et al. 1993). If we assume a protein:DNA ratio for chromatin of 1.5 the elemental composition of chromosome becomes 40.8% C, 16.5% N, and 3.6% P (C:N:P = 29:10:1; Fig. 2.4). Note that the storage of genetic information is associated with a very low N:P. In procaryotes the ratio is even more extreme; there, the DNA strand is naked and thus has N:P of 3.7, the N:P of nucleic acids. Eucaryotic chromosomes of course are isolated in the cell within the nucleus, which is comprised of the nuclear membrane itself, a quantity of externally situated ribosomes, an internal proteinaceous matrix, and various enzymes involved in nucleotide processing. As we saw for mitochondria, this complexity makes it difficult to calculate directly the elemental composition of the nucleus, the largest single organelle found in eucaryotic cells. Once again, however, we can turn to cell fracture studies (Bowen 1979). These data are for mammalian nuclei and support the idea that storage of genetic information is associated with low N:P material demands (12.8% N, 2.3% P, N:P = 11; Fig. 2.4), a reflection of the high phosphorus content of DNA (Fig. 2.2). In both procaryotes and eucaryotes and no matter whether one considers just the genetic material or the whole nucleus, storage of genetic information requires P-rich, low-N:P material investment.

Ribosomes

Genetic information is acted upon via communication between genes on the chromosomes and ribosomes in the cytoplasm. Ribosomes are the organelles of protein synthesis, where enzymes that catalyze various cellular activities as well as structural proteins that contribute to cellular apparatus are formed. Ribosomes therefore are associated with synthetic activity and biomass growth. Happily for us, the structure of ribosomes has been thoroughly documented. The eucaryotic ribosome consists of two parts called the 40S and 60S subunits. Within each, 82 unique ribosomal proteins and four ribosomal RNAs of differing size are found: one 18S rRNA molecule is incorporated into the 40S subunit while one molecule each of 5.8S, 5S, and 28S rRNA is found in the 60S subunit. Eucaryotic ribosomes have an RNA:protein ratio of 1.2 while procaryotic ribosomes are more RNA rich (RNA:protein ratio of 1.8; Campana and Schwartz 1981). These biochemical constituents indicate a particularly P-rich and low-N:P elemental composition (41.8% C, 16.3% N, 5.0% P, C:N:P = 21:7.2:1 for eucaryotes; 40% C, 16.1% N, 5.6% P, C:N:P = 18:6:1 for procaryotes; Fig. 2.4). In fact, ribosomes are the most P rich and lowest-N:P organelles in cells. The major stoichiometric implication (to which we will continue to refer throughout this book) is that biosynthetic functions of cells require investment in low-N:P materials. To the extent that cells and organisms differ in their biosynthetic capacities and thus in the complement of ribosomes they contain, we would expect to see differences in elemental composition, especially in P content, C:P, and N:P.

Observations of Intracellular Elements

The stoichiometric ratios we have presented so far have largely been obtained by calculations from known characteristic biochemical composition. In addition, several cell fractionation measurements have been used. Although these methods should be reliable for structures of highly characteristic composition (e.g., the RNA molecule) or which are readily purified (e.g., mitochondria), they are less useful for structures where either the biochemical composition is not easily characterisized (e.g., cytoplasm) or clean fractionations are not easily obtained (e.g., nucleolus).

An alternative method for estimating intracellular stoichiometry uses quantitative image analysis coupled to electron microscopy. These imaging techniques take advantage of the differences in interactions of different elements with energized electrons in the microscope to produce very high-resolution maps of C, N, and P atoms, allowing for observations of these and other high-P structures within the cell (Hendzel and Bazett-Jones

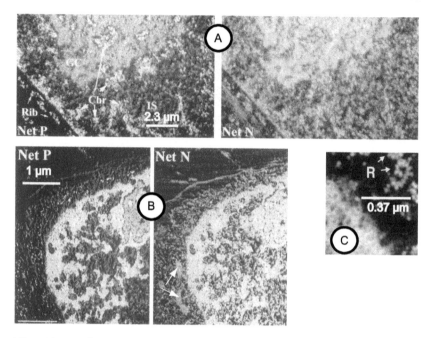

Fig. 2.5. Visualization of intracellular N and P using electron spectroscopic imaging (ESI). A. Phosphorus (left) and nitrogen (right) maps of thin sections of Indian muntjac fibroblasts, showing nuclear (upper right) and cytoplasmic (lower left) regions. Rib: ribosomes arranged along the nuclear envelope; Chr: chromatin; and IS: interchromatin space. B. Thin section of mouse 10T½ cell showing nucleus and cytoplasm. Note the mitochondria (white arrows) that are visible in the N image (right) but not the P image (left). C. Phosphorus image showing ribosomes (R) from a 10T½ fibroblast cell. Images A and B are reprinted from portions of micrographs in Hendzel et al. (1999) and image C is reprinted from a portion of a micrograph in Bazett-Jones et al. (1999).

1996; Olins et al. 1996; Wasser et al. 1996; Beniac et al. 1997; Quintana et al. 1998). One such approach is electron spectroscopic imaging (ESI), which couples a conventional transmission electron microscope with an analytical imaging spectrophotometer (Hendzel and Bazett-Jones 1996; Hendzel et al. 1999). Use of ESI to estimate patterns of intracellular element concentration in the cell involves a comparison of images collected with different ionization energies. Because different elements ionize at different energy levels, concentrations of individual elements in thin sections of cells can be measured, effectively mapping their location in the cell.

Figure 2.5 presents several images using ESI. Comparison of phosphorus maps (left) with nitrogen maps (right) in panels A and B shows,

TABLE 2.3

Ranking of intracellular N:P ratios as determined by electron spectroscopic imaging (ESI). Structures are arranged from top to bottom as high to low N:P (Hendzel et al. 1999)

Structure Type
Mitochondrion
Ribosome-depleted cytoplasm
Nuclear pores
Ribonucleoprotein-depleted nuclear matrix
Interchromatin granule cluster
Nucleolus (granular component)
Chromatin

first, that N is more homogeneously distributed in the cell than is P, as we would expect from our biochemical calculations (Figs. 2.2 and 2.4). The N images have noticeably less contrast than do the P images. Second, one can see that structures that have lower N:P ratios (e.g., chromatin or ribosomes, panel A, left, labeled) are more easily visible in the P maps than in the N maps. Third, structures with high N:P ratio (e.g., mitochondria, panel B, right, arrows) are more apparent in the N maps than the P maps. At high magnification (panel C), individual ribosomes (R) are easily viewed in intracellular P maps. Quantitative analysis of similar images then can be used to estimate the N:P ratios of different cellular structures (Table 2.3). Here, we are presenting structures in terms of their relative position on an N:P gradient. Note from this table that structures with high nucleic acid content, such as chromatin, have low N:P ratio while structures high in protein, such as mitochondria, have high N:P ratio. The relative order of these structures obtained by ESI provides independent verification of the trends in N:P ratios in cellular components that we have discussed.

The stoichiometry of ribosomes as well as chromatin in the nucleus has been directly measured using ESI (Bazett-Jones et al. 1999). In this study, the C:N:P of ribosomes was found to be 4.97:1.68:1 and of chromatin was found to be 2.84:1.40:1. The high-P, low-N:P composition of these nucleic-acid-rich structures is obvious. Extremely small structures, down to single DNA molecules, were able to be visualized!

Advances in techniques like these will be a great boon to ecological stoichiometry. To date, these techniques have been used primarily to ask questions about ultrastructure. It would be very interesting to use these methods to explore the dynamics of intracellular structures of contrasting

N:P under different resource regimes, for instance, or to examine the distribution of P in the cells of organisms selected for rapid or slow growth.

Statistical-Empirical Patterns

From this chapter's opening quote through this extensive consideration of biochemical and cellular structures, we have seen associations between particular elements and certain biological functions. For example C, Ca, and Si were associated primarily with structure and P was associated with biosynthesis. Later, when we consider multicellular organisms, we will see even more examples concerning the relation of structural materials (bone, wood, chitin) to stoichiometric formulas.

An entirely different approach to the same set of issues is to consider statistical-empirical associations among chemical elements in biomass across taxonomic categories. If particular sets of elements are associated with particular functions, species emphasizing those functions should have high levels of those elements, and there should be a signal in the multivariate patterns of element distributions across taxa. Garten (1978) performed a statistical analysis on the content of multiple elements (Al, B, Ca, Cu, Cs, Fe, K, N, Na, P, Mg, Mo, S, and Sr) of 100 plant species. Certain subsets of these elements had good statistical associations. Interestingly, the first principal component in this analysis was associated with the elements P, N, Cu, S, and Fe, elements that play a role in the construction and processing of nucleic acids and proteins (Tables 2.1 and 2.2; Fraústo da Silva and Williams 1991). Similarly, Krivtsov et al. (1999) used X-ray microanalysis to measure the content of eight elements (Ca, Cl, K, Mg, Na, P, S, and Si) in the cyanobacterium *Anabaena* during a bloom. They found several associations: a grouping of P, Mg, and K, a second grouping of Na and S, and a third grouping of Ca, Cl, and Si.

These studies indicate multivariate elemental signatures of differing biological functions. They show that changes in concentration in one element are often associated with changes in other elements that also participate in the same biochemical functioning. Correlations such as these indicate stoichiometric constraints: element concentrations in living biomass are correlated.

Functioning Mixtures of Molecules and Organelles

By examining the elemental composition of major biomolecules and cellular organelles we have begun to assemble chemical stoichiometric patterns from the bottom up and to associate them with biological functions such as growth, structure, and storage. From atoms to molecules to organelles (and, as we will see, to cells), as the system of interest has become more complex there have been new constraints put on elemental composition.

Perhaps you have noted that our method of analysis does not consider the turnover rate of different biochemical pools. We did not take account, for example, of the enormous differences between turnover rate of P within ATP compared to bone. Those turnover rates differ by many orders of magnitude (minuscule fractions of a second for ATP vs. months to years for bone). This was not an oversight. Ecological stoichiometry relates organism composition to its external environment, and hence it is the rates of demand and exchange of the whole living organism that we must focus on. The extraordinarily high intracellular rate of turnover of ATP-P does not, by itself, increase P demand of the entire organism. For the questions we pursue, turnover within the organism becomes an unimportant issue, except for how it might influence the gain or loss from the whole organism (such as by changed leakiness through membranes). This simplification is one reason why ecological stoichiometry can potentially do so much with so little information.

Outside of living systems, a single chemical element may exist in its "native" elemental state. However, biomolecules have more restricted stoichiometry than individual elements; for example most contain C, H, and O. Organelles require multiple functions and hence multiple biomolecules. They must combine biomolecules in such a way that they are more stoichiometrically restricted than individual molecules. Further, when all the functions of a single living thing are found in one place, the complex mixture must have even narrower stoichiometric bounds. The properties of C, H, and O, for example, are sufficient for imparting structure to plant cells, but other elements must be added to allow for other functions. The upper range of P content illustrates nicely this progression with complexity. The highest P content in major biomolecules is greater than 15% (in ATP; Fig. 2.2). In organelles, however, the maximum P content is between 5 and 6% (in ribosomes). Later, we will see complilations of P content for individual species, but for now it suffices to say that they are in nearly all cases lower still. Somewhere between the level of individual biomolecules (Fig. 2.2) and the level of intact, functioning, multicellular organisms (Fig. 1.2), numerous stoichiometric constraints, resulting from the fundamental chemistry of the mixed functions of life, are imposed.

Living systems bring elements and their individual properties together to generate mixtures having particular biological function. Mixtures must have more restricted stoichiometries than the individual components making them up. All major biological functions must be present in any one living cell. All cells must combine the individual stoichiometries of molecules (Fig. 2.2) and organelles (Fig. 2.4) into functioning complex mixtures. The cell must metabolize, grow, retain shape, and react to its environment. It must have ribosomes; it must have an outer covering; it must encode genetic information. Although cells emphasize these different strategies to different degrees, they must all be present. The stoichiometry of free-living unicellular microorganisms ("protoplasmic life," close to the Redfield ratio; Chapter 1) shows us the bounds to the stoichiometry of growth and metabolism in living cells. We saw several examples of this in the last chapter (Figs. 1.5, 1.6, and 1.10), and unicellular autotrophs will be covered in detail in the next chapter.

SUMMARY AND SYNTHESIS

This chapter went from atoms to biochemicals to organelles within cells, showing that the elemental composition of biochemicals is directly linked to their biological function. First, we briefly reviewed the chemical properties of some of the elements in the periodic table, emphasizing in particular the specific properties of C, N, and P that relate to their biological functions. We also emphasized the fact that the elemental composition of biological material does not represent a random sample of the periodic table nor of the chemical elements available on Earth. We saw how the physical chemistry of C, N, and P dictates the nature of the primary chemical effects of which those elements are capable.

We then surveyed the elemental composition of major classes of biological materials likely to contribute substantively to organismal elemental composition. We found that different classes of biochemicals differ considerably in elemental composition. Carbohydrates and lipids are both poor in N and P but differ somewhat in terms of C content: carbohydrates are relatively C poor (32% C) while lipids are the most C rich of the major biochemicals (65–70% C). We saw that two biologically critical molecules, proteins and nucleic acids, differ little in terms of C and N content. However, the P content of major molecules is extremely variable and thus differences in C:P and N:P in organisms are likely to be generated primarily by variation in P content rather than by differences in C or N content. Particularly striking among these differences is the high P content of nucleic acids, a fact that suggests that variable nucleic acid content of cellular structures and organisms should generate considerable variation in structural and organismal C:P and N:P ratios.

We next evaluated the elemental composition of the subcellular structures of organisms. We saw that major cellular structures contrast strongly in C:N:P stoichiometry. Plant cell walls are nutrient poor and carbon rich. The N content of many biological structures is relatively uniform, consistent with the fact that proteins and nucleic acids have similar N content. We saw that organelles involved in cellular energetics (chloroplasts and mitochondria) are N-rich but low-P structures due to the predominance of enzymes and structural proteins involved in energy processing. However, organelles involved in storage and execution of genetic information (nucleus, chromosomes, and ribosomes), while similarly N rich, are also high in P due to the central role of P-rich nucleic acids. We did not emphasize nutrient storage in this chapter, but will cover it in more detail soon. In sum, we see that basic biological functions (mechanical support, energetic metabolism, storage and processing of genetic information via protein synthesis) are associated with contrasting biochemical and elemental demands.

The end result of the efforts in this chapter is that we now have a foundation on which to build a framework linking the characteristics of an organism (such as its size, mode of nutrition, major life history characteristics, and/or ecological function) with its elemental composition. This may be possible because major contrasts in the lifestyles of particular taxa are inextricably linked to the mix of cellular and biochemical machines involved in pursuing those lifestyles. In Chapters 3 and 4 we will use some of the basic tools developed in this chapter to connect important biochemical and cellular features to organismal C:N:P stoichiometry.

3

The Stoichiometry of Autotroph Growth:
Variation at the Base of Food Webs

Receiving a new truth is like adding a new sense. —*Justus Liebig (1803–1873)*

Photoautotrophs are the globally dominant interface between nonliving and living systems, where elements are combined to form living biomass. The elemental stoichiometry at the base of food webs is established when primary producers use light to fix carbon dioxide and simultaneously assimilate inorganic nutrients, thus creating biological systems with a biochemical and elemental mixture related to structure and function as we discussed in the previous chapter.

Autotroph elemental content is determined by the net difference between uptake and losses due to such processes as respiration, exudation, and (in higher plants) leaf and root excision. A key concept in this chapter is that the acquisition and losses of different elements (C, N, P, and others) are not perfectly coupled to each other in autotrophs. In other words, autotrophs need not exhibit balanced growth. As we saw in the first chapter, they are not constrained to be strictly homeostatic. Their uncoupling of carbon fixation and nutrient acquisition, along with their ability to reorganize cellular and tissue allocation patterns in response to environmental conditions, allows them to exploit variable resource supplies in what seem to be highly adaptive ways (Berman-Frank and Dubinsky 1999). The net result of decoupling of individual uptake rates is large variation in autotroph C:N:P stoichiometry. Interspecific variation in autotroph stoichiometry is also important, particularly in terrestrial plants where large stoichiometric differences among groups such as forbs, shrubs and evergreens play a large role. We will deal more with interspecific patterns in Chapters 7 and 8, although a major influence of autotroph size and growth form will be covered here. As we will see in later chapters, this stoichiometric variation has ramifications for herbivore feeding and growth, the dynamics of food webs, consumer-driven nutrient recycling, detrital breakdown, and long-term changes in the global carbon cycle and Earth's climate. It holds the key to much of interest in studying food webs and biogeochemical cycling.

It is entirely appropriate that we draw inspiration in the opening page of this chapter from Justus Liebig. Liebig was a skilled chemist and, turning to what he regarded as the chief chemical challenges of his day, namely, living plants and animals (Holmes 1964), he made many seminal contributions. Liebig is widely remembered for proposing the "law of the minimum," which is all about stoichiometry. It says that organisms will become limited by whatever resource is in lowest supply compared to their needs. Trained as a chemist, when he approached biological problems, he already would have had stoichiometric yield and limiting reagents in his mind. Liebig also discovered a means to measure C and H from the combustion products of CO_2 and H_2O (those of us with CHN analyzers can thank him). He was a strong proponent of the central importance of chemistry to both agriculture and animal physiology. With a nod to Liebig, let us see what "new truths" this chapter has in store.

In this chapter, we will focus on the processes affecting the elemental composition of autotroph biomass. We will briefly touch on some basic cellular mechanisms and then we will provide a synthesis of laboratory and greenhouse data, exploring the factors that determine autotroph C:N:P stoichiometry. We will also review some of the main theoretical approaches regarding autotroph growth rate and nutrient content. Finally, with insights from the lab, greenhouse, and microcomputer in mind, we will examine field data on autotroph C:N:P and describe a set of hypotheses that may help explain C:N:P variation under the contrasting light and nutrient conditions that characterize a variety of natural ecosystems.

CELLULAR AND PHYSIOLOGICAL BASES

The nutritional physiology of photoautotrophs ("autotrophs" hereafter) is a large field (for vascular plants, see Marschner 1995; Aerts and Chapin 2000) that has grown enormously since Liebig first articulated his law of the minimum. To us, this great mass of work provides a firm foundation for understanding autotroph-based food webs in nature. However, there are advances that still need to be made. As pointed out by Chapin (1980), many patterns that are well established in applied plant studies (agronomic research) have not always been widely applied to wild plants. Chapin emphasized the need for a greater flow of information from agronomy to terrestrial plant ecology. A similar information gap may exist for aquatic systems, where information from aquaculture and algal physiology moves somewhat slowly into limnology and oceanography. However, we believe there is an even more significant sociological gulf in this subject: namely, the unrecognized similarities and differences in the nutritional physiology of lower plants (e.g., algae, mainly aquatic) and higher plants (vascular, mainly terrestrial). Much to our surprise, the nutritional physiologies of these two groups have rarely been discussed together. In reviewing these two sets of literature under this single roof, we have been struck by the fundamental similarities of many of the processes that are, for no good reason that we can see, always separately discussed.

All biology students are exposed to the structures of archetypal eucaryotic plant and animal cells. What they generally do not hear about are the distinct stoichiometric patterns in these two groups. As we discussed in Chapter 2, the anatomical similarities are obvious: each has a nucleus, some mitochondria, various membranous organelles, and many ribosomes. However, the contrasts are also striking, such as the presence of chloroplasts in the plant cell. Two other important aspects of cell structure absent in animals but present in archetypal autotrophs are cell walls and large central vacuoles. In this regard, cell anatomy has a distinct stoichiometric imprint. The cellular basis of contrasts in the biological stoichiometry of autotrophs versus heterotrophs relates both to support and storage. First, cell walls add C but no P and little N to cellular construction (Fig. 2.4). From this fact alone, it is easy to see why autotroph cells have higher C:nutrient ratios than animal cells. Second, the large central vacuole creates the potential for strongly varying elemental composition due to storage of organic compounds and inorganic nutrients not immediately used in metabolism and structure. Throughout the rest of this book we will see how these two contrasts have profound ecological consequences.

In the next chapter we will establish the stoichiometric connections between biochemical investment and growth rate in animals. This is possible because in animal cells it is not generally necessary to account for large and variable cellular pools of storage nutrients not immediately linked to cellular function. In autotrophs, nutrient elements such as N (e.g., nitrate, amino acids, or other organic-N compounds) and P (e.g., phytic acid, inorganic phosphate, or polyphosphate) as well as nutrient-free organic compounds (e.g., sugars) can be present in very high concentrations in the vacuole or in the cytoplasm.

"Luxury consumption" refers to increases in organismal nutrients over and above what is immediately required for growth. Luxury consumption of one nutrient occurs when a different limiting factor (such as light or another nutrient) begins to limit growth. At this point, the uptake of non-limiting nutrients may continue and thus growth rate and net nutrient uptake are not equal. Luxury consumption results in unbalanced growth and variable chemical composition. Accumulation of nutrients from luxury consumption sometimes occurs in the form of special storage compounds. For example, if exposed to phosphate levels above immediate growth requirements set by other resources, algae (as well as bacteria, fungi, and some higher plants) may accumulate P as polyphosphate (Rhee 1973; Siderius et al. 1996; Jaeger et al. 1997), a salt or ester of polyphosphoric acid. In many higher plants, P is stored in seeds and grains in the form of phytate, a phosphorylated alcohol (with a C:P of one!). Through P storage autotrophs can have low C:P and N:P when they have access to abundant P under conditions of (for example) N-limited growth. Autotrophs similarly acquire N in excess of immediate needs under conditions of P-limited growth (Greenwood 1976; Miyashita and Miyazaki 1992). Cyanobacteria may accumulate N in the form of cyanophycin, a polymer of the two N-rich amino acids aspartate (C:N = 4) and glutamate (C:N = 5). Polyphosphate and cyanophycin typically accumulate in granuoles in the cytoplasm, not in a central vacuole. However, the central vacuole of plant cells plays a very large role in storage of other inorganic and organic compounds.

For eucaryotic autotrophs the large central vacuole, with its wildly variable chemical composition (Leigh and Wyn-Jones 1985), complicates the direct connection between nutrient content and growth. The key role of the vacuole in the variability of the stoichiometric patterns in plants is apparent when one considers that the concentrations of free inorganic and organic constituents are tightly regulated in the cytoplasm (Bieleski 1973; Leigh and Wyn-Jones 1985), just as they are in animal cells. Thus, the vacuole can be a major factor in decoupling nutrient dynamics from biomass growth as autotrophs exhibit wide intraspecific variation in elemental composition in response to immediate environmental conditions. As we will examine in detail below, the variation in nutrient content with respect to the limiting nutrient will be directly associated with growth rate but variation in terms of nonlimiting nutrients will not. Thus, understanding the coupling of nutrient elemental composition and autotroph growth requires knowledge of the particular factors controlling growth at a given point in time.

The dichotomy in chemical variability between vacuole and cytoplasm led Leigh and Wyn-Jones (1985) to refer to the "selective cytoplasm and the promiscuous vacuole." They proposed a conceptual model describing the cellular basis of changes in the elemental composition of eucaryotic autotroph cells. Consider an example involving potassium, an element occasionally limiting to primary producers in terrestrial systems and one for which relationships between biomass production and tissue percentage have been long established (Fig. 3.1). These authors proposed that the shape of the relationship shown in Figure 3.1A reflects relative changes in the concentrations of a limiting element in the vacuole and cytoplasm. We begin at high nutrient supply and consider what happens as nutrient supply is reduced. As nutrient supply to the plant declines, tissue content declines; however, this decline in whole-tissue content is initially at the expense of nutrients only in the vacuole (Fig. 3.1B, region B). Concentrations in the cytoplasm are maintained at fixed levels as nutrient supply

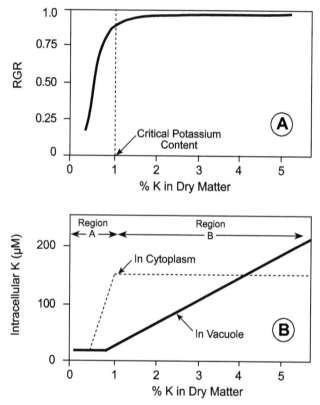

Fig. 3.1. The cellular basis of variation in biomass production as a function of tissue nutrient content in oats. A. Production (as measured by RGR) increases hyperbolically with increasing tissue potassium content. B. During the onset of K limitation (moving from the right to the left), first vacuolar K is depleted, resulting in little change in plant production (region B). However, at a critical K content, vacuolar K is totally depleted and cytoplasmic content then begins to decline (region A). Reduction of cytoplasmic K results in a reduction in production, as K is critically involved as a cofactor in enzyme activation. Based on Leigh and Wyn-Jones (1985).

declines. With further tissue reductions, at a critical tissue nutrient level, vacuolar nutrients reach a minimum and cytoplasmic concentrations then also begin to decrease (Fig. 3.1B; region A). This model relates stoichiometry to growth in that the reduction of cytoplasmic concentration is the direct cause of the growth reduction shown in Figure 3.1A. Similar patterns have been shown for P content of root tips of pea plants (Lee and Ratcliffe 1983). During the onset of P starvation, tissue inorganic P concentration (P_i) in root tips declined by a factor of 4 entirely due to changes in vacuolar P_i with no change in cytoplasmic P_i. Reflecting the constancy of cytoplasmic P_i, growth decreased by only 2%.

Thus, interpretation of changes in tissue elemental composition in organisms with specialized storage locations must be made with two distinct regions in mind. Below a critical level, changes in tissue nutrient content are closely coupled to autotroph growth rate (Region A). Above that critical value, large changes in elemental composition can occur with no change in growth rate (Region B). In most of this chapter we will focus on variation in elemental composition that is linked to autotroph growth rate under conditions of nutrient limitation. However, with respect to the consequences of variation in autotroph stoichiometry (e.g., implications for food quality or nutrient recycling), all variation of C:nutrient is of interest, whether or not the variation is directly coupled to growth or associated with luxury consumption.

The vacuole and other cell storage mechanisms buffer autotroph growth against the vagaries of the environmental supply regime and in so doing they contribute to the variability in the elemental composition of autotroph biomass. The vacuole also provides a potential cellular mechanism for differentiation among autotroph species in the physiological integration of the external nutrient regime. Consider the trade-offs involved. The vacuole increases cell size while providing no capabilities for energy acquisition (chloroplasts), energy processing (mitochondria), or growth (ribosomes). For pelagic autotrophs, large cell size can be disadvantageous. In phytoplankton, increased cell size increases sinking rate (Reynolds 1984), increasing population losses from the water column. Increased cell size also lowers the surface area:volume ratio and thus decreases nutrient acquisition ability. Thus, if we consider unicellular autotrophs, some taxa might evolve a strategy of low allocation of cellular space to storage. The cellular stoichiometry of such plant cells with minimal vacuoles would thus generally lie within region A of Figure 3.1B, with elemental composition directly coupled to growth. Such cells should, however, achieve higher absolute growth rates due to higher surface area:volume ratios and lower allocation of cellular resources to uses not immediately directed toward growth. The trade-off is that they would be extremely sensitive to a varying nutrient regime, unable to take advantage of brief pulses in nutrient concentrations. Conversely, some taxa might evolve a strategy of increased allocation to vacuoles, displacing cell space available for use in energy acquisition and growth but providing considerable capacity for storage of potentially limiting nutrients. Such a strategy would be advantageous under conditions of variable nutrient supply. In such cells growth and elemental composition would frequently be decoupled.

Alternative nutrient storage strategies based on allocation of cellular volume to cytoplasm versus vacuole have interesting parallels with other schemes categorizing growth strategies for phytoplankton and higher plants. Grime (1977) proposed a tripartite set of evolutionary strategies and higher-plant trait syndromes. The first group of the three are the competitive strategists. These are plant species adapted to relatively fertile conditions and characterized by high growth rates and high degrees of phenotypic plasticity in response to competitive interactions with adjacent plants. The second group are the ruderal strategists, which are adapted to habitats with high levels of resources but frequent incidence of physical disturbance and thus a low degree of interaction with adjacent plants; these too often have relatively high growth rates. Finally, stress tolerators are those plants specializing in habitats characterized by inhospitable physical conditions (e.g., low temperature) or chronically low availability of key resources (shaded habitats, mineral-deficient soils). These have a growth physiology in which, should excess resources be supplied, uptake of nutrients and biomass gain are decoupled, leading to extensive accumulation of reserves. For phytoplankton, Sommer (1985) proposed a similar scheme involving three strategies for phytoplankton in confronting a variable nutrient regime. The first was affinity strategists with steep uptake and growth kinetics curves (i.e., relatively high growth rates on low external resource concentrations); these are well adapted to chronically low nutrient levels. The second was growth strategists, which have high growth capacities that permit them to respond quickly to temporal or spatial nutrient pulses. Finally, storage strategists have large intracellular storage pools and rates of maximum nutrient uptake much higher than that of maximum growth.

Figure 3.1B helps place the frameworks of Grime and Sommer in a stoichiometric context. Competitive (Grime) or affinity (Sommer) strategists would commonly have cellular nutrient content near their optimal level while ruderal (Grime) or growth (Sommer) strategists would frequently find themselves experiencing depleted internal concentrations (Region A) as nutrient levels fall and their growth rate slows. In contrast, stress-tolerant (Grime) or storage (Sommer) strategists would exhibit variable biomass nutrient content but relatively constant, though low, growth rates (Region B). Thus, categories of alternative plant strategies identified by previous workers in plant evolutionary ecology appear to have distinct stoichiometric consequences due to cellular allocation syndromes. Understanding the ecological conditions that favor dominance by members of these groups may help in developing expectations about how much variation in autotroph C:N:P might be expected in particular habitats. We will return to these ideas in Chapter 7 when we discuss the role of resource competition and stoichiometry in the context of community interactions.

TABLE 3.1
N content and C:N of different tissues in apple trees. C:N was calculated from % N assuming 48% C in biomass. Information is also presented for wood and roots as a function of age. Data from Murneek (1942) as reported in Kramer and Kozlowski (1979)

Tissue	Age (y)	% N	C:N
Leaves		1.23	45
Spurs		1.04	54
Wood aged	1	0.93	60
	2	0.67	84
	3	0.54	104
	4–6	0.35	160
	7–10	0.27	207
	11–18	0.16	350
Main stem		0.14	400
Total above ground		0.33	170
Root stump		0.26	215
Roots aged	1–6	1.24	45
	7–13	0.6	93
	14–18	0.32	175
Total below ground		0.4	140
Entire tree		0.34	165

C:N:P STOICHIOMETRY OF ENTIRE HIGHER PLANTS

Appreciating the promiscuity of the vacuole and the selectivity of the cytoplasm helps us understand the nutritional status of individual cells. However, additional considerations must be brought to bear on multicellular, tissue-differentiated higher plants. New complexities arise because of tissue specialization and transport within the plant (see, e.g., papers in Roy and Garnier 1994). Within a single large plant, leaves, stems, and roots have highly contrasting elemental composition (Table 3.1). Specifically, leaves generally contain more N per unit dry mass than either stems or roots (Epstein 1972; Kramer and Kozlowski 1979). In addition, this table indicates that tissue age affects nutrient content; old wood and old roots are extremely nutrient poor. Similar patterns exist for other major elements. Leaves, young roots, and flowers are generally much richer in P, K, and S compared to stems and older roots (Kramer and Kozlowski 1979; Abrahamson and Caswell 1982).

Nutrient content at the whole-plant scale depends on the distribution of plant biomass among plant organs (Lawlor et al. 1981) as well as their separate nutrient contents. The patterns of allocation to organ systems are an important aspect of plant architectural "strategy" (Tilman 1988; Körner 1993). Across a range of plant life forms (trees, shrubs, perennial forbs, annuals, floating plants), the allocation of biomass to leaves (so-called leaf weight ratios) varies from 2–3% (in trees) to 80–98% (floating plants) (Körner 1993). Allocation is also very plastic. The allocation to different plant parts responds to environmental conditions such as light intensity and nutrient availability (Tilman 1988; Aerts et al. 1991; Marschner 1995). Differences in the C:N:P stoichiometry of various plant tissues and relative allocation patterns to those tissues combine to determine overall allocations of C, N, and P during growth at the scale of the entire plant (e.g., Abrahamson and Caswell 1982). Nutrient levels in leaves are strongly influenced by plant subsidies from other organs (roots, older leaves) and by withdrawal of nutrients, as occurs prior to leaf abscission (Berendse and Elberse 1989; Killingbeck 1993). Below ground, the most active tissues are fine roots. These have variable C:N:P stoichiometry, with generally higher nutrient content in the smallest diameter roots (Gordon and Jackson 2000). A huge literature is devoted to understanding how patterns of plant allocation, tissue composition, and nutrient balance are affected by environmental conditions and species traits (Lambers et al. 1989; Roy and Garnier 1994; Marschner 1995; Aerts and Chapin 2000).

The plasticity of higher-plant responses to abiotic nutrient supplies includes changes in allocation. It has been suggested that to optimize growth a plant should reallocate biomass and other resources toward acquisition of a limiting resource at the expense of acquisition of other resources (Bloom et al. 1985). A simple rule of allocating new biomass to the organs that acquire the most strongly limiting resources can achieve this result (Chapin 1991b). The net result would be to reduce the effect of limitation by one resource but make limitation by other resources more likely. Growth models indicate that optimal allocation occurs when all potentially limiting resources equally limit the plant (Rastetter and Shaver 1992). In practice, genetic or physical limits can keep this from happening. We will see a good example of this when we discuss responses of plants to N limitation below.

We cannot fully explore these issues here, although we will return to some of them later in the book when we discuss the role of growth stoichiometry in biogeochemical cycling at the ecosystem scale. For the time being, however, we will take a "herbivore-centric" view, and focus primarily on the elemental composition of leaves, as these will be on the menu of most terrestrial herbivores.

AUTOTROPHS IN CAPTIVITY

Agronomists, physiologists, and ecologists have subjected a variety of higher and lower plants to varying regimes of nutrient supply and light intensity to better understand their mineral nutrition. In this section we will present examples of both algae and vascular plants to illustrate general principles of how autotroph elemental composition responds to key environmental parameters. This exercise will equip us to understand and interpret patterns from field settings where environmental conditions vary enormously.

Questions we will address include the following: In what ways do higher and lower plants alter their elemental composition in response to changes in nutrient or light regimes? What are the limits of response to such changes? What characteristics of an autotroph influence the way that it alters its elemental composition in response to changing nutrient and light regimes? What do controlled laboratory and greenhouse studies tell us about how static measures of autotroph elemental composition are linked to the dynamic processes of growth?

The chemostat is a basic tool of the phytoplankton physiologist (Fogg 1975). In this continuous culture device, a "reaction vessel" contains the growing algae, and a given fraction is continuously replaced with fresh medium, the new medium displacing an equal volume of culture. Changing the inflow concentration of nutrients along with the flow rate regulates nutrient supply rate to the reaction vessel. The rate of loss experienced by the algae also is set by the rate of flow of medium through the vessel. By adjusting the rate of inflow of supply medium relative to the size of the vessel, the experimenter can control the rate of algal growth at equilibrium, which must match the rate of loss via outflow. The researcher can collect algae from the outflow of the vessel and analyze them for elemental composition. The experimenter can also regulate the temperature and light intensity experienced as well as the identity of the nutrient limiting growth (for example, assure that P is limiting by supplying N at high N:P and other micronutrients at saturating levels) and adjust the equilibrium biomass attained (by altering the concentration of limiting nutrient in the medium). Using this tool, algal physiologists have elucidated many details of the growth physiology of a variety of algal species.

Vascular plants, on the other hand, do not like being grown in chemostats. Their large ontogenetic changes also mean that nutrient requirements are often strongly related to size and age structure. Experimental systems that can achieve equilibria of vascular plant growth and resource concentration are difficult to obtain. Thus, it is challenging to evaluate changes in the nutritional status of vascular plants under equilibrial conditions of nutrient supply, light intensity, etc. Nevertheless, a variety of plant culture techniques have permitted detailed studies of the growth and compositional responses of plant tissue to changes in environmental conditions. We will highlight a few studies here.

We focus primarily on the elemental composition of leaves. There are several reasons for this. First, there have simply been more data collected for leaves than for stems and roots. More importantly, however, leaves (along with secondary roots) are a major form of annual production by most terrestrial plants. Thus, understanding the factors regulating the C:N:P stoichiometry of primary production in an ecosystem context includes understanding those factors affecting leaf nutrient content (Aerts and Chapin 2000). Second, as mentioned earlier, leaves are the targets of consumption by many terrestrial herbivores and thus consideration of the stoichiometry of terrestrial grazing food webs will often start with the elemental composition of leaves. However, there is no reason we know of that similar stoichiometric perspectives should not be brought to bear on the dynamics of below-ground food webs, which are becoming increasingly recognized as key regulators of terrestrial ecosystem dynamics (Strong et al. 1996). Finally, leaves respond more to changes in fertility than do stems and roots (Marschner 1995); thus, we will have a good idea of the range and causes of variation in C:N:P by focusing on leaves. While morphological complexity and nutrient translocation among plant parts complicate matters, we will try to evaluate the extent to which leaf elemental composition is directly coupled to growth rate and assess how elemental composition is affected by nutrient supplies, light intensity, and atmospheric CO_2 concentrations (pCO_2). A major goal is to evaluate the potential similarities between patterns documented for higher plants and algae.

Effects of Growth Rate and Identity of Limiting Nutrient

One of the strongest determinants of C:nutrient ratios in autotroph biomass is the severity of nutrient limitation of growth. Recall that we will gauge nutrient limitation in terms of relative growth rate (RGR = μ/μ_m; Chapter 1). This measure of nutrient limitation compares the realized growth of the existing community to the maximum growth that the same species assemblage could obtain given sufficient resources. There are other possible definitions of nutrient limitation (Howarth 1988). Chapin et al. (1986) pointed out that changes in species abundance given changes in fertility mean that, with sufficient time elapsed, maximal growth at a site may differ from maximal growth of an extant community. Our quantitative measure of nutrient limitation is concerned only with the extant community. The RGR is lower with increasing severity of nutrient limitation. We consider μ_m to be a function of light intensity and thus RGR measures nutrient, but not light, limitation.

We will first consider algae. Droop (1973) first demonstrated the close linkage between intracellular nutrients and growth rate. We will describe his theoretical approach in detail in a moment. Goldman et al. (1979) presented illustrative data of this type from studies of several taxa growing under P or N limitation. Consider the photosynthetic flagellate *Monochrysis lutheri* growing under P limitation (medium N:P > 85). At high RGR, cell composition was close to the Redfield ratio (Figs. 3.2A,B, dashed lines). Under P limitation (Fig. 3.2A), as growth rate declined, cellular C:P and N:P ratios increasingly deviated from Redfield proportions, reaching extremely high values (C:P ~1,000, N:P ~100) when the RGR was below 0.2. In contrast, cellular C:N did not deviate much from Redfield proportions (~7) across this gradient of P limitation. Cellular adjustments were different when the limiting nutrient was N; data for the diatom *Thallassiosira pseudonana* grown at a medium N:P of 5 are shown in Figure 3.2B. Here, as N limitation became increasingly severe with decreasing dilution rate, cellular C:P and N:P declined strongly, with larger decreases in N:P than in C:P. Thus, C:N increased with decreasing RGR. These results illustrate a general feature of limitation by a single nutrient: as limitation of growth rate by nutrient X becomes increasingly severe (RGR declines), the biomass C:X increases. Carbon:Y ratios (where Y is a different nutrient) may or may not respond to variations in limitation by X. Mathematical formalizations of this relationship (Droop 1973; Ågren 1988) will be described later in this chapter.

Nutrient limitation of vascular plants also decreases nutrient content (and elevates C:nutrient ratio). Illustrative data from Greenwood (1976) for leaves of wheat are shown in Figure 3.2C. Changes in relative growth rate under N limitation are clearly linked to the N content of the biomass. Plant N content (~N:C) increased with plant RGR under N limitation, just as we saw for algae (Fig. 3.2B). However, tissue-differentiated whole plants have responses not possible in unicellular autotrophs. For example, the plant may draw upon other tissues to prevent leaf nutrient levels (especially in the cytoplasm) from falling below those necessary for normal growth. Nutrients may be translocated out of old leaves and into growing leaves (Williams 1955; Mimura et al. 1996). Thus, the decline in nutrient content in response to declining nutrient supply shown in Figure 3.2 may not always occur.

Furthermore, as we discussed earlier, reallocation of resources to increase acquisition of a limiting resource at the expense of acquisition of other, nonlimiting, resources is favored. Under N limitation, roots may develop enhanced absorptive capacities for the limiting nutrient. Specific root length, which is equal to root length per unit root mass, is one trait that has been shown to be adjusted in this way (Aerts and Chapin 2000). Root:shoot ratios often increase as nutrient supply declines or light intensity increases (Marschner 1995). In terms of physiological mechanisms,

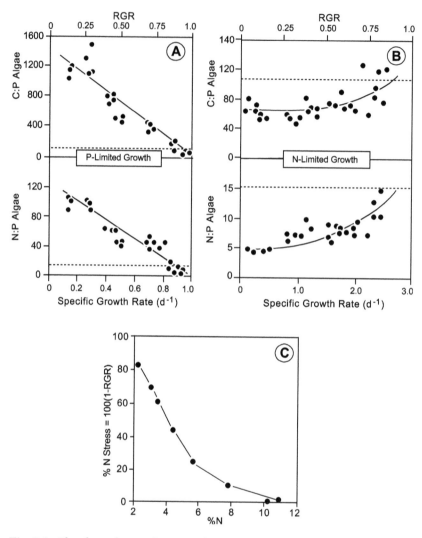

Fig. 3.2. The dependence of autotroph C:N:P stoichiometry on the severity of nutrient limitation. Panels A and B show data for the phytoflagellate *Monochrysis lutheri* and the diatom *Thalassiosira pseudonana* while panel C shows data for N-limited growth in wheat. Panel A shows how algal C:P and N:P are similar to the Redfield ratio (dashed lines) at high growth rate (RGR ≈ 1). Panel B shows responses for C:P and N:P under N limitation. N:P declines more substantially than C:P as N limitation increases in severity; thus, cellular C:N increases with more severe N limitation. Panel C shows how foliar N content decreases with increasing nitrogen stress (degree of N limitation of growth calculated as shown). Given that plant % C is relatively constant, declining leaf % N indicates a decrease in N:C

these responses may simply reflect the fact that root meristems are closer to the source of limiting nutrient supply (the soil). Thus, by having the first chance to intercept nutrients coming in from the soil they could grow more rapidly than aboveground tissues until the point at which their growth rate is constrained by the rate of carbohydrate supply from the stems, which itself is a function of leaf growth and carbon fixation (Chapin 1980; Rastetter and Shaver 1992) and N content (Chapter 8). Thus, the root:shoot ratio could be set mechanistically by the balance of carbon and nutrient acquisition linked to the intensity of light relative to soil fertility. One way to interpret these patterns is that the plant grows in such a way as to achieve an optimal light:nutrient balance that maximizes growth rate integrated over the whole plant. We present more on the effects of light:nutrient ratios on autotrophs below, and effects on food webs are considered in Chapter 7.

Effects of Inorganic N:P

The severity of nutrient limitation is not the only determinant of autotroph C:N:P stoichiometry. The ratio of nutrient supply also plays a role. We saw this already for algae in Figure 1.5 and now we will look at this effect in greater detail both in algae and in higher plants.

Rhee's work from which Figure 1.5 was made also reported the biochemical forms of accumulation of luxury consumption products. Under high N supply (N:P > 30), surplus intracellular N accumulated primarily as protein and free amino acids. Under high P supply (N:P < 30), surplus P accumulated largely as polyphosphate. The close match between cellular stoichiometry and supply ratios indicates that essentially all N and P were taken up from the dissolved pool at all N:P ratios. Although one or the other of these elements limits growth, luxury consumption brings the extracellular levels of the other, nonlimiting element, also down to near zero. One could say these algae not only "are what they eat," they "are what they have available to eat." This is a large contrast with the homeostatic metazoans to be discussed in Chapters 4 and 5.

Rhee's study was performed at a single growth rate. The effect of the resource supply N:P ratio on algal C:N:P stoichiometry at different growth rates is illustrated in Figure 3.3A where biomass stoichiometry in the phy-

Fig. 3.2. (*Continued*)

(increase in C:N). Note that foliar % N appears to be approaching a minimum value of 2% at zero growth rate (i.e., 100% "N stress"), suggesting that the minimal content of N in wheat is 2%, or, in stoichiometric terms, its maximum C:N is ~28. Panels A and B are based on Goldman et al. (1979) and panel C is based on Greenwood (1976).

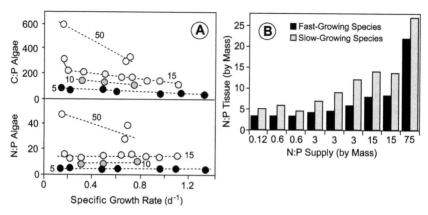

Fig. 3.3. Growth-related variation in C:P and N:P depends on supply N:P. A. In the alga *Dunaliella tertiolecta*, when nutrients are supplied in relatively balanced proportions (e.g., N:P of 15; see values on the figure panel), variation in C:P and especially N:P with growth rate is muted. When nutrients are supplied at or below Redfield proportions, cells have close to Redfield proportions even at slow growth rates. B. The effect of supply N:P on biomass N:P in terrestrial grasses. Foliar N:P is shown for *Brachypodium pinnatum* (a species with characteristically slow growth rate) and *Dactylis glomerata* (a fast grower) grown in the greenhouse with soil containing different levels of nutrients at different N:P ratios. Note that most variation in foliar N:P is associated with soil N:P; multiple values for foliar N:P at particular soil N:P indicate different levels of nutrient concentrations added. Under all soil conditions, the fast-growing grasses have lower foliar N:P than the slow-growing species. Panel A is based on Goldman et al. (1979) and panel B is based on Ryser and Lambers (1995).

toflagellate *Dunaliella tertiolecta* grown at medium N:P from 5 to 50 is shown. At any given growth rate, cellular C:P and N:P were positively related to inflow N:P. Further, biomass C:P increased most dramatically with declining growth rate (decreasing nutrient supply rate in the chemostat) when medium N:P was high. Note also that cellular N:P remained at approximately Redfield proportions for all growth rates when inorganic nutrients were supplied at Redfield proportions. Thus, a key implication here is that, to generate unbalanced C:N:P ratios in algae, nutrient supply ratios must also be unbalanced. It is not enough just to have slowly growing algae, as such algae will have Redfield stoichiometry if inorganic nutrients are provided at the Redfield ratio.

The effects of supply N:P on cellular composition have important implications for interpretation of elemental composition data from natural environments. Goldman et al. (1979) used the relationships shown in Figure

3.2A to infer something about phytoplankton growth in the oceans. Noting the "canonical" Redfield ratio in particulate matter from oceanic surface waters (Chapter 1) and that cultured algae achieve the Redfield C:N:P when growing at their physiological maximum (Figs. 3.2A,B), they concluded that algae in the oceans must be growing without strong nutrient limitation. This interpretation has been accepted by some (e.g., Harris 1986; Elser and Hassett 1994). Curiously, though, this interpretation overlooks some of the results in the original paper. Figure 3.3A shows that the cellular stoichiometry of algae is actually insensitive to growth rate when N and P are supplied in the inflow medium in Redfield N:P proportions or lower, even when the algae are growing at rates less than 20% of their physiological maximum. In other words, an observation of Redfield C:N:P in particulate matter in the oceans or elsewhere is consistent with rapid, nutrient-saturated growth rate but it does not rule out the possibility that algae are experiencing an inadequate but balanced (close to Redfield) nutrient supply. Further, as we will see, Redfield stoichiometry is also consistent with a light-limited, slow growth rate (see next section).

Examination of data for the nonlimiting nutrient indicates that, as the RGR declines with increasing limitation by X, quantities of the alternative nutrient (call it Y) remain relatively constant or increase due to luxury consumption if that nutrient is in considerable excess. Thus, cellular C:Y often declines as limitation by nutrient X becomes more severe (for example, the data for C:P under N limitation shown in Fig. 3.2B). Since the concentrations of nutrients X and Y change in opposite directions, cellular X:Y (e.g., N:P) can be particularly dynamic during the onset of nutrient limitation, especially if luxury consumption is occurring (Figs. 1.5, 3.2, and 3.3).

Now, let's again compare algae to vascular plants. Due to the difficulty of growing higher plants to equilibrium with their resources in completely defined growth substrate, less is known about how higher plants respond to changes in the relative supply of N and P in the environment. Nevertheless, evidence indicates that foliar N:P also is positively correlated with the N:P supply ratio. In two grasses differing in maximum growth rates (slow-growing *Brachypodium pinnatum* and fast-growing *Dactylis glomerata*), foliar N:P increased strongly with inorganic N:P in artificial soil (Fig. 3.3B). Note also that, in what we will eventually see to be a recurring pattern, tissue N:P was related to growth strategy: N:P was lower in the faster- than in the slower-growing species.

With good stoichiometry-growth couplings, one might be able to ascertain nutrient limitation patterns in the field from measurements of nutrient content. This has been reported for some heterotrophs (Heldal et al. 1996), but it has been examined primarily for autotrophs because of their widely variable stoichiometry (Fourqurean et al. 1992; Koerselman and Meuleman 1996; Bedford et al. 1999; Hillebrand and Sommer 1999; Lockaby and Conner 1999). Recent evidence from field experiments also suggests that biomass N:P of both terrestrial plants and algae may be a reliable indicator of whether autotroph growth is potentially limited by N or by P. Verhoeven and colleagues reported on the results of experiments where growth and tissue nutrient content of herbaceous vegetation in wet habitats in The Netherlands were assessed in conjunction with fertilization with N, P, and K singly and in combinations (Verhoeven et al. 1996). They found that initial tissue N:P was a reliable indicator of autotroph response to nutrient addition. That is, plants that had high initial tissue N:P (>33) generally responded to P fertilization (i.e., had P-limited growth) while plants with low leaf N:P (<33) generally responded to N. Verhoeven and colleagues noted that an N:P ratio of 33 is considerably above the Redfield ratio, which has sometimes been considered to be the threshold for differences between N and P limitation for phytoplankton algae, and attributed this difference to increased structural investment in vascular plants relative to algae. Whether it is appropriate to regard the Redfield ratio as a fundamental breakpoint between N and P limitation is debatable. As we will see below, different autotroph species have different optimal N:P ratios that separate their responses to N versus P. Further, we have seen that plant structural material has extremely low N and P content. Increasing allocation to cellulose compounds in structure would add only C to biomass and, similar to the effect of adding storage lipid discussed in Chapter 2, would proportionally raise C:N and C:P without affecting biomass N:P. There is no obvious reason to suspect a shift in N:P ratio for nutrient limitation when one adds or subtracts C.

Furthermore, it is not at all clear that the thresholds separating N and P limitation are in fact different for phytoplankton and vascular plants. Recent analyses of algal nutrient limitation in lakes suggest a close correspondence between growth limitation and N versus P availability in algae and higher plants. The responses observed by Verhoeven and colleagues bear a remarkable resemblance to responses of N versus P limitation of phytoplankton that have been observed in lakes (Elser et al. 1990; Downing and McCauley 1992). Elser and colleagues summarized published studies of algal nutrient limitation in which investigators examined algal growth response to enrichment by N, P, or both. Downing and McCauley then combined the enrichment response data for each lake from that study with corresponding data for total N (TN) and total P (TP) concentrations obtained from other sources. They found that algae generally responded to N enrichment when TN:TP was less than 31 and to P when TN:TP was greater than 31. This value is very similar to the ratio of 33 observed for the same transition in wetland plants. These two examples support the use of biomass N:P ratios for diagnosing nutrient supply rates and N:P balance across a range of ecosystem types.

Comparison of the results of Verhoeven et al. (1996) and Downing and McCauley (1992) is not, however, free of difficulty. One potential problem is that the lake enrichment responses were for whole communities while the terrestrial responses were at the level of individual species. Another is that the algal study measured total nutrient pools (including nutrients both in algae and in the surrounding water) while the wetland plant study was based on tissue nutrients. Finally, allometries of plant growth rates and N and P requirements (discussed in more detail below) suggest, at least in theory, that the N:P threshold for N versus P limitation of growth of vascular plants should generally be higher than that for algae because vascular plants generally have lower μ_m values. Nevertheless, the two similar cutoffs are a remarkable congruence worthy of further contemplation.

Effects of Light Intensity, Trace Elements, and pCO_2

The investigations discussed above were performed under experimental conditions of optimal or near optimal light, carbon, and micronutrients. However, differences in one or more of these factors can influence autotroph C:N:P ratios. Light intensity should alter autotroph C:N:P because light intensity is directly linked to carbon acquisition via photosynthesis. Similar arguments might be made about pCO_2, since increasing the concentration of CO_2 at the enzymatic sites of C fixation often will increase photosynthetic rates. In other words, as for N and P, autotroph C content might respond to differential C availability by increasing quota. We also discuss the effects of trace metals, especially iron, in this context. We will see that cellular elemental composition is not simply a function of the absolute growth rate under nutrient limitation. And once again we will begin with algae.

Tett et al. (1985) pointed out that algae have C:N:P close to Redfield proportions when growing slowly under light-limited conditions. However, even under conditions of moderate to high light coupled to nutrient limitation, differences in light intensity have large effects on cellular elemental composition. The experiments of Healey (1985) illustrate the general nature of algal response to light under nutrient limitation. In this study the cyanobacterium *Synechococcus linearis* was grown at various dilution (growth) rates under P or N limitation and at different light intensities. The responses (Fig. 3.4A) clearly show the joint dependence of stoichiometry on nutrient limitation and light intensity. At a given light intensity, cellular nutrient:C increased with increasing growth rate, patterns similar to those we have already described. However, it is also true that, at a given growth rate, cellular nutrient:C increased (C:nutrient decreased) with decreased light intensity. Others (Rhee and Gotham 1981; Smith 1983a; Goldman 1986) have reported similar findings with eucaryotic algae. Clearly, light intensity influences nutrient content. Goldman (1986)

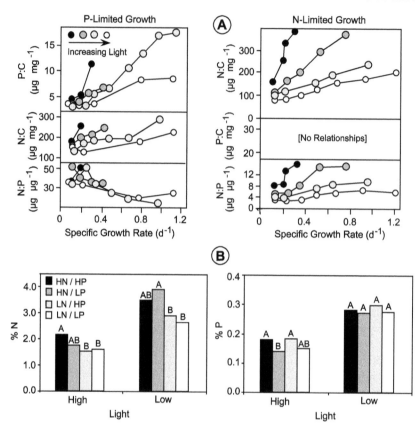

Fig. 3.4. Effects of light intensity on autotroph C:N:P stoichiometry. A. Effects on the cyanobacterium *Synechococcus linearis* under P limitation (left) or N limitation (right). Note that as light intensity increases, the cyanobacterium has higher maximum growth rates (μ_m) but its nutrient content is consistently decreased. At a given growth rate, increasing light intensity leads to decreasing cellular P:C and N:C (increasing C:P and C:N). There is little effect on N:P. Under N limitation, increasing light intensity at a given light intensity leads to decreasing N:C (increasing C:N) and N:P. B. Effects of nutrient supply and light intensity on the N and P contents of one-year-old needles in red pine seedlings (*Pinus resinosa*). Seedlings were grown in four combinations of high and low N and P at low and high light intensity. Different letters corresponding to each bar within each light intensity treatment indicate values that were significantly different according to multiple comparison tests. The statistical main effects of light on %N and %P were highly significant. Note that in this study the effects of light on nutrient content were generally larger than the effects of fertilization. Panel A is based on Healey (1985); Panel B is based on Elliot and White (1994).

pointed out that a C:nutrient-RGR relationship holds as long as one defines μ_m as a nutrient-saturated growth rate at ambient light and not under saturating light. Recognition of the effect of light on nutrient content may permit a better understanding of factors regulating autotroph elemental composition under field conditions where both nutrient supply and light intensity vary. This will be discussed in more detail below.

The role of trace nutrients, such as iron, in limiting pelagic productivity is under increasing scrutiny. As we saw in Chapter 2 (Table 2.1), the main role of metals in cell physiology is catalytic, in either acid-base or redox reactions. Hence, metals impact heavily on energy transformation processes in the cell. Specifically, Fe has a key role in the energy transduction pathways of photosynthesis and respiration (Williams and Fraústo da Silva 1996; Falkowski and Raven 1997). These observations lead us to a hypothesis that the effects of metal limitation on macronutrient (C:N:P) stoichiometry may resemble the effects of light limitation (both may limit photosynthetic C acquisition). A role of metals in macroelement stoichiometry could have major implications for large-scale biogeochemical cycling, as limitation by Fe availability appears to be an important factor regulating oceanic nutrient cycling at globally significant scales (Chapter 8; Chisholm and Morel 1991).

Surprisingly little experimental work has been done to examine the stoichiometry of algal growth under Fe-limited or colimited conditions and the data are somewhat equivocal at the moment. Greene et al. (1991) grew the marine diatom *Phaeodactylum tricornatum* under Fe-replete and Fe-limited ($\sim$20% of maximum growth rate) conditions; the N:P of the inorganic nutrient medium was at the Redfield ratio. Detailed physiological data indicated a deterioration of photosynthetic performance under Fe-limited conditions, consistent with the idea that Fe limitation is manifested primarily in energy metabolism and thus may have a similar impact on C:N:P stoichiometry as does light limitation. Meanwhile, cellular C:N:P composition remained close to Redfield proportions in both cultures (109:18:1 for +Fe, 103:16:1 for $-$Fe), which is not entirely surprising given the evidence presented above that algal C:N:P ratios remain near Redfield proportions when inorganic nutrient supply is balanced. Similar data showing modest effects of Fe status on C:N:P have been obtained by others (Rueter and Ades 1987; Takeda 1998). In addition, the response to Fe may depend on the specific form of inorganic N supplied (NO_3 vs. NH_4) due to the requirement for Fe in nitrate reductase (Muggli and Harrison 1996; Muggli et al. 1996). There appear to be stronger effects of Fe supply on Si:C and Si:N ratios (Takeda 1998; De La Rocha et al. 2000) than on C:N:P ratios. Much more work is needed in this important area, but perhaps the general idea that iron impacts heavily on energy metabolism gives us reason to expect low C:nutrient ratios in Fe-limited phytoplankton, as with light limitation.

Because of concerns over the effects of increased CO_2 concentrations in the atmosphere, considerable research on the effect of atmospheric pCO_2 on carbon flux is underway. Photosynthesis involves the enzymatic fixation of free CO_2 into carbohydrates. Thus, CO_2 availability should logically affect algal C:N:P. Carbon limitation may resemble limitation by N or P in that there can be a positive relationship between C:N ratios and growth rate under C limitation (Clark 2001). As we will see below, vascular plant biologists are actively investigating the effects of pCO_2 on plant nutritional status. In contrast, only a small number of studies have looked for CO_2 effects on algal C:N:P ratios, perhaps because it has long been argued that CO_2 is not limiting in aquatic systems due to rapid diffusion from the atmosphere (Schindler 1977). Regardless of the long-time-scale effects at the ecosystem scale, it remains possible that, at the physiological or cellular scale, CO_2 availability limits algal photosynthesis and thus can alter cellular physiology and affect cellular C:N:P.

A few recent studies have examined this possibility and indicate that there are, in fact, effects of pCO_2 on algal stoichiometry. Burkhardt and Riebesell (1997) grew a marine diatom across a range of pCO_2 values and found that cellular C:P and N:P varied by up to 65%. However, they cautioned that more algal species needed to be studied. Indeed, when they followed up with a more extensive study in which seven additional species were grown under high-light, nutrient-replete conditions across a range of pCO_2 levels (Burkhardt et al. 1999), they found that some of the species showed increased cellular C:nutrient ratios and increased N:P but other species responded with lowered C:N or C:P with increased pCO_2! They further noted that algal C:N:P mainly responded to altered CO_2 at pCO_2 levels considerably lower than current values of pCO_2 in the oceans. From these findings they questioned the relevancy of algal C:N:P adjustments to pCO_2 in the field. However, we note that Burkhardt and Riebesell performed their experiments under nutrient-replete conditions. The C:N and C:P ratios in Burkhardt and Riebesell's study were generally low and not particularly divergent from Redfield proportions. This is not surprising based on the physiological principles discussed in this chapter, i.e., algal C:N:P ratios do not diverge strongly from Redfield proportions under conditions replete in macronutrients or when nutrients are supplied at the Redfield ratio. Thus, effects of increased pCO_2 on algal C:N:P might be better gauged by examining responses to pCO_2 under conditions of nutrient limitation, the prevailing situation in most pelagic ecosystems. Additionally, the effects of pCO_2 might be stronger if inorganic nutrients were supplied in unbalanced proportions.

New experiments shed some light on these possibilities but the evidence remains somewhat contradictory. Gervais and Riebesell (2001) grew the marine diatom *Skeletonema costatum* in P-limited batch mode. Algae grown under P-replete conditions were transferred to low-P medium (0.14 μmol L^{-1}, 100 μmol L^{-1} NO_3) at two CO_2 levels (20.6 and 4.5 μmol L^{-1}) in closed bottles. While strong effects were observed for carbon isotopic discrimination, little or no effect of pCO_2 on algal C:P and N:P was observed. Instead, algae developed very high C:P (1,000) and N:P (60) ratios in both CO_2 environments. The authors concluded that "CO_2-related variation in phytoplankton stoichiometry appears to play a minor role in the ocean carbon cycle under nutrient-limiting conditions." Perhaps such a wide extrapolation is premature, as these findings may not generalize to other species or other conditions. It is also not certain if the results would hold for algae grown at equilibrium in chemostats where possible carryover effects of internal algal P quota would be avoided. In addition, it is important to know if the identity of the limiting nutrient (N vs. P) influences the response of algal stoichiometry to pCO_2, especially since pCO_2 effects are likely mediated through photosynthetic physiology which appears to involve especially N-intensive, high-N:P investments (Chapter 2). Recent experiments by Caraco and colleagues in fact indicate that effects of pCO_2 on algal C:nutrient occur when N is limiting (N. Caraco, personal communication). In their experiments, algae were grown in batch mode under N limitation at different levels of pCO_2. The effects of pCO_2 were greatest for large cells, with the cellular C:N of the large diatom *Ditylum* sp. increasing three-fold in response to an increase in pCO_2 similar to that expected during the coming century. They suggest that this is a biotic response to global CO_2 meaningful at the global scale. According to their calculations, even a 20% increase in phytoplankton C:nutrient ratios (from 106 to only 127 C:P) could increase export production (the removal of C from surface oceans to bottom waters by sinking of organic matter) by 1 billion metric tons per year, a value equivalent to the estimated postindustrial increase in the terrestrial C sink (Schimel 1995). Although the notion of direct CO_2 limitation of photosynthesis in phytoplankton is controversial (Falkowski and Raven 1997) and is contrary to the results of Gervais and Riebesell (2001), the stakes are high on this question. In our view, more work, including equilibrium culturing in chemostats, is needed to better establish the responses of freshwater and marine phytoplankton to pCO_2 under nutrient-deficient conditions.

We just saw how the nutrient content of algal biomass decreased with increasing light intensity at a given growth rate (which corresponds to the nutrient supply rate at equilibrium in chemostats). Let us turn again to vascular plants for a comparison, where similar changes have been observed in leaf elemental composition. Illustrative data come from the study by Elliot and White (1994), who grew red pine (*Pinus resinosa*) for four months in pots under high light and low light and under N or P limitation (all eight possible combinations of light, N, and P were studied). Fertilization increased foliar nutrient content (Fig. 3.4B), consistent with similar patterns discussed above. In addition, light intensity had a major effect on foliar N and P contents (Fig. 3.4B), which were dramatically lower in the high-light treatment, especially in low-nutrient pots. Similar decreases in nutrient content with increased light were also observed in stems and roots. Greenwood (1976) also showed that increased light intensity lowered leaf % N (raised C:N) in wheat tillers at a given degree of N deficiency. An increase in C:nutrient ratios in response to increased light seems to be general in autotrophs.

Numerous studies have examined the effect of pCO_2 on terrestrial plant nutrition and several review articles have summarized these studies (Schulze and Mooney 1994; Strain and Cure 1994; McGuire et al. 1995). Here, we use the meta-analysis of Curtis and Wang (1998) concerning data from 41 studies reporting leaf nitrogen content under ambient and elevated CO_2 conditions. They also examined changes in whole-plant allocation (root:shoot ratios) and leaf starch concentration. Leaf N content (N:C) was consistently decreased by exposure of plants to high-CO_2 atmospheres. On average, leaf N content decreased by a factor of more than 16% under high CO_2 while leaf starch content increased by factors of 24% (in gymnosperms) to 68% (in angiosperms). Although several individual studies have reported effects of CO_2 on root:shoot ratios, the meta-analysis did not find any overall significant trend of altered plant allocation patterns. It seems clear that elevated CO_2 increases C:nutrient in vascular plant production. As has been pointed out previously (Zak et al. 1994; McGuire et al. 1995; Koch and Mooney 1996; Cotrufo et al. 1998), increased leaf C:nutrient under elevated CO_2 supply has a variety of ramifications for ecosystem processes, including potential effects on food quality and trophic dynamics and litter decomposition (see Chapter 8). While it is sometimes claimed that increasing pCO_2 will be a boon for humankind's agricultural production, this production will be less nutritious unless yet more fertilizer is applied to agricultural fields, likely exacerbating runoff problems. That is, increased CO_2 may well help us produce 20% more lettuce but our salads will have to be around 20% bigger for us to achieve the same N intake in our diet!

Species-Specific Differences

Not all autotroph species alter their elemental composition in response to nutrient limitation in the same way, even when grown at similar growth rate, light intensity, and supply N:P. In fact, such differences form much of the basis of nutrient competition among autotroph species: taxa that are good competitors for element X are generally those that require less of element X to maintain a given growth rate (see discussion in Chapter 7).

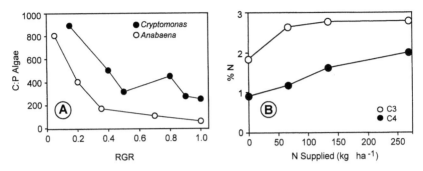

Fig. 3.5. Species dependence of autotroph C:N:P stoichiometry. A. Response of biomass C:P to increasing P limitation (moving from right to left on graph) for two phytoplankton species (*Cryptomonas erosa* and *Anabaena variabilis*). For any given P-limited growth rate, *Cryptomonas* has higher C:P than *Anabaena*. At zero growth rate, *Anabaena* has a lower biomass C:P (higher minimal P quota), indicating that it is a relatively poor competitor for P (see Chapter 7). B. Dependence of tissue N content on fertilizer N supply in graminoid species with contrasting photosynthetic physiology (C_3 vs. C_4 photosynthesis). At each N level applied, the C_3 grasses have higher N than the C_4 grasses. Panel A is based on Healey and Hendzel (1979) and panel B is based on Colman and Lazemby (1970).

Such taxa will generally have inherently high C:X, especially as growth rates decline.

As an illustration, consider species-specific algal C:N:P ratios from Healey and Hendzel (1979) who grew five species of algae under differing degrees of N or P limitation (Fig. 3.5A). At any given P-limited growth rate, the cyanobacterium *Anabaena variabilis* generally had lower C:P (higher P content) than the cryptophyte *Cryptomonas erosa*. Effects of cell-specific stoichiometry on competition between these two algae are discussed in Chapter 7. Interspecific variation in C:N:P ratios means that, under field conditions with multiple species present, we need to consider not only physiological responses of algal species to local conditions but also the possibility of species shifts among taxa with different characteristic C:N:P traits.

Large differences in N content of C_3 and C_4 grasses are a further example of the taxon dependence of autotroph stoichiometry. Marschner (1995) discusses the study of Colman and Lazemby (1970), who examined the dry matter production and tissue N content of various grasses under several levels of N fertilization. As we discussed in Chapter 2, C_3 grasses have a larger N allocation to photosynthesis (RUBISCO enzyme) than C_4 grasses. The stoichiometric consequence is large: for each level of added fertilizer, C_3 grasses had significantly higher N content (lower C:N) than did C_4

grasses (Fig. 3.5B). Hence, carbon acquisition in C_3 grasses requires relatively more N than in C_4 grasses.

Resource competition theory (Tilman 1982) can be used to understand these species shifts. An "optimal N:P" for a species can be defined as the N:P ratio where its growth is equally limited by N and P (we'll make this a little more complicated shortly). For those familiar with resource competition theory, this would be the N:P at the "kink point" of a zero net growth isocline of a species growing on these two resources. It is often calculated as the ratio of the species' N and P contents at zero growth rate under limitation by that particular nutrient (N content under N limitation and P content under P limitation) (Andersen 1997). Although some expect shifts from N to P limitation in autotrophs always to occur at the Redfield ratio, this is not necessarily the case. Optimal N:P varies considerably among algal species, ranging from 7 to 87 in one review (Hecky and Kilham 1988). Species-specific differences in optimal N:P for terrestrial plants also occur (Ingestad 1979a).

A more complex view of optimal N:P ratios considers the influence of growth rate. Elrifi and Turpin (1985) explored the effect of growth rate on shifts between N and P limitation and showed that optimal N:P is not a fixed parameter dependent only on the minimal cell quotas for N and P for that taxon. They argued for consistent shifts in optimal N:P with growth; specifically, that optimal N:P declines as growth rate increases (Elrifi and Turpin 1985; Terry et al. 1985; Turpin 1986). The implication is that algae are more easily limited by P at high growth rates and more easily limited by N at slow growth. This shift is consistent with disproportionate increases in P demand with increased growth rate, driven by increased allocation to P-rich ribosomal RNA, which we will discuss further in Chapter 4. As pointed out by Elrifi and Turpin (1985), the growth rate dependence of algal optimal N:P is of "great ecological importance in determining the outcome of competitive interactions in nutrient-limited habitats." However, to our knowledge no empirical or theoretical studies have followed up on this insight.

One measure of the species dependence of stoichiometry is interspecific variation in minimal P:C ratio (the P:C of the organism at zero growth rate under P-limited growth, $P:C_{min}$). The range of these values can give some idea of the range in algal elemental composition we might expect to see under extremely severe P limitation. Andersen (1997) reviewed chemostat data on $P:C_{min}$ for 30 species of phytoplankton. Median $P:C_{min}$ was 1.4×10^{-3} (C:P $\approx$ 700) and the range was from 2.3×10^{-4} (C:P $\approx$ 4350) for *Asterionella formosa* to 4.5×10^{-3} (C:P $\approx$ 221) for *Chlorella pyrenoidosa*. This is a striking degree of variation, and it says that under environmental conditions of strong P limitation we can expect autotroph

C:P in the pelagic zone to generally be ~700, but not higher than ~4000 and no lower than ~200. This variation is meaningful to the autotroph communities as it has a heavy bearing on competitive dynamics (Chapter 7), and it is meaningful to various ecosystem processes like herbivore growth (Chapter 5) and nutrient recycling (Chapter 6). Nevertheless, there is little information on the genetic and physiological determinants and evolutionary pressures on minimal cell quotas in algae.

Interspecific variation in C:N:P is also pronounced in the photosynthetic tissues of vascular plants (Field and Mooney 1986). Nielsen et al. (1996) compiled a large array of data for a wide variety of autotroph taxa from algae to trees to assess the allometric scaling of growth rate and % N and % P in autotroph biomass. They showed some differences in % N or % P among plant functional groups (trees, shrubs, herbs, aquatic angiosperms) (Fig. 3.6A). Their review suggested that growth form (single- vs. multicellular, small vs. large) rather than habitat type (aquatic vs. terrestrial) is a key determinant distinguishing pelagic from terrestrial autotrophs, as the elemental composition of aquatic angiosperms differed little from terrestrial plants. In their allometric analyses, they noted that as "plant size" (thickness of photosynthetic unit) increased from algae to aquatic plants to grasses to conifers to cactus, plant maximum growth rate, P content, and N content all generally decreased. Combining their allometric equations we can examine how N content and P content scaled with growth rate (Fig. 3.6B): both N content and P content increased with increasing growth rate. Slow-growing (large) autotrophs had high N:P while rapidly growing (small) taxa had lower N:P (Fig. 3.6C), an identical trend to the one suggested for algae by Elrifi and Turpin (1985), and yet another suggestion of a P-rich signature of rapid growth to be emphasized in Chapter 4.

In summary, autotroph elemental composition is associated with taxon-specific traits such as size, characteristic biochemical pathways, and maximal growth rate.

Generalizations

The preceding sections have given an overview of the factors affecting autotroph C:N:P ratios, a major driving force in ecological stoichiometry. Certain patterns hold for autotrophs from cyanobacteria to trees. We propose here a set of factors that affect autotroph C:nutrient ratios. For any given nutrient element X, the C:X of autotroph biomass

1. Increases with the severity of growth limitation by nutrient X;
2. Increases with decreased availability of X compared to alternative limiting nutrients (i.e., Y:X);

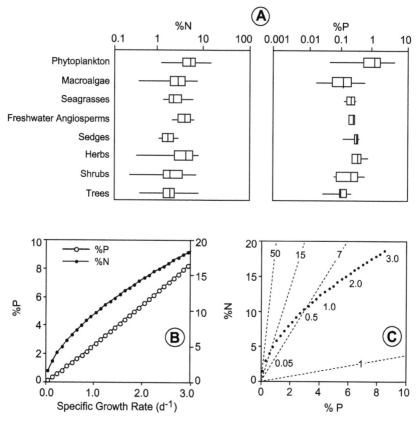

Fig. 3.6. Growth form and allometric dependence of growth rate and elemental composition for photosynthetic organs in autotrophs from algae to trees. A. Box and whisker plots for % N and % P for various autotroph groups. B. Scaling of P content and N content with maximum growth rate for autotrophs. The curve fits come from the original source (Fig. 2 from Nielsen et al. 1996). C. Stoichiometric diagram for the autotroph data compiled by Nielsen and colleagues. Values near the curves indicate growth rates for any given % N and % P combination. Dotted lines indicate different N:P ratios. At low growth rates, autotrophs generally have low % N and % P and high N:P while high-growth-rate autotrophs have high % N but especially high % P and thus have low N:P. Based on Nielsen et al. (1996).

3. Increases with increasing light intensity;
4. Increases with increasing $p\mathrm{CO}_2$ (for terrestrial plants, possibly for aquatic plants); and
5. Is a function of plant species, with good competitors for element X and larger taxa having high C:X.

Further, due to luxury consumption, in both aquatic and terrestrial realms we expect autotroph N:P to reflect supply N:P. Finally, there is likely to be important variation among species in the degree of phenotypic plasticity in C:N:P in response to various environmental factors.

THEORIES OF AUTOTROPH STOICHIOMETRY

We have seen a general, positive relationship between autotroph growth rate and content of the growth-limiting nutrient. This relationship means that autotroph growth and biomass stoichiometry are tightly coupled. The consistency of this relationship across numerous taxa lends itself to mathematical analysis and in the following we present an overview of some of the main theoretical frameworks designed to characterize these relationships.

Aquatic Approaches: A Cell Quota Model

We have mentioned the concept of "cell quota" frequently already; it is one way to refer to the nutrient content of an organism. Now let us define and analyze "quota" more formally. Droop defined the cell quota (Q) as the quantity of nutrients in a cell (typical units are moles per individual cell but different units are also commonly used). The major advance in coupling phytoplankton growth rate to intracellular nutrients came when it was noticed that growth (μ) was linearly related to $1/Q$ (Droop 1974). The resulting two-parameter "Droop model" linking growth to quota,

$$\mu = \mu'_m \overbrace{\left(1 - \frac{Q_{\min}}{Q}\right)}^{A},\qquad (3.1)$$

has had a long and successful history in phytoplankton ecology (Droop 1983; Morel 1987). In this model, μ as always is the specific biomass growth rate (time^{-1}), μ'_m is an asymptotic maximum growth rate at infinite cell quota (time^{-1}), $Q_{\min}$ is the minimum quota of nutrient needed to support life (moles per cell at zero growth), and Q is the actual cell quota for any finite growth rate (moles per cell). Note that the term marked "A" has a value of zero when Q is at its minimal value and a value of 1 when Q is infinite. The theoretical growth rate μ'_m is greater than the true maximal growth rate μ_m. Depending upon the nutrient and species under consideration, μ'_m may or may not be closely approached as μ approaches μ_m.

Growth rate is frequently expressed in relative terms to express the severity of growth limitation by a nutrient (relative growth rate, RGR; Chapter 1). Using the theoretical maximum growth at infinite cell quota gives

$$\text{RGR} = \frac{\mu}{\mu'_m} = 1 - \frac{Q_{\min}}{Q}. \tag{3.2}$$

The fit of this model to example data is illustrated in Figure 3.7. In this case, growth rate asymptotically closely approaches μ'_m as cell quota increases to high levels from an X intercept (where growth rate is zero) equivalent to $Q_{\min}$.

The Droop model links growth to intracellular nutrients. In contrast, the Monod expression relates growth to external resources:

$$\mu = \mu_m \left(\frac{R}{K_\mu + R} \right), \tag{3.3}$$

where μ_m is the maximum specific growth rate, R is the concentration of the limiting resource in the external medium, and K_μ is the half-saturation constant for growth (the resource concentration at which growth is one-half of its maximum). The Monod model is mathematically identical to Michaelis-Menten enzyme kinetics. Under equilibrium conditions, Equations (3.2) and (3.3) are consistent with each other but additional information is necessary to model non-steady-state conditions properly (Burmaster 1979; Morel 1987).

While most studies of algal growth kinetics using the cell quota approach have expressed Q in terms of quantity of nutrient per cell, the cell quota concept is just as readily viewed in stoichiometric terms. It is simple to convert cell quota to quantity of nutrient per unit carbon biomass of the cell (as was done in our discussion of the Andersen compilation of minimal cell quotas in the section "Species-Specific Differences" above). Doing so facilitates comparison of various species as different species generally differ in cell size. One can rearrange expression (3.2) to derive a relationship between cellular elemental composition for a limiting nutrient X and relative growth rate of the autotroph,

$$\text{RGR} = \frac{\mu}{\mu'_m} = 1 - \frac{(X:C)_{\min}}{(X:C)_A}, \tag{3.4}$$

where $(X:C)_{\min}$ is the minimum quota in stoichiometric terms and $(X:C)_A$ is the algal $X:C$ at a given positive growth rate. Alternatively, we can just as readily express the Droop relationship in terms of algal $C:X$,

$$\text{RGR} = \frac{\mu}{\mu'_m} = 1 - \frac{(C:X)_A}{(C:X)_{\max}}, \tag{3.5}$$

where $(C:X)_A$ is the algal $C:X$ at growth rate μ and $(C:X)_{\max}$ is the maximum $C:X$, which is achieved at zero growth rate under X-limited conditions. Equation (3.5) predicts a linear, negative relationship between RGR and algal $C:X$ with a slope of $-1/(C:X)_{\max}$. This expression is predicted to

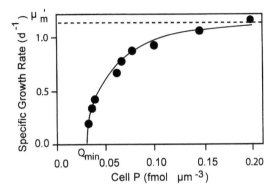

Fig. 3.7. Relationship between growth rate (μ) and cellular P content (cell quota, Q) under P-limited growth for the alga *Scenedesmus*. The curve indicates a fit to the data using the Droop model [Eq. (3.1)]. The parameters of the Droop equation are shown on the axes. Based on Rhee (1973).

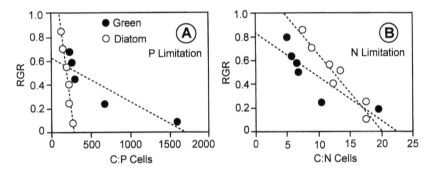

Fig. 3.8. RGR (μ/μ_m) and cellular C:nutrient are inversely related in two species of algae. A. The association of RGR and cellular C:P under P limitation is shown for a diatom (*Cyclotella meneghiniana*, open circles) and a green alga (*Scenedesmus acutus*, dark circles). B. The same two species but under N limitation. Based on Sterner (1995).

hold as long as nutrient X remains the nutrient limiting growth, even as the RGR approaches 1. The stoichiometric version of the Droop model [Eq. (3.5)] gives dynamical meaning to the static measure of elemental composition.

Support for the Droop Model

A considerable amount of data from chemostats has been analyzed using the Droop model, including tests made by Droop (1983) himself. We already saw that the Droop model [Eq. (3.1)] provided a good fit to the data in Figure 3.7. Looking back at our other previous examples, we can find several cases where the expected linearity between cell stoichiometry of the limiting element and growth rate predicted from the Droop equation was seen (Figs. 3.2A for P, 3.4A for both N and P). There was one case where there was apparent nonlinearity and thus some departure from expectations based on the Droop model (Fig. 3.5A for P).

Sterner (1995) examined the growth–elemental composition relationships for a diatom (*Cyclotella meneghiniana*) and a green alga (*Scenedesmus acutus*) when grown under P or N limitation (Fig. 3.8). Under P limitation, the RGR declined strongly with increasing C:P for both species but the decreases differed for the two species, further illustrating the species dependence of growth-C:N:P relationships discussed previously. The data fitted the linear decline predicted by the Droop expression very well for the diatom but the green alga had a nonlinear response (Fig. 3.8A, note departures from dashed line). Similar changes of growth rate with C:N were observed under conditions of N limitation (Fig. 3.8B), where again the predicted linear relationship was seen for the diatom but a somewhat nonlinear response was seen for *Scenedesmus*. Note in this plot that growth rate is plotted on the y axis and elemental composition on the x axis, while in previous plots related to chemostat studies the authors presented data with growth rate on the x axis and elemental composition on the y. This raises the question of what is causing what here. Does nutrient content cause growth change or does growth cause change in nutrient content?

From different perspectives and for different purposes, both views are correct. In chemostats, the experimenter sets the growth rate; cells develop a characteristic biochemical and elemental signature in response to that regime. However, as should be obvious from basic biology and from discussions we will develop in Chapter 4, it is the biochemical machinery that causes growth itself. In other situations, as in batch culture mode, one can think about nutrients being supplied and then algae taking them up and processing them to produce a certain growth rate, leading to an association of growth rate and elemental composition. It is interesting to note that in some situations in nature we might find scenarios more like the chemostat (for example, when grazers impose high loss rates on algae, requiring surviving algal populations to grow fast to keep up) or more like a batch culture (for example, when spring mixing introduces large amounts of nutrients into the surface layers and a small inoculum of algae is present). In summary, we have seen that the Droop relationship connects stoichiometry to growth conceptually and, at least in some instances, precisely and mathematically as well.

Allocation Patterns and Cell Quota

The Droop model connects growth to a single total cellular pool of nutrients. By itself it does not predict changes with light or temperature, though it can be used to fit data at different light or temperatures. The importance of establishing a satisfactory mechanistic understanding of the determinants of phytoplankton elemental composition has motivated a variety of additional theoretical developments (Shuter 1979; Laws et al. 1983; Geider et al. 1998). These have moved beyond the phenomenological approach of the Droop equation (3.1) to consider the plasticity of cellular allocation in different conditions. These models seek to understand changes in algal C:N:P stoichiometry as reflecting adaptive responses of functional biochemical pools to environmental forcing factors.

One of the first of these models builds from basic biochemical and cellular principles to make predictions about allocation patterns and autotroph growth under contrasting environmental conditions. Shuter (1979) developed this model for unicellular algal growth. In the model, biomass (carbon) is allocated into four functional pools (storage, structure, photosynthetis, and synthetis). The structural pool includes cell wall, cell and internal membranes and associated proteins, and genetic material. The photosynthetic pool includes all components involved in conversion of inorganic C to stored carbon, including pigments, chloroplast membranes, and photosynthetic enzymes such as RUBISCO. Cellular components involved in conversion of stored carbon into functional cellular material (ribosomes, mitochondria) are in the synthetic pool. A variety of assumptions based on established physiological principles regarding effects of temperature, light, and nutrients on cellular metabolism and growth are incorporated. Of particular importance to us, the nutrient contents of the three functional (i.e., nonstorage) compartments are assumed to be equal, while the storage compartment material is assumed to lack N and P. The analysis involves determining the optimal allocation of biomass to various compartments as a function of changing conditions of nutrient supply, light intensity, or temperature. The optimal solution is the allocation pattern producing the highest growth rate under a given set of environmental conditions. The model predicts changes in cellular allocation patterns as a function of growth rate under different patterns of limitation by light and nutrients and under different temperature regimes.

Several of Shuter's model's predictions about growth and optimal cellular allocation under differing light, nutrients, and temperature are shown in Figure 3.9. The figure indicates the underlying assumption of a fixed component of structural biomass (F). Under conditions of constant temperature and nutrient supply (Fig. 3.9A), increasing growth due to increas-

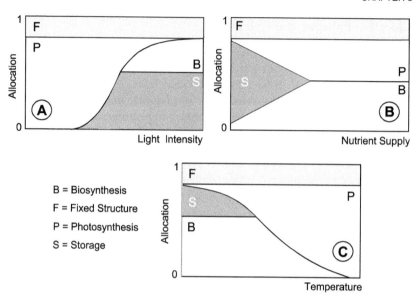

Fig. 3.9. Shifts in algal cellular allocation in response to changes in environmental conditions, as predicted by the Shuter model. A. With increasing light intensity, the model predicts decreased allocation to photosynthetic apparatus (P) and increasing allocation to biosynthetic constituents (B) and storage (S). Such a change should result in increased cellular C:nutrient because the storage component is assumed to lack nutrients. Allocation to structural materials (F) is fixed in the model. B. With increasing nutrient supply, the model predicts decreased allocation to storage and increased allocation to photosynthetic and biosynthetic components. This should result in decreased cellular C:nutrient. C. With increasing temperature, a shift of allocation between biosynthetic and photosynthetic compartments is predicted, along with a shift away from storage. This should result in a modest decline in cellular C:nutrient with increasing temperature. Based on Shuter (1979).

ing light intensity is predicted to first shift allocation away from the photosynthetic apparatus (P) and toward the biosynthetic compartment (B). Allocation to the C-rich storage compartment (S) does not increase until nutrient demands for growth exceed the ability of the cell's uptake and transfer systems to acquire and process nutrients at a pace to match increased carbon acquisition as light intensity increases. Under constant temperature and light intensity (Fig. 3.9B), increasing growth due to increased nutrient supply occurs in concert with proportional increases in the contributions of photosynthetic (P) and biosynthetic (B) compartments and at the expense of the storage compartment (S). Increased temperature affects growth and allocation (Fig. 3.9C) via effects on growth maxima, half-saturation constants, synthetic efficiency, and maintenance metabolism. This leads to an increase in allocation to photosynthetic compartments (P) relative to biosynthetic compartments (B) at higher temperature, a direct result of the assumption that temperature affects the operation of the synthetic but not the photosynthetic apparatus.

Shuter tested the model with data from cultured phytoplankton to see whether it accurately predicted cellular allocation with changed growth rate. He found reasonable agreement between data and theory. He did not explicitly consider cellular C:nutrient in his tests but we will show how his tests can be used to make stoichiometric predictions. Recall that photosynthetic, biosynthetic, and structural components were all considered to have the same elemental composition while only the storage compartment differed in nutrient content (it contained only C and no nutrients). Thus, cellular C:nutrient should track the proportion of total cell biomass within the storage compartment: high cellular C:nutrient will be associated with relatively large storage pools and low allocation to synthetic and photosynthetic compartments. Conditions where other functional components (structural, photosynthetic, or synthetic) dominate will have lower cellular C:nutrient ratios. Shuter's model captures the qualitative nature of changes in C:X under light and nutrient limitation seen in chemostats (Figs. 3.2 and 3.4). Specifically, his model predicts decreased C:X with decreased growth under light limitation, but higher C:X with decreased growth under nutrient limitation. This correspondence suggests that stoichiometric responses to light and nutrients result from optimal allocation of biomass to key subcellular pools that differ in elemental composition. Shuter's model demonstrates that adaptive reasoning from first principles of cellular allocation can yield qualitatively realistic predictions of stoichiometric response.

A more difficult proposition would be to see if cellular allocation models can also quantitatively predict algal elemental composition in laboratory studies. Laws and colleagues (1983, 1985) tested predictions of a modified version of the Shuter model against existing and new chemostat data. To do so, they needed to estimate some parameters of the model via least-squares regression using chemostat results, and thus growth rate predictions were not built exclusively on mechanistic relationships independent of the data. Nevertheless, the results provide some insight into the utility of focusing on cellular processes in understanding variation in elemental composition. Predicted changes in C:N with growth rate under N limitation agreed well with actual observations (Fig. 3.10A). High N:C (low C:N) occurred under conditions of high growth rate as cellular allocation moved away from storage pools lacking N to N-rich pools involved in active growth. However, the match between model and data was somewhat weaker for P (Fig. 3.10B). It is interesting to note that the model did not successfully

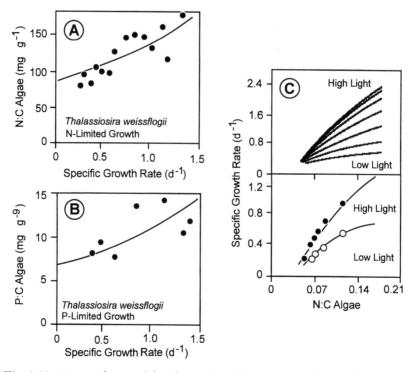

Fig. 3.10. Empirical tests of the Shuter (A and B) and Geider (C) cellular allocation models. A. Predicted and observed cellular N:C as a function of growth rate
under N-limited conditions. B. Predicted and observed cellular P:C as a function of
growth rate under P-limited conditions. C. Predicted (top) and observed (bottom)
relationships between algal growth rate and N:C as a function of light intensity
under N-limited conditions. Note in the top panel of C that cells grown at a given
growth rate at high light have lower N:C than those grown at lower light. The
bottom panel illustrates the close correspondence between model predictions and
chemostat data for the chrysophyte *Pavlova lutheri* grown under N limitation at
high (189 μmol photons m^{-2} s^{-1}) and low (63 μmol photons m^{-2} s^{-1}) light
intensity. Panels A and B are based on Laws et al. (1983) and panel C is based on
Geider et al. (1998).

account for observed variations in RNA and ATP allocations (data not
shown). The poor ability of the theory to work for P-intensive biochemicals
might be traced to Shuter's assumption that functional compartments were
uniform in elemental composition. Our analysis in Chapter 2 showing that
major molecules and organelles are similar in N content but differ in P
content (Figs. 2.2 and 2.4) indicates that an assumption of uniform chemical content within major pools would be more valid for N than for P. More

specifically, the machinery of energy metabolism (mitochondria, chloroplasts) is N rich but low in P (high N:P) but the machinery of biomass production (ribosomes) is similarly N rich but also disproportionately P rich (low N:P). Thus, the assumption that the P content of the synthetic compartment is similar to that of the photosynthetic and structural compartments is probably invalid. Improved predictions of P:C and allocation to P-rich molecules might result if more accurate assessments of the differing P demands of these compartments were incorporated into the formulation.

In a more recent analysis, Geider et al. (1998) constructed a model based on several features of phytoplankton physiological acclimation. These were (1) down-regulation of pigment content at high light, low nutrients, or low temperature, (2) accumulation of energy-storage polymers under conditions of high light and low nutrients, and (3) feedback between nitrogen and carbon metabolism. Their model explicitly calculated relationships between cellular elemental composition (N:C) and growth rate as a function of light, nutrients, and temperature. They also compared their model's predictions with data from cultured algae. One of their major results was that Droop plots were very light dependent, with greater growth at a given N quota under increased irradiance (as in Fig. 3.4A). These relationships also say that increased light at a given algal growth rate lowers cell N quota (and thus increases algal C:N). Model and data were in good agreement (Fig. 3.10C).

Studies such as these demonstrate the power of focusing on the C:N:P of functionally contrasting molecules and organelles (Chapter 2) in developing mechanistic principles regulating biomass stoichiometry under various growth conditions. We will see another detailed application of this approach when we consider the question of how to build an animal in Chapter 4.

Terrestrial Approaches: The Productivity Equation

Theoretical connections between tissue nutrient concentrations and plant growth rate have also been advanced in the terrestrial plant literature. In this section we consider the approach of Ågren (1988), whose models begin from a different starting point, but, as we will see, converge upon solutions very similar to the Droop model.

Ågren (1988) argued that a theory of plant nutrition should be built not on the basis of plant specific growth rates and tissue nutrient concentrations but instead on absolute growth and absolute amounts of biomass and nutrients. His reasoning was that a plant interacts with its environment through its absolute size and absolute growth rate. The derivation is as follows. Assume that a plant's absolute rate of growth of biomass (dM/dt) is proportional to the total mass of nutrient X in the plant (Q') with a constant of proportionality (Φ'_M) equal to a species-specific character referred to as the "nutrient productivity." We will use the symbol Φ to refer to such proportionalities that relate growth according to one substance based on the amount of another substance. As always, the prime indicates mass units and the subscript refers to the substance beting gained, in this case, total mass (M). Nutrient productivity Φ'_M has units of biomass/(nutrient mass time)$^{-1}$. We will encounter this parameter again in Chapter 8. The proportionality means

$$\frac{dM}{dt} = \Phi'_M Q'. \tag{3.6}$$

Not all nutrients within the plant will be used for growth and thus there is a minimal amount of nutrient that must be present before growth can occur. Let this amount, expressed as a concentration, be referred to as $\Gamma_{x,min}$ (recall from Chapter 1 that Γ generally refers to a nutrient content; see also the Appendix). The parameter $\Gamma_{x,min}$ is conceptually identical to the $(X{:}C)_{min}$ we saw in the stoichiometric version of the Droop model in Equation (3.4). As before, growth rate increases with nutrient content. However, let there be an upper concentration $\Gamma_{x,opt}$ above which no increased growth is observed with further increases in tissue nutrient content. Under such assumptions, absolute plant growth rate is given by

$$\frac{dM}{dt} = \Phi'_M(Q' - \Gamma_{x,min}M) = \Phi'_M(\Gamma_x - \Gamma_{x,min})M, \tag{3.7a}$$

$$\text{where } \Gamma_{x,min} < \Gamma_x < \Gamma_{x,opt},$$

or

$$\frac{dM}{dt} = \Phi'_M(\Gamma_{x,opt} - \Gamma_{x,min})M, \text{ where } \Gamma_x \geq \Gamma_{x,opt}, \tag{3.7b}$$

and where Γ_x is the actual tissue nutrient concentration. These expressions are easily transformed into expressions for specific growth rate by dividing both sides by M:

$$\mu = \Phi'_M(\min[\Gamma_x,\Gamma_{x,opt}] - \Gamma_{x,min}). \tag{3.8}$$

The minimum function in Equation (3.8) takes the place of the two versions of Equation (3.7). Ågren (1988) referred to Equation (3.8) as the "productivity equation." Equation (3.8) predicts a rectilinear relationship between specific growth rate and internal nutrient content as depicted in Figure 3.11. Depending on the parameters $\Gamma_{x,min}$, $\Gamma_{x,opt}$, and Φ'_M, there is a variably broad range of tissue nutrient across which growth rate and nutrient content are directly coupled. Further, we can reexpress Equation (3.8) in terms of biomass stoichiometry. We do this by rescaling nutrient content as the ratio of nutrient X per unit carbon (X:C), rather than in its original biomass terms, by expressing growth in terms of C mass instead of dry mass, and by defining Φ as the rate of C production per unit plant nutrient X (and hence it is called Φ_C'). If we are in the region of subsaturating internal nutrient concentration ($\Gamma_{x,min} < \Gamma_x < \Gamma_{x,opt}$) then we can express Equation (3.8) in stoichiometric terms as

$$\mu = \Phi_C'[(X{:}C)_A - (X{:}C)_{min}], \tag{3.9}$$

where $(X{:}C)_A$ is plant X:C and $(X{:}C)_{min}$ is plant X:C at zero productivity. Thus, in the limiting region where $(X{:}C)_A$ is less than $(X{:}C)_{min}$, growth rate is a positive function of the X:C ratio with a slope of Φ_C' and an X intercept (where growth is zero) of $(X{:}C)_{min}$.

Ågren noted structural similarities between his approach and that of Droop, showing that Equation (3.1) can be rewritten as

$$\mu = \mu_m\left(1 - \frac{\Gamma_{x,min}}{\Gamma_x}\right). \tag{3.10}$$

Here, μ_m is specific growth rate at $\Gamma_x = \Gamma_{x,opt}$. Expressing this now in terms of relative growth rates and stoichiometric parameters,

$$\mathrm{RGR} = \frac{\mu}{\mu_m} = \left(1 - \frac{(X{:}C)_{min}}{(X{:}C)_A}\right), \tag{3.11}$$

or, in terms of plant C:X,

$$\mathrm{RGR} = \frac{\mu}{\mu_m} = \left(1 - \frac{(C{:}X)_A}{(C{:}X)_{max}}\right). \tag{3.12}$$

Note that Equations (3.11) and (3.12) are equivalent to Equations (3.4) and (3.5) (stoichiometric versions of the Droop equations), but only in the region where internal nutrient concentrations in Ågren's approach are less than optimal. Also note that the parameters are not formally equivalent. In the Droop model, maximum growth rate (μ_m') is defined as a theoretical maximum that is only approached in the limit while Ågren's maximum growth rate is somewhat more pragmatically defined. Despite these somewhat technical differences, the two approaches make qualitatively similar predictions about the relationship between autotroph elemental composition and growth rate. This is obvious from comparing Figure 3.11 with Figure 3.7. Ågren (1988) has explicitly considered the similarities and differences between his and Droop's approaches. He compared the fits of the two models to data on growth and internal nutrient content for two strongly divergent autotrophs: a tree (birch, *Betula pendula*) and a photosynthetic flagellate (*Monochrysis lutheri*). Data for experiments where these organisms were grown under N, P, and vitamin B_{12} (*M. lutheri* only)

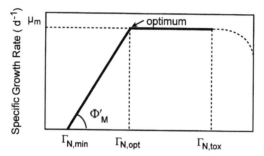

Fig. 3.11. Predicted relationship between plant growth rate and nutrient content based on the Ågren productivity equation [Eq. (3.8)]. The growth rate μ_m is achieved when plant nutrient content has reached or exceeds the optimal concentration ($\Gamma_{N,opt}$). The minimum nutrient content ($\Gamma_{N,min}$) needed for growth is shown, as is the "nutrient productivity" (Φ_M', the rate of biomass production achieved per unit plant nutrient). A point at which the level of internal nutrients becomes toxic is also shown but will not be considered further. An empirical example is discussed later (Fig. 8.7).

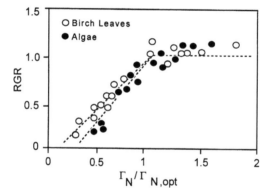

Fig. 3.12. Similar couplings of growth and nutrient content in "higher" and "lower" autotrophs. Here, for birch (*Betula pendula*) or an alga (a photosynthetic flagellate, *Monochrysis lutheri*), growth rate for each species is normalized against its maximum growth capacity (RGR = μ/μ_m) and its internal nutrient content is normalized against its optimal value ($\Gamma_N/\Gamma_{N,opt}$). The obvious similarity of the relationships for the two autotrophs indicates that the same degree of reduction in nutrient content results in the same degree of reduction of organism growth, even if the autotrophs are as different as trees and algae. Based on Ågren (1988).

limitation were analyzed. The relationships between specific growth rate (normalized to each species' maximum) and internal nutrient content (normalized to each species' $\Gamma_{x,opt}$ for each nutrient) were examined (Fig. 3.12).

There are two remarkable aspects of Figure 3.12. First, Ågren (1988) noted that both his productivity equation and the Droop equation fitted the data well. However, he preferred the productivity equation because it is built on basic principles of plant performance from which specific growth rate (μ) and nutrient content (Γ) are derived. In contrast, in the Droop expression μ and Q are primary variables and thus, he argues, the cell quota model of Droop is more descriptive and less mechanistic. To us, this difference is somewhat philosophical. Of greater importance in our view is the second remarkable feature of the data. There is a striking similarity in the growth-nutrition response for autotrophs as strongly divergent as a tree and a photosynthetic flagellate. When scaled to each taxon's specific values for $\Gamma_{x,opt}$ and μ_m, the relationships between growth and nutrient content for the two autotrophs are extremely similar. That is, a similar degree of reduction in the internal content of a limiting nutrient (relative to its optimal level) produces a similar relative reduction in growth rate in an autotroph, regardless of whether that autotroph is as complex as a tree or as "simple" as a unicellular alga. Note that Ågren is not saying that all species of autotrophs have the same elemental composition, nor does this mean that they have the same growth–nutrient content relationship. Rather, when expressed relative to each taxon's particular parameters, relative changes in growth and nutrient content have a similar form and magnitude across these two autotroph taxa. As we have seen repeatedly already, rapidly growing autotrophs are nutrient rich (low C:nutrient) and nutrient-limited autotrophs have elevated C:nutrient. If this result were found to be general across numerous species, this similarity would expose a general quantitative relationship coupling elemental stoichiometry and autotroph growth. Ecologists searching for unifying principles that hold across ecosystem types and major phylogenetic boundaries might wish to give serious consideration to Figure 3.12.

We have seen that a variety of expressions for the relationship between autotroph growth and elemental composition have been proposed. Regardless of the precise configuration of the mathematical expression, the key, we believe, is that stoichiometry provides direct information about a critical dynamic variable: growth rate. The insight that biomass stoichiometry and growth rate are strongly coupled has been applied frequently in interpreting differences in C:N:P for phytoplankton (e.g., Goldman et al. 1979; Harris 1986; Sommer 1988; Elser and Hassett 1994), in agronomy and forestry science (e.g., Greenwood 1976), and to a considerable extent in the study of unmanaged terrestrial ecosystems (e.g., Körner 1989; Vitousek et al. 1993a; Reich et al. 1999; Aerts and Chapin 2000). However, applications in several other important areas of ecological research, such as benthic ecology, are still scanty (for exceptions, see Kahlert 1998; Hillebrand and Sommer 1999). In the following sections we leave laboratory chambers and equations behind to compile and interpret collections of field data for autotroph C:N:P in oceanic, lake, and terrestrial ecosystems.

AUTOTROPHS IN THE WILD: OCEANS, LAKES, AND LAND

We have seen that the elemental composition of individual autotroph species is coupled to growth rate as modulated by the supply of limiting nutrients relative to the demand set by available energy inputs. This coupling suggests that autotroph C:N:P in the field might indicate the identity of a limiting nutrient and the severity of growth limitation (Goldman et al. 1979; Healey and Hendzel 1980; Vitousek and Howarth 1991; Tanner et al. 1998). Although the various relationships with light, temperature, and supply N:P ratios, etc., need to be considered, C:N:P has commonly been used to diagnose nutrient limitation in both terrestrial and aquatic ecosystems. We discussed possible N:P thresholds delimiting N versus P limitation in autotrophs earlier (Downing and McCauley 1992; Verhoeven et al. 1996). Partly for this reason, there have been numerous studies of autotroph elemental composition in the wild. Here we consider some of these efforts, examining how three different major ecosystem types (terrestrial, freshwater, and marine) are based on autotrophic biomass differing in means and variances of elemental composition. At the same time, we will also find some surprising similarities in N:P ratios among these different ecosystems. We emphasize stoichiometric patterns in autotrophs here; we will reexamine a few of these contrasts and patterns in the context of the stoichiometry of food webs and whole ecosystems in Chapter 8.

 Comparing autotroph stoichiometry across ecosystems presents some methodological and conceptual challenges. The nature of the information that can be collected on autotroph C:N:P in pelagic systems differs in several important ways from information that can be collected on terrestrial autotrophs. Pelagic autotrophs are microscopic, multispecies assemblages with overlapping cell sizes, mixed to varying degrees with detritus, bacteria, Protozoa, and small metazoan zooplankton. One cannot pluck sufficiently large species-specific samples of autotrophs from natural communities like these. Instead, as we saw in Chapter 1, the elemental composition of the base of pelagic food webs is measured on bulk samples of particles ("seston") on filters. Water samples are generally first passed through screens (50–100 μm mesh) to remove larger metazoan zooplankton prior to concentrating particulate matter onto fine mesh filters (which often collect most, but usually not all, heterotrophic bacteria). Seston can be size fractionated to assess the contribution of small, bacteria-sized particles to elemental signatures (e.g., Elser et al. 1995a) or in some cases to obtain approximately monospecific, detritus-free fractions for elemental analysis (e.g., Sommer 1988). In some situations (particularly at very low productivity: del Giorgio and Peters 1994; Biddanda et al. 2001) perhaps a quarter or as much as half of seston biomass is comprised of heterotrophic bacteria or zooplankton (Sterner et al. 2001). Seston includes all particles, living and dead; however, several lines of evidence indicate that measurements of the elemental composition of particulate matter in stratified lakes are not greatly influenced by detrital contamination (Harris 1986; Sommer 1988; Hecky et al. 1993; Urabe 1993b; Elser et al. 1995a). Nevertheless, there are situations where detrital contamination might be significant (e.g., lakes of low residence time, with considerable external litter inputs or extensive resuspension of sediments). The data examined here contain few if any such environments. Collections of particles will also average over any intraspecific differences in nutrient content or ratios, which are known to occur (Gisselson and Granéli 2001). Finally, seston stoichiometry in many environments is a function of depth; considered here are data based on samples from the relatively homogeneous mixed upper layer in stratified waters. In general, the chemical content of suspended particulate matter in most cases should be a reliable indicator of the composition of bulk autotroph biomass.

Considering the nature of seston, truly equivalent data for terrestrial ecosystems would require a sample consisting of biomass-weighted collections of all autotroph and small herbivore species comprising the community, with the addition of some surface leaf litter and soil microbes. To get a measurement equivalent to "seston" in a forest, one can imagine a giant ecologist collecting a forest sample with a very large cork borer! Most terrestrial data involve information only for leaves (and to a much lesser extent stems) but the biomass present in terrestrial systems can include a large amount of wood and bark as well as roots. We know that wood has elemental composition differing greatly from leaves (Table 3.1) but there seems to be little information available on the stoichiometry of roots. An example of measurement of total aboveground stoichiometry in a deciduous forest ecosystem will be given in Figure 8.6D; such studies are rare. Finally, another difference is that because terrestrial data are collected at the species level they include variability arising from site-specific factors as well as from species-specific characteristics. Species-level variability in pelagic ecosystems would be lost in the bulk measurement. In spite of these challenges, comparisons of stoichiometric data for aquatic and terrestrial systems are informative if we desire insight into broad patterns like differences in the quality of potential food available to generalist herbivores; the composition of actively photosynthesizing biomass; the parent material potentially entering detrital food webs; or the nature of the coupling between mineral cycles and the carbon cycle on annual time scales.

Elemental analysis of "wild" plant tissues in terrestrial ecosystems has been somewhat of a cottage industry for more than 20 years. Impressive amounts of data for the nutrient content of field-collected leaves are available. A working group at the National Center for Ecological Synthesis and Analysis (NCEAS) has compiled a considerable quantity of these data in characterizing the C:N:P stoichiometry of autotroph biomass in terrestrial ecosystems (Elser et al. 2000b). Most studies that were used reported elemental composition as a percentage of tissue dry mass. Values for % N and % P were converted to C:N and C:P by assuming that C was 46% of mass (the mean value when reported). The data encompassed observations for 406 plant species compiled from dozens of different studies and represents the largest such compilation to date.

The NCEAS working group also added to existing compilations of data on freshwater particulate matter in stratified lakes (Hecky et al. 1993; Elser and Hassett 1994). The new lake data set also is the largest to date. It includes measurements for ~270 lakes. While this represents a large number of study sites, it has sampling biases. In particular, the data are strongly influenced by the small lakes of temperate North America.

The third major ecosystem type we consider is oceans. As we have discussed in Chapter 1, characterization of oceanic particulate matter has been a common theme in oceanographic research ever since Redfield's work decades ago. These studies have expanded on Redfield's concept of a biogeochemically balanced ecosystem and identified a number of spatial and temporal patterns. For example, work by Karl and colleagues (Karl et al. 1995; Karl 1999) shows increases in C:P and N:P of particulate matter during El Niño conditions in the central Pacific gyre, as phytoplankton growth apparently shifted from N to P limitation under El Niño Southern oscillation (ENSO) forcing (see Chapter 8). For comparison with lake particulate matter and terrestrial vegetation from the NCEAS analysis, we use data from the study of Elser and Hassett (1994), who reported particulate matter stoichiometry for marine surface waters at 17 sites that included estuaries, embayments, and coastal ocean systems around North America. This data set is quite limited in scope but considering the relative uniformity of C:N:P in particulate matter in the world's oceans, inclusion of additional data would not alter the contrasts we discuss below.

So, what are the patterns in C:N:P stoichiometry "in the wild," i.e., at the base of the food webs of oceans, lakes, and terrestrial ecosystems? The data reveal differences among these three ecosystems both in mean values and in variation around those means. Given what we have already seen in

Chapter 1 regarding marine systems, it is not surprising that oceanic particulate matter was relatively homogeneous across sites and was nutrient rich (relatively low C:nutrient; Fig. 3.13). The C:N of oceanic particulate matter ranged from 4.5 to 10 with a mean value of 7.7, close to Redfield's classic proportion (6.6). Similarly, particulate C:P varied over a comparatively narrow range (90–250) with a Redfield-like mean of 143 (recall that Redfield C:P is 106). The N:P stoichiometry also fell within a narrow range (12–29) with a mean value (19) close to the Redfield value of 16. Although particles in the offshore ocean may be sparsely distributed, on a per biomass basis, they are nutrient rich.

The stoichiometrically uniform base of oceanic food webs contrasts strongly with the situation in lakes (Fig. 3.13). In lakes, particulate matter was often very phosphorus poor. The mean C:P was 307 and the range was 55–1630. The mean N:P was 30 and the range was 6.5–125. However, lake C:N was only slightly elevated relative to typical marine particulate matter (mean 10.2; range 2–24). These ratios indicate that lake phytoplankton are commonly phosphorus deficient (we discuss N vs. P limitation in lake phytoplankton more below, and also in Chapter 8).

Let us now compare aquatic and terrestrial ecosystems. In comparison to seston, terrestrial autotrophs are distinguished by extremely low N and P content (Fig. 3.13). Foliar C:N and C:P both typically greatly exceeded ratios in aquatic particulate matter, even for freshwaters. For example, leaf C:P ranged from Redfield-like proportions of 115 to very high values up to 5990. Mean C:P for leaves was 970, making C more than eight times more abundant compared to P than in the Redfield ratio. Similarly, C:N was high and variable, ranging from 7.5 to 225. Its mean was 36. The contrast between aquatic and terrestrial stoichiometry is highlighted by noting that the most nutrient-rich leaf on land is no higher in nutrients than the average particle in the ocean! Further, the contrasts based on Figure 3.13 certainly underestimate what we would expect for equivalent samples of entire autotroph biomass, as the terrestrial data are for leaves and do not include low-nutrient woody tissues. Even in relatively nutrient-rich secondary growth situations, wood and bark contain approximately half of the aboveground N and P and by implication the great majority of carbon in biomass (see Fig. 8.6D).

Different ratios varied to different degrees and habitats differed in the degree of variability of particular ratios. In each habitat, C:P was generally the most variable (as measured by coefficient of variation) of the ratios (Fig. 3.14A, gray bars). For each stoichiometric ratio, variability in autotroph stoichiometry was lowest in oceanic systems and greatest for terrestrial autotrophs, although variability of freshwater seston approached variability of terrestrial leaves (Fig. 3.14B). However, the contrast in variability between terrestrial and aquatic autotrophs is not necessarily a sign of

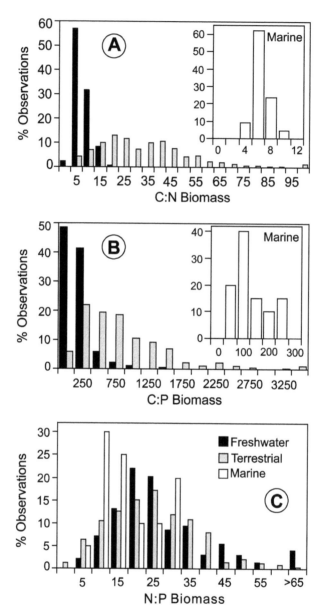

Fig. 3.13. Variation in C:N:P stoichiometry in autotroph biomass at the base of marine (white), freshwater (black), and terrestrial (gray) food webs. Data for lakes and oceans reflect values for suspended particulate matter samples ("seston") while data for terrestrial habitats reflect values for individual leaves of various species from a variety of habitats. A. C:N B. C:P C. N:P. For estimates of variability, see Figure 3.14. Terrestrial and freshwater data from Elser et al. (2000b) and marine data from Elser and Hassett (1994).

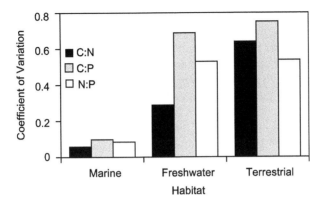

Fig. 3.14. Degree of stoichiometric variation at the base of aquatic (marine and freshwater) and terrestrial food webs (for means, see Fig. 3.13) using coefficients of variation (the standard deviation divided by the mean). C:P is the most variable of the three ratios and C:nutrient was most variable in terrestrial systems but the degree of variation in N:P was essentially identical for lake and terrestrial data.

greater between-site variability in the mineral nutrition of terrestrial autotrophs. Recall that the terrestrial data are at the species level. Thus, these data include variability arising from site-specific factors as well as from species-specific characteristics. If it were possible to obtain data for autotroph C:N:P stoichiometry at the species level for phytoplankton we might also observe variation in the lake data closer to that observed for the terrestrial autotrophs.

Recall that Nielsen et al. (1996) indicated that stoichiometric differences in autotrophs related more to organism size than habitat or functional group. However, microscopic plankton dominates both freshwater and marine seston, so differences between those two groups cannot be attributed to size. For a further comparison that would allow us to explore stoichiometric patterns in autotrophs related to size, habitat, and other factors, we can expand our analysis to include large aquatic plants. Atkinson and Smith (1983) reported a median C:N:P ratio for benthic marine macroalgae and seagrasses of 550:30:1. This high-carbon stoichiometry could reflect relatively low, nutrient-limited growth rates, or it could be based on a contribution of structural carbohydrates. Fernández-Aláez et al. (1999) looked at C, N, and P content in three functional groups of large aquatic plants (macroalgae, emergent angiosperms, and floating or submerged angiosperms). Their results do not readily relate to structural investment; the macroalgae had lowest P and C contents but the emergent angiosperms had the lowest N content.

Further considering the stoichiometry of large aquatic plants, Duarte

(1992) compared the N and P content of 46 macroalgal species, 27 sea-grass species, 11 species of freshwater angiosperms, and several mixed phytoplankton and macroalgal communities. The N and P content of all these groups overlapped, but the range for phytoplankton included higher N and P content than for the larger autotrophs. Generally, values for species from all the groups were consistent with single stoichiometric trends across all the data, for example, in N versus P contents. From this information, Duarte concluded that differences among the groups reflected "a greater degree of P and N limitation of growth of natural macrophyte populations, rather than an intrinsic difference in their chemical composition relative to that of phytoplankton." Such a conclusion may work for aquatic plants with reduced need for C-rich structural biomass, but would be inappropriate in comparing aquatic to terrestrial autotrophs, the latter having often huge structural investment. The nutrient-poor stoichiometry of terrestrial autotrophs compared to aquatic species is clearly a manifestation of physical factors that necessitate investment in C-rich structural carbohydrates in cell walls and for construction of support structures (Chapter 2) to maintain the elevation, shape, and orientation of leaves and stems in air. The extra support afforded by the aquatic medium and their small size seems to indicate that aquatic plants do not require such heavy structural investments and thus they incorporate significantly less carbon for every unit of N or P assimilated than terrestrial plants (Fig. 3.6A). The importance of this additional carbon in terrestrial autotroph stoichiometry is underscored by observing that in the NCEAS data foliar N:P did not differ between aquatic and terrestrial autotrophs. While terrestrial C:N and C:P were many times higher than in lakes and oceans, the N:P of terrestrial foliage (mean 28; range 3–110) was indistinguishable from the N:P of lake particulate matter.

We conclude that—despite differences in the relative contribution of C-rich, low-nutrient materials to terrestrial autotroph biomass—the central "core" of actively metabolizing biomass of all autotrophs is fundamentally the same. This similarity is striking and, based on the logic used in the previous chapter, it suggests that the biochemical evolutionary pressures on all autotrophs in all habitats are similar except for differing demands for structure. The similarity in composition also reminds us of the nearly identical thresholds for N versus P limitation in terrestrial vegetation (33; Verhoeven et al. 1996) and lake phytoplankton (31; Downing and McCauley 1992) that we noted earlier. Further, the fact that the mean N:P values of autotrophs in lakes and on land fall remarkably close to these threshold values is truly intriguing. A number of fundamental questions are raised by the similarity of all these N:P ratios and thresholds. For example, do they mean that existing paradigms of a predominance of P limitation in lakes and N limitation on land are incorrect? Further, are there widespread factors that maintain autotroph nutrient limitation evenly balanced between N and P limitation? We will come back to the general issue of N:P balance in ecosystems in Chapter 8.

CAUSES OF VARIATION IN AUTOTROPH C:N:P IN NATURE

Now that we have described the stoichiometric variability of autotroph composition in the field, let us examine the question of what causes this variation. Specifically, we will ask the question, "What environmental factors affect autotroph growth status and thus contribute to variation in C:N:P stoichiometry in the field?"

Pelagic Systems

We have recently developed some ideas about the key mechanisms behind the strong contrasts between particulate C:N:P stoichiometry in lakes versus oceans (Elser and Hassett 1994), in large versus small lakes (Sterner et al. 1997), and in lakes in different regions (Hassett et al. 1997). In thinking about nutrient limitation, it is easy to focus on an inadequate nutrient supply ("Hey, there aren't enough nutrients around here!"). Unfortunately, it is equally easy to overlook the question that should naturally follow, "Not enough for *what*?" "What" in this case refers to autotroph demand, and if we are to understand nutrient limitation our attention should focus not only on factors related to nutrient supply but also on those associated with demand. Nutrient limitation occurs when the available nutrient supply falls short of demand.

Earlier we reviewed considerable evidence from laboratory studies that phytoplankton elemental composition is a function of nutrient-limited relative growth rate (Figs. 3.2–3.4, 3.7). In nature, μ is often a function of nutrient supply rate and thus of factors such as external loading, internal returns (such as diffusion of nutrients across the thermocline from nutrient-rich bottom waters to nutrient-limited productive layers), and recycling by heterotrophs (e.g., by zooplankton; see Chapter 6). Factors influencing demand (μ_m) are less obvious. One major factor is light. We saw earlier that increased light intensity leads to increased algal growth capacity (μ_m) (Shuter 1979) and to increased algal C:nutrient (Fig. 3.4). Thus, environmental conditions that increase light intensity may make it more difficult for nutrient supply rates to keep up with nutrient demand of algae with high μ_m; more and more C is incorporated per unit nutrient as light intensity and algal growth capacity increase. This line of reasoning led Sterner et al. (1997) to propose a "light:nutrient hypothesis," which says that particulate C:nutrient in lakes is regulated by the ratio of light to nutrients

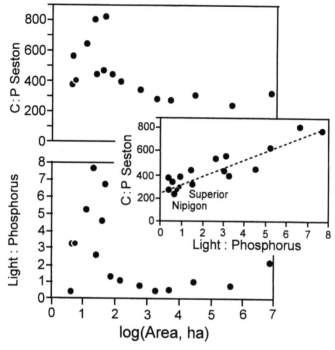

Fig. 3.15. Variation in elemental composition and growth conditions at the base of the food web in pelagic ecosystems as a function of lake size (surface area). The top graph shows that particulate matter has variable and sometimes high C:P in small lakes and the bottom graph shows that algae growing in the mixed layer of small lakes experience variable and often disproportionately high illumination (mean light intensity in the mixed layer, I_m) relative to nutrient availability (as gauged by total P concentration, TP). Lakes greater than about 100 ha surface area are consistently low light:nutrient environments and have consistently low seston C:P. The inset shows that particulate C:P and light:nutrient ratio (I_m:TP) are highly correlated in this set of lakes. Data come from a series of lakes on the Canadian Shield. Based on Sterner et al. (1997).

supplied to phytoplankton. In lake environments with a high light:nutrient ratio, high light intensity elevates μ_m while low nutrient supplies prevent μ from keeping pace, producing low RGR and elevated C:nutrient. In contrast, in situations of low light:nutrient in the environment, low light intensity would produce low algal μ_m, putting less demand on nutrients and maintaining the RGR at high values, leading to low C:nutrient in algae.

Such a mechanism is logical, but is it in fact a major factor in the real world? In other words, is the light:nutrient balance really responsible for the variation in phytoplankton C:nutrient in the wild? To test whether this was the case, Sterner and colleagues calculated an index of the light:nutrient ratio. "Light" was taken to be the mean light intensity in the mixed layer (I), a function of both the depth of that mixed layer and the degree of light attenuation. Nutrient supply is much more difficult to assess; Sterner and colleagues used the concentration of total phosphorus (TP) as an index of P supply. Whether this concentration truly measures a supply rate is a rather subtle and not yet answered question, but the logic is based on the assertion that large pools will generally have large fluxes in and out of them. Using these measures they calculated the I:TP ratio as an index of light:nutrient balance for lake systems. Having done so, they found a strong positive relationship between particulate C:P and I:TP (Fig. 3.15, inset). However, here they were treading in the dangerous waters of correlations with ratios (see Chapter 1). Thus, they also used multiple regression to evaluate relationships between particulate C:P and I, TP, and I:TP in a larger set of 89 lakes. They found that only I:TP was a significant term while individual components (I,TP) were nonsignificant predictors. This result emphasizes the importance of the balance of light and nutrients for autotroph stoichiometry (rather than the absolute values of one factor or the other) and suggests that the relationship between C:P and I:TP is real and not simply a statistical artifact of some kind. (If autocorrelation were strong, then there would have been a significant relationship between C:P and TP in the multiple regression.) A similar correlation between particulate C:P and I:TP was also seen in an independent data set of small lakes (Hassett et al. 1997). In that case, differences in particulate C:P for a set of Wisconsin lakes (low particulate C:P) and a set of Ontario lakes on the Canadian Shield (high particulate C:P) were associated with differences in both nutrients (TP was higher in Wisconsin) and light (I was higher in Canada).

So, if the light:nutrient balance causes differences in seston C:P ratios across lakes, what, then, causes variation in light:nutrient balance? Likely explanations are somewhat specific to limnology and so we will not discuss them extensively here. For the curious, we provide the following summary of arguments presented elsewhere (Fee et al. 1994, 1996; Hassett et al. 1997; Sterner et al. 1997). Mean light in the mixed layer is set by two primary parameters (Kirk 1983; Sterner 1990a): light extinction and mixed-layer depth. Differences in light extinction can arise from algal self-shading but also from turbidity due to suspended inorganic particles as well as from colored substances (dissolved organic carbon, or DOC). In the case of the Wisconsin versus Ontario contrast mentioned above, Ontario lakes had higher values of I primarily due to lower levels of colored organic matter (i.e., the Wisconsin lakes had browner water). Mixed-layer depth is set by a complex combination of local climatic conditions, lake orientation relative to the wind, lake size, and water clarity (Gorham and Boyce 1989; Mazumder and Taylor 1994; Fee et al. 1996). In general, mixed layers are deep (and thus have lower mean light) in larger lakes with greater exposure to the kinetic energy input of wind (for an example, see Urabe et al. 1999). In the contrast between large and small lakes in Figure 3.15, large lakes generally had deep mixed layers (and thus lower I) while small lakes generally had shallow mixed layers (and thus higher I). Nutrient supplies to epilimnetic phytoplankton are determined by many parameters, including watershed runoff, anthropogenic inputs, internal supplies from the sediments, cross-thermocline diffusion of nutrients, and internal recycling by bacteria and zooplankton. Thus, gauging the denominator of light:nutrient balance is an exceedingly complex task to which enormous effort in pelagic ecology has been devoted. In any case, the light:nutrient hypothesis calls attention to the possibility that understanding the impact of ambient nutrient supplies on pelagic ecosystems also requires an assessment of the light environment because of its influence on phytoplankton nutrient demand.

The impression of widespread, strong nutrient limitation of lake phytoplankton that we gained from Figure 3.13, while consistent with the conclusions of a number of previous studies (Sommer 1989; Elser et al. 1990), also contradicts some others. Harris (1986) noted that C:N and C:P for several of the Laurentian Great Lakes were relatively low and similar to the Redfield ratio. Harris concluded, "Nutrient limitation of growth cannot be a frequent occurrence in the surface waters of lakes and oceans." The difference between Harris's assessment and one drawn from Figure 3.13B can be resolved now that we see how lake size and stoichiometry are connected. This, therefore, is another stoichiometric ecosystem contrast: nutrient limitation of phytoplankton growth and resultant skewed elemental ratios are common in small lakes, but diminish with increasing lake size. Strong nutrient limitation, in fact, may not occur at all in large lakes. Harris (1986) considered seston data primarily from the large Laurentian Great Lakes and thus his view of the frequency of algal nutrient limitation likely applies to conditions in large but not small lakes. By this perspective, in terms of seston stoichiometry, the ocean is the world's "largest lake" (Sterner et al. 1997).

Focusing on the light:nutrient balance and its relationship to the laws of supply and demand permits the following paradoxical viewpoint: while one might attribute strong nutrient limitation and high phytoplankton C:nutrient to low nutrient supply, it may be just as correct to say that phytoplankton become strongly nutrient limited because they have too much light! This way of thinking leads one eventually to a concept of the "paradox of energy enrichment" (Loladze et al. 2000) to be discussed further in Chapter 7. The point emphasized here is that the elemental stoichiometry in lakes may be connected to geomorphic and watershed factors such as lake size and the influx of colored organic matter. Via such mechanisms, physical and geologic factors combine to influence autotroph stoichiometry in pelagic ecosystems and—as we will see in pages to come—affect the entire food web in many ways.

Terrestrial Systems

Is there any evidence that the light and nutrient balance also regulates autotroph elemental composition and ecosystem processes in other habitats? While we are aware of no studies that directly address this question, several conceptual approaches bear a close resemblance to these same ideas. For example, some considerations of plant functional morphology (Tilman 1988; Chapin 1993) have emphasized the opposing demands placed on terrestrial plants by vertical gradients of light and nutrients: plants must balance their resource allocations to extend upward for light and downward for nutrients. These trade-offs establish the well-known relationships between root:shoot ratios and plant nutritional status (Aerts and Chapin 2000). Thus, environmental conditions of contrasting light:nutrient balance should generate changes in the root:shoot ratio of individual plants (high light:nutrient should be associated with high root:shoot ratio) and, possibly, over the longer term, in the relative dominance of species differing in root:shoot allocation (but see various caveats on this extrapolation described by Aerts and Chapin 2000).

Hobbie et al. (1993) arrayed terrestrial plant functional groups (deciduous trees, shrubs, herbs, mosses, etc.) along a gradient of nutrient availability and argued that under high fertility, light becomes limiting due to self-shading. According to this view, light and nutrients vary inversely across terrestrial plant communities. In their scheme, high-nutrient environments support communities dominated by large deciduous trees, autotrophs with high specific growth rates (using our terminology) and high rates of leaf and root turnover. They also pointed out that under fertile conditions plants have high tissue N content (low C:N). In contrast, under low nutrient availability, light intensities are high and plant community composition is dominated by slow-growing herbs, mosses, and lichens with low rates of tissue turnover, high nutrient retention, and low tissue N content (high C:N). All of this sounds vaguely familiar to us.

Thus, it appears that, in both aquatic and terrestrial systems, autotroph functional morphology and physiology, evolutionary ecology, light:nutrient balance, and C:N:P stoichiometry are intimately inter-twined. The elemental composition of autotroph biomass in an ecosystem is simultaneously established by the mix of species, with their set of physiological responses to growth conditions, and by the current physiological status of those dominants. A complex set of dynamics is established: the relative supplies of light and nutrients influence the types of species present in the autotroph community, whose effects on light penetration and nutrient availability simultaneously alter environmental conditions experienced in the habitat and thus stimulate physiological and competitive adjustments. We will argue in upcoming chapters that these responses of autotroph C:N:P then have further feedbacks on growth conditions via stoichiometric effects on nutrient cycling by herbivores, detritivores, and decomposers.

Given the potential complexity of these reciprocal interactions, it is not particularly surprising that few ecological models have yet been produced that incorporate the full set of connections we have sketched in this chapter. Nevertheless, we will present evidence that the consequences of variable autotroph stoichiometry for ecosystem dynamics are profound, and therefore it is imperative that we make these effects a more explicit part of ecological theory. These consequences include major effects on plant-microbe interactions, rates of mineralization of detrital-bound nutrients, herbivore success and consumer-driven nutrient cycling, and food-web dynamics. Later, we will also discuss the findings of some new models of ecosystem function incorporating stoichiometric processes, showing that such models need not be overly complex and that they predict a rich array of dynamics not previously incorporated into our thinking about the workings of ecosystems.

CATALYSTS FOR ECOLOGICAL STOICHIOMETRY

Beginning with this chapter, we will be presenting our thoughts about the critical data, theory, or conceptual advances needed for ecological stoichiometry to progress. Like an enzyme lowering activation energy, we want to hasten the rate of formation of product, which in this reaction is knowledge.

In this chapter we considered how C, N, and P are coupled to each other during primary production. The stoichiometry of autotroph growth is one of the best-defined aspects of ecological stoichiometry due to the extensive literature on the mineral nutrition of algae and higher plants and well-developed theories of autotroph competition, such as resource ratio competition theory. Nevertheless, some important aspects of autotroph C:N:P stoichiometry remain unresolved.

- At the subcellular and genetic levels, what biochemical allocations and genetic mechanisms are responsible for patterns of autotroph elemental composition at low growth rate? In particular, we are interested in the physiology and evolution of minimal nutrient content in autotrophs, as this has major implications for competitive outcomes and, ultimately, for ecological dynamics as we will see in upcoming chapters.

- More refinements of models of biochemical allocation in autotrophs are needed. In particular, information about differential nutrient investments in various subcellular apparatus, such as that presented in Chapter 2, might improve the predictive ability of models attempting to model biochemical and elemental composition of autotrophs under different light and nutrient regimes.
- Autotroph C:N:P stoichiometry is not well documented in many important habitats. For example, what is the range of variation in elemental composition of periphyton in benthic systems (streams or littoral zones)? What environmental factors are associated with that variation?
- What abiotic or biotic factors regulate the C:N:P stoichiometry of aquatic autotrophs under natural conditions? Sterner et al. (1997) argue for a key role of the balance of light and nutrients in determining the elemental composition of the base of the food web in pelagic environments. Does this hypothesis hold in other habitats? How do atmospheric pCO_2 or water availability or temperature alter predicted patterns in terrestrial environments? Does pCO_2 play a yet unappreciated role in aquatic systems?
- How much variation in autotroph C:N:P stoichiometry among ecosystems is due to intraspecific variation (physiological adjustment) versus interspecific differences (shifts in species composition among taxa having different characteristic C:N:P ratios)? Analyses by Aerts and Chapin (2000) suggest that phenotypic plasticity is the primary factor responsible for variation in foliar nutrient content in terrestrial systems; Duarte (1992) suggested something similar for aquatic autotrophs. On the other hand, Nielsen et al. (1996) argued for an overriding importance of autotroph size.
- How do tissue-scale differences in C:N:P stoichiometry in higher plants scale up to whole-organism composition under various environmental conditions and allocation strategies?

SUMMARY AND SYNTHESIS

Photoautotrophy is arguably the dominant biogeochemical mode of organism nutrition in the living world. In stoichiometric terms, the growth of autotrophs is characterized by its extreme variability linked to fundamental growth processes. This variation has a basis in cellular structure (cell walls, vacuoles for excess energy and nutrient storage) and in autotroph physiology (the ability to flexibly adjust growth rate and cellular allocation to fit local conditions). This extreme plasticity, which is in strong contrast to the homeostasis of C:N:P maintained by animals (see Chapters 4 and 5), makes sense in light of the way that autotrophs make their living: unlike animals, which obtain their energy and elements together, preformed in food items, autotrophs must obtain their energy and materials from disparate sources (solar radiation and uptake of inorganic forms from the external medium). Since there is no necessity for supplies of solar radiation and various inorganic elements to be coupled in the environment, autotrophs must have strategies for capture and storage of these resources whenever and wherever they are made available. Thus, the physiologies of energy and nutrient acquisition are decoupled to a large degree in autotrophs. This, then, is another hypothesis for the evolutionary basis of the degree of homeostasis (we saw some other hypotheses in Chapter 1).

As a result, autotroph elemental composition varies considerably among ecosystems and this variation seems to be associated with the relative balance of key resources such as CO_2, solar energy, and mineral nutrients in the environment. Such plasticity generates the potential for considerable variation in C:N:P stoichiometry at the base of food webs. As we come to appreciate the nutritional requirements and strategies of other organisms in coming chapters, we will perhaps begin to recognize the overarching influence that autotroph C:N:P variation has for ecological dynamics. The fundamental contrast in nutritional strategy of autotrophs and animals implies that differences in autotroph and animal C:N:P should be commonplace and potentially extreme. Upcoming chapters (Chapters 5 and 7) focus on the frequency and magnitude of these differences and their implications for trophic dynamics and we will also consider their ramifications for large-scale ecosystem processes, such as nutrient recycling (Chapter 6) and C sequestration (Chapter 8). Next we consider the factors determining the C:N:P stoichiometry of animals and begin to glimpse the magnitude of elemental imbalance at the bottom of food webs.

4

How to Build an Animal:
The Stoichiometry of Metazoans

The time has come to stop talking about entropy in the biosphere as if it were the dominating feature and focus on kinetics, complexity, and molecular, cellular, and organismic hardware.—Morowitz (1992)

In the history of biology, the dominant way of analyzing animal growth is with biochemical (Lehninger 1971), physiological (Pandian and Vernberg 1987; Jobling 1994), or ecological (Wiegert 1976) energetics. Within ecological energetics, animals are abstracted as local thermodynamic perturbations—repositories of chemical energy—and whose feeding, metabolism, growth, and reproduction are analyzed by application of the energy concepts of thermodynamics: calories, free energy, work, entropy, heat, efficiency, and productivity. The basic assumption is that flows of energy organize thermodynamic systems, including biological ones (Morowitz 1968). Calories of food are ingested and expended via metabolism or captured in covalent bonds during organism growth. In ecological energetics, the key apparatus is the bomb calorimeter. However, as Morowitz implies above, a focus on "organismic hardware" may be a profitable approach. Here we use the foundation of Chapter 2 to build a view where animals are repositories of more than one thing, specifically, of multiple chemical elements. In the stoichiometric view, they are mixtures of multiple substances, as for example in the "human molecule" we encountered in Chapter 1. This "organismic hardware" must be assembled in proper proportions from the wide array of elements available in the environment via foraging, ingestion, digestion, assimilation, excretion, and egestion. For the field of ecological stoichiometry, the key apparatus is the elemental analyzer.

This chapter has a central question of, "What determines the elemental compositions of different animals?" For an answer we will build from the bottom up, relying heavily on information we presented in Chapter 2 on the biological chemistry of the elements. Since biological machinery is built with molecules of contrasting elemental composition, an animal's elemental composition is linked to its evolved structure and life history. Much of the thinking behind this chapter can be traced to the work of Reiners (1986) that we introduced in Chapter 1. Reiners dichotomy between "mechanical structures" and "protoplasmic life" will be important. Recall that the former, which for animals includes spines or shells for defense and bones for support, differ greatly in elemental composition. Reiners considered the latter, which is made up of the "soft" cellular constituents of animals, to be relatively uniform in elemental composition. However, building on the work of Elser et al. (1996), we will show why "protoplasmic life" also can vary considerably in its stoichiometry. We will try to demonstrate that how you build an animal depends on what kind of animal you want to build and how fast you want to build it (or, more accurately, how fast it wants to build itself). In Chapter 5, we will examine the implications of animal elemental composition for the stoichiometry of animal growth processes.

BIOCHEMICAL AND BIOLOGICAL DETERMINANTS OF BODY ELEMENTAL COMPOSITION

The primary biomolecules involved in the protoplasmic components of animal biomass fall into five main categories: carbohydrates, proteins, nucleic acids, lipids, and nucleotides involved in energy transformations (e.g., ATP, GTP, etc.). In Chapter 2 we showed that these molecules differ in C, N, and P content in a way that is consistent with their biological chemistry. Some important biomolecules contain neither N nor P (e.g., storage lipids), and increases in the contribution of such molecules to biomass will increase C content and reduce N and P content, leaving body N:P unchanged while increasing body C:N and C:P (Fig. 4.1). On the other hand, other important biomolecules contain considerable N and P and thus influence the elemental composition of the body in different ways. Increasing protein content (high N, low to zero P) will generally increase body N content (approaching the ~17% N value of the average protein, Fig. 2.2). It will also lower C:N and raise N:P (Fig. 4.1). In contrast, increased nucleic acid content (N- and especially P-rich, 9.5% P, Fig. 2.2) will generally raise body P content and lower body C:P and N:P (Fig. 4.1).

Biochemical constituents that vary over wide ranges of percent of biomass in animals include storage lipids in many taxa (for example, see the discussion of wax esters in copepods in Chapter 2) and nitrogenous wastes in some arthropods. We will also consider below the role of the vertebrate skeleton as a repository for the mineral elements. Sterner and Schwalbach (2001) considered growth patterns as a function of relatively short-term nutrient storage in zooplankton. In animals, homeostasis is often not

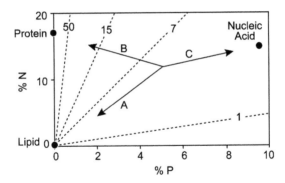

Fig. 4.1. Stoichiometric diagram illustrating how altering storage lipid allocation proportionately affects % N and % P, leaving N:P constant but changing C:P and C:N (arrow A); how altering protein allocation increases % N while lowering % P, increasing N:P (arrow B); and how altering nucleic acid allocation disproportionately affects % P, lowering N:P (arrow C). Dashed lines indicate particular N:P ratios.

entirely strict, but, in comparison with the variation in C:N:P in autotrophs we just encountered, animal stoichiometry is much less variable. Hence, while it is interesting to contemplate the consequences of large alterations in major biochemical pools for body elemental composition, individual animals generally do not store and deplete large percentages of most elements within their bodies. Instead, molecules are actively deployed in specific cellular and extracellular structures as the organism regulates its biochemical and elemental constitution while executing its developmental program. Overall, we tend to think of animals as homeostatic. Because specialized storage pools such as polyphosphates or nitrate in vacuoles are deemphasized, when animals differ strongly in their elemental composition, two things must be true. First, different cellular and extracellular structures must differ in elemental composition. Second, animals must have different contributions of such structures to their body make-up.

Chapter 2 was about the first of these points: biochemical and cellular structures do differ considerably in elemental composition. For example, mitochondria and ribosomes have contrasting elemental composition (Fig. 2.2) primarily because of the P-rich RNA in ribosomes. Different allocations to energy metabolism versus growth should therefore affect organismal stoichiometry. Extracellular materials also differ considerably in their elemental composition. As we have seen, chitin, the structural carbohydrate molecule in many invertebrates, lacks P but contains N (N:P = ∞) while bone has both N and P but is particularly rich in the latter (N:P of

~0.8). Thus, evolutionary trends toward larger sizes necessitate stiffer structural support, which has qualitatively different stoichiometric effects in invertebrates and vertebrates. Increased structure should raise body C:P and N:P in invertebrates but lower them in vertebrates. We will discuss this contrast in more detail later in this chapter.

   In the following, we will focus on metazoan animals, exploring several interlinked questions. Does the relative contribution of organelles or extracellular materials differ significantly among animals? (It does.) Is such variation linked in intelligible ways to organism traits likely to be important for an organism's survival and reproduction? (It is.) Much of the rest of the book will then be devoted to showing that the resulting contrasts in body stoichiometry of animals have wide-ranging ecological repercussions.

INVERTEBRATE STOICHIOMETRY:
C:N:P IN ZOOPLANKTON AND INSECTS

Compared to differences across, say, microbial taxa, or when compared to plants, all metazoan animals have similar biochemical functioning. Chapter 2 explored the linkages between biochemistry and the stoichiometry of living cells. Using similar reasoning as we used there, with a basically similar biochemical functioning, one might presume that all metazoans within a single basic body plan should have similar stoichiometry. However, this would be wrong. As we will now see, animals do differ strongly in their elemental composition. These differences go way beyond the obvious factors of differing investments in extracellular structural material. Even closely related species of similar size and built on common body architecture contrast in their content of N and especially P. To demonstrate this point, we will draw on the previously published and often discussed stoichiometric variation within freshwater zooplankton. Then we will compare those results with one of the first attempts to characterize body C:N:P in terrestrial insects, a group with much greater taxonomic diversity.

   The freshwater zooplankton are not taxonomically diverse. They consist mainly of rotifers and two types of crustaceans: cladocerans (or according to some taxonomic schemes, anomopods) and copepods. Andersen and Hessen (1991) and Hessen and Lyche (1991) determined the element content of various species of common crustacean zooplankton from Norwegian lakes. These field studies showed that the elemental composition of individual taxa was remarkably stable (Andersen and Hessen 1991) even in the face of strong seasonal variation in abundance and elemental composition of their food. It was also found that zooplankton had constant chemical content across different experimental food manipulations in the laboratory (starvation, food supplementation, P enrichment) (e.g., Hessen 1990).

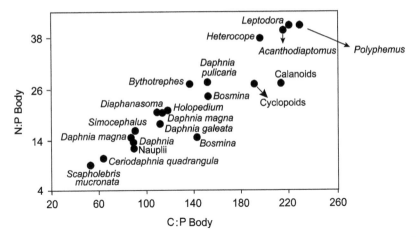

Fig. 4.2. Interspecific variation in the elemental composition of various taxa of freshwater zooplankton (cladocerans: *Bosmina, Bythotrephes, Ceriodaphnia, Daphnia, Diaphanasoma, Holopedium, Polyphemus, Scapholeberis, Simocephalus*; copepods: others).

Thus, a homeostatic model of C:N:P in these organisms was adopted. More recently, variation in P content within single zooplankton species has been scrutinized (Fig. 1.9; DeMott et al. 1998). However, the range in P content shown in Figure 1.9 for a single taxon is more limited than the full range of P content from all zooplankton taxa, which extends from about 0.5% to over 2.5% (details on this shortly).

The same study of Hessen and colleagues mentioned above also showed for the first time that zooplankton taxa had contrasting nutrient content. This variation was surprising because all species were crustaceans, all were within a size range of one to several millimeters, and many potentially coinhabited the same lakes. These data (presented here along with additional measurements by others since then, Fig. 4.2) reveal striking stoichiometric variation (C:P ranges more than fourfold; N:P ranges more than fivefold) in spite of the overall taxonomic homogeneity of this group. One major pattern was that certain mature copepods (*Acanthodiaptomus, Heterocope*) had high N:P and C:P (i.e., were low in P) while certain cladocerans (*Ceriodaphnia, Daphnia, Scapholeberis*) had low N:P and C:P (i.e., had high P content). Note from this figure how most stoichiometric variation in whole zooplankton is associated with differences in P (C:P and N:P vary but C:N is more constrained). This pattern of strong variation in P content but narrow variation in N is consistent with what we saw for many biochemicals (Fig. 2.2). Also (with the exception of cell walls, which are not a factor in animal stoichiometry) the pattern is very reminiscent of what we saw for major cellular structures and organelles (Fig. 2.4). It will also reappear when we consider patterns within vertebrates in a later chapter (see Figs. 6.6 and 6.7). It seems thus to be generally true that P is more variable within animals than is N. In the next chapter we will discuss the growth limitation implications of this interspecific variation in chemical content.

Although we can often use the simplification of a characteristic species-specific nutrient content for zooplankton, population structure and maturation patterns sometimes must be considered too. Cladocerans have simple life histories, with modest size-based variation in their stoichiometry (we will see several instances of this in the coming pages). Copepods have more complex life histories and there is large ontogenetic variation in elemental composition in these taxa. Schulz (1996) observed extreme differences in body elemental composition of copepod nauplii (early instars in the complex life cycle) and adults. She found that nauplii had a P content (~1.4%) that was among the highest of all taxa and was far higher than that of mature copepods (~0.5%) (see Fig. 4.2 for C:P and N:P). Carillo et al. (2001) also observed a wide ontogenetic range in stoichiometry in the copepod *Mixodiaptomus laciniatus*, with maximal P content in intermediate life stages. Others have also observed large ontogenetic shifts in copepod stoichiometry (Villar-Argaiz et al. 2000). Large ontogenetic variation in stoichiometry is a new and potentially important aspect to this set of issues. For example, it suggests that effects of stoichiometric food quality on copepods may be greater than previously appreciated and may impose as yet poorly understood life history bottlenecks (Sterner and Schulz 1998). Villar-Argaiz et al. (2002) suggested that copepod life history and stoichiometry are linked via differential emphasis on survival versus reproduction at different stages.

The NCEAS working group we referred to in the last chapter also explored the range of variation in elemental composition of zooplankton taxa by combining unpublished data with a thorough compilation of published data for freshwater crustacean zooplankton (Elser et al. 2000b). The data included values for 39 zooplankton species from 22 genera, 13 families, 8 orders, and 3 classes and were obtained from studies published as early as 1953. Zooplankton stoichiometry was compared to the stoichiometric patterns in zooplankton with insect herbivores from terrestrial ecosystems.

Comparison of the frequency distributions of the two groups reveals very similar means and variances in C:N and N:P (Fig. 4.3). There was a

---

Fig. 4.3. Frequency distributions for C:N (A), C:P (B), and N:P (C) of freshwater zooplankton (dark bars) and herbivorous insects (gray bars). The variation in C:N is less than the variation in C:P and N:P (note coefficients of variation, c.v.). This difference in variability is consistent with the observation that, within the major

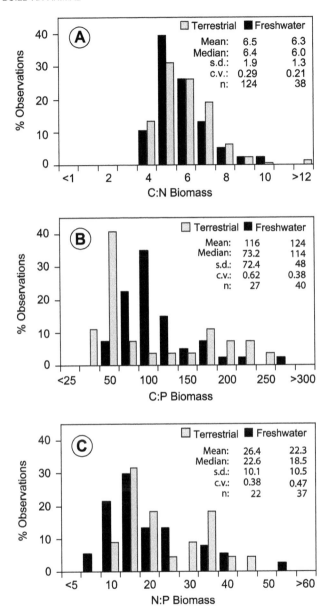

biomolecules that vary substantially in organisms, P content is more variable than
N content (see Fig. 2.2). The data further show that herbivorous insects in terrestrial ecosystems and herbivorous crustaceans in lakes exhibit similar but not identical distributions in C:N:P stoichiometry. The symbol n indicates the number of
different species in each group involved in the analysis; s.d. is the standard deviation. Based on Elser et al. (2000b).

suggestion that insects were somewhat higher in P (lower C:P) than zoo-plankton but this was not statistically significant. Zooplankton N:P differed considerably among species, ranging from 7.5 for the cladoceran *Scapholeberis mucronata* to 55 for the calanoid copepod *Eudiaptomus gracilis*, a more than sevenfold range. Similarly, among insect taxa N:P ranged from 13 for the beetle *Tenebrio molitor* to 47 for the grasshopper *Mermeria bivittata*, a fivefold range. Once again we see in these data that for both groups of animals C:N had a relatively constrained range of variation (coefficient of variation: ~0.25) while C:P varied greatly (coefficient of variation: ~0.5).

This comparison leads to the somewhat surprising conclusion that these distantly related animals (Insecta vs. Crustacea), occupying drastically different habitats (air, water), exhibit essentially the same basic set of relationships among these three major elements. They have similar means and similar variances. In both groups strong variation in C:N:P stoichiometry occurred among species. Furthermore, whatever evolutionary pressures have produced these chemical signatures in the two groups, they have produced not just similar means but also similar interspecific variation around that mean. This seems to be a challenging set of problems for comparative evolutionary ecology between aquatic and terrestrial systems and may in the end relate to fundamental limits to the ability to create functioning mixtures of biochemicals that we discussed in Chapter 2.

We have established what appear to be some highly general stoichiometric patterns within animal groups. Let us now consider what causes these patterns. The reasons are quite interesting and it will take several sections to get somewhere near the bottom of things.

DETERMINANTS OF C:N:P IN INVERTEBRATES: THE GROWTH RATE HYPOTHESIS

The previous section showed that invertebrate animals differ in their nutrient content, and that P is particularly variable among taxa. How could this variation arise? What makes one animal high P and another low P? Candidate molecules that could drive substantial variation in N:P are proteins and nucleotides (nucleic acids, ATP; Fig. 2.2). As discussed earlier, increasing protein content would increase N content without contributing any P, thus increasing N:P (Fig. 4.1). In contrast, an increased nucleic acid allocation would lower C:P and N:P (Fig. 4.1), moving the whole-organism stoichiometric signature more toward the composition of nucleic acids themselves (Fig. 2.2). Which of these macromolecules vary enough among organisms to substantively alter body elemental composition? In answering this question we will describe a central idea of current stoichiometric investigation, the "growth rate hypothesis" (GRH), an approach that may provide a mechanistic framework for integration of ideas of life history evolution, cell biology, population dynamics, and ecosystem function.

To explain variation in P but not N content, we will confine ourselves to molecules containing P (phospholipids, ATP, and nucleic acids). Recall from Chapter 2 that some of these classes of biomolecules either vary little between organisms or contribute insufficiently to the biomass of the animal to alter its elemental composition to a significant degree. For example, phospholipids contribute less than 5% to total body mass in most organisms and therefore cannot explain the whole-organism patterns we are trying to understand. Similarly, ATP is present in animal tissue at very low concentrations, far too low to account for a significant proportion of body P in any but the lowest-P animals.

It has been suggested that RNA is the most likely biochemical contributor to variable C:P and N:P in zooplankton (Hessen and Lyche 1991; Sterner 1995; Elser et al. 1996). Evidence for that conclusion was based on reasoning as follows. The contrasts in body C:P and N:P between the cladoceran *Daphnia* and adult calanoid copepods shown in Figure 4.2 reflect the fact that *Daphnia* has a P content that is 0.8% greater than that for these copepods (1.4% vs. 0.6%). Fortunately, some early data for RNA content for *Daphnia* and calanoid copepods can be used to estimate the contribution of RNA to this difference (Dagg and Littlepage 1972; Baudouin and Scoppa 1975; Båmstedt 1986; McKee and Knowles 1987). These data indicate that calanoid copepods contain ~2% RNA while *Daphnia* contains ~10% RNA (Table 4.1). Since RNA is ~9% P by mass (Fig. 2.2), we can calculate RNA-P in these two groups. Doing this, we find that RNA-P differs between calanoid copepods and *Daphnia* by ~0.8%. An excellent match!

TABLE 4.1
Accounting of % P, % RNA, and % RNA-P for *Daphnia* and copepods. Data for elemental composition are from Andersen and Hessen (1991); data for biochemical composition are from McKee and Knowles (1987) and Båmstedt (1986)

	N and P Content			*RNA Content*	
	% N	% P	N:P	% RNA	% RNA-P
Calanoid copepods	8–10	~0.6%	~35	~2%	~0.2%
Daphnia	8–10	1.2–1.6%	~12	~10%	~1%
Difference:		~0.8% "extra" P			~0.8% RNA-P

These rough calculations are intriguing, but they are based on data from multiple sources and thus must be considered provisional. Recently, Vrede et al. (1999) measured the P content of bodies, carapaces, and eggs. Within the animal bodies, P was divided into pools consisting of nucleic acids, phospholipids, and "other." The authors examined two species of *Daphnia* and one species of calanoid copepod. They found that nucleic acids were indeed major pools of P, and they also found a relatively uniform contribution of phospholipids to animal P budgets. Unfortunately, nearly 50% of the animal P was in undetermined pools (other) in the high-P daphnids, so it is unwise to place much quantitative emphasis on the exact figures in this study. However, as emphasized by the authors, the results suggest that pools of P involved in zooplankton phosphorus homeostasis are still not known for certain.

The conclusion that RNA-P is an important contributor to interspecific variation in P content in zooplankton has more recently been supported by direct measurements of RNA and P on the same animals in a single study. Dobberfuhl (1999) found a very strong relationship between total body % P and % RNA-P with a slope close to 1 across several taxa and two size classes of one of the taxa (Fig. 4.4). The close correspondence between the differences in RNA-P and total body P and the slope near unity indicate that RNA indeed may cause substantial variation in whole-body % P. Extrapolation to zero RNA suggests that these organisms have 0.66% of their biomass in non-RNA P pools such as phospholipids, DNA, and ATP. We do not imagine that this degree of match will hold for all cases in which P and RNA content data are compared. However, we think it is reasonable to suggest that the search for explanation of differences among invertebrates in C:N:P stoichiometry should focus on RNA, which is where we will go now.

In growing organisms ~85% of cellular RNA is associated with the rRNA in ribosomes (Lehninger et al. 1993). Messenger RNA (mRNA) and transfer RNA (tRNA) comprise the rest. tRNA and especially mRNA molecules turn over rapidly in response to cellular demands while rRNA is relatively stable on physiological time scales (Maaløe and Kjeldgaard 1966). Since ribosomes are the machinery of protein synthesis, it seems natural to wonder whether differences between animals in P content and thus C:N:P stoichiometry are associated with differences in maximum biomass growth rate (Elser et al. 1996) when ribosomes are all being used at capacity. Correspondences between P content and maximal growth rates in zooplankton have in fact been reported (Main et al. 1997; Elser et al. 2000a). The growth rate hypothesis suggests a strong linkage between evolved life histories and patterns of body nutrient content:

> The Growth Rate Hypothesis states that differences in organismal C:N:P ratios are caused by differential allocations to RNA necessary to meet the protein synthesis demands of rapid rates of biomass growth and development.

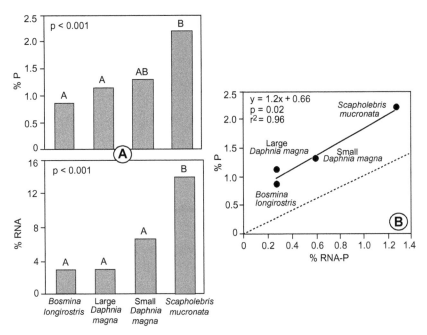

Fig. 4.4. Variation in body P content is associated with RNA content in crustacean zooplankton. P and RNA were measured on the same populations of animals. A. Results by species and size. B. Correlation. While the number of taxa in the comparison is small, the strong correlation and the fact that the slope of the relationship is close to 1 (compare to dotted line in panel B) indicate that nearly all of the variation in total body P content in these animals is attributable to variation in RNA content. Based on Dobberfuhl (1999).

The GRH connects cellular biochemical allocation, animal growth, and C:N:P stoichiometry within an evolutionary framework. We will see that it already has a lot of supporting evidence going for it. As will be discussed in upcoming chapters, body C:N:P stoichiometry has major effects on ecological processes such as food-web dynamics and nutrient release; thus the GRH potentially links the cellular biology of growth to nutrient cycling during trophic interactions in food webs. The GRH suggests a possible biogeochemical signature of major evolved life history traits in animals and thus it also potentially connects evolutionary biology to ecosystem processes. These two fields of biological investigation have seldom been brought into such close partnership within a strongly mechanistic framework. Let us now examine the GRH in more detail.

Relationships between protein synthesis rate (a key process in growth) and RNA levels are widely known for diverse organisms (Thorpe 1984). Considerable data on cellular RNA levels and growth rate in animals have

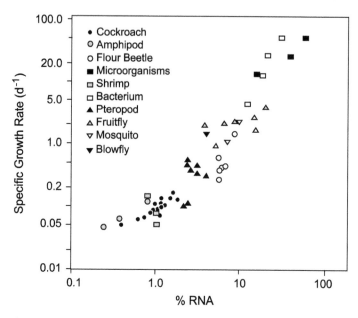

Fig. 4.5. Positive association between growth and RNA content for a variety of arthropods and microorganisms. Based on Sutcliffe (1970).

existed for some time. For example, Sutcliffe (1970) showed that there is a good correlation between growth rate and RNA content for a variety of invertebrate taxa (terrestrial and aquatic) (Fig. 4.5). He further showed that the relationship was largely contiguous with one previously demonstrated for microorganisms by Maaløe and Kjeldgaard (1966).

At the level of generalized life history strategies of *Daphnia* and copepods, the GRH seems to work well. We have already mentioned some of their similarities, but these taxa do differ considerably in their life history strategies (Allan 1976). Copepods can be characterized as "K strategists" due to their long life spans, sexual reproduction, complex metamorphosis, and low predator vulnerability. Daphnids on the other hard are relatively "r selected" with short life spans, asexual reproduction, direct development, and high predator vulnerability. General life history analysis usually confines itself to phenotypic traits such as birth and death schedules, time to reproduction, and others (Stearns 1976, 1992). The GRH is a different kind of approach, explicitly considering the material basis of life history phenomena. It hypothesizes that rapid growth cannot just happen: it requires a certain pattern of cellular investment.

Several examinations of stoichiometry-growth linkages have now been completed. Elser et al. (1996) considered published data for maximum growth rates (μ_m) and biochemical composition of *Daphnia magna* (McKee and Knowles 1987) and *Drosophila melanogaster* (Church and Robertson 1966) at different stages. Calculating elemental content from biochemical pools, they estimated that during development P content changed in both species and N content changed for *Drosophila* (Figs. 4.6A,B). Calculated body N:P declined with increasing μ_m for each species, primarily because of increases in P-containing biomolecules at higher growth rates. A more explicit test of the GRH was performed by Main et al. (1997), who simultaneously measured elemental composition and μ_m for six species of freshwater zooplankton. To assess ontogenetic variation, populations of several species were subdivided according to size. Their data show a clear association between elemental composition and μ_m (Fig. 4.7). With increasing growth rate, % N increased slightly and % P increased fivefold so that N:P declined. Carrillo et al. (2001) also related ontogenetic patterns in growth rate and stoichiometry in *Mixodiaptomus laciniatus* and found, for certain life stages of the copepod, a negative relationship between growth rate and N:P ratio and a positive relationship between P content and growth rate, trends supportive of the GRH.

The most direct test of the GRH in an evolutionary context was by Elser et al. (2000a), who performed a comparative geographic study within the *Daphnia pulex* species complex. They showed that individuals from the arctic where the growing season is short and selection on growth rate should be severe had higher μ_m and higher body % P than individuals from the same species complex from temperate habitats (Figs. 4.8A,B; differences in % N and N:P were nonsignificant). Further, the arctic animals were inefficient recyclers of P but recycled N with equal efficiency. Arctic *Daphnia* also were more susceptible to P-deficient diets (Figs. 4.D). Another test of the GRH is to see if stoichiometry can be predicted from known life history traits. For example, if the GRH is correct, most small, rapidly growing animals should have high P and low N:P. Aphids, for example, have a high potential growth rate (Van Hook et al. 1980; Risebrow and Dixon 1987) and it is interesting therefore to note that aphids appear to have body P content similar to their fellow parthenogens *Daphnia* (preliminary data indicate that aphids are ~1.2% P; J. Elser, unpublished). Food quality implications for aphids are discussed in Chapter 5. We will see many more connections between stoichiometry, growth, and nutrient fluxes in coming chapters.

Thus, diverse evidence indicates that body elemental composition is linked to growth rate in zooplankton. Insects (and, as we will see below, bacteria) also appear to show similar relationships, but the data are scarcer. Thus, there seems to be a general association in heterotrophic organisms of rapid growth rates with high P content and low N:P. In Chapter 3, we saw that autotrophs also exhibit positive associations between P content

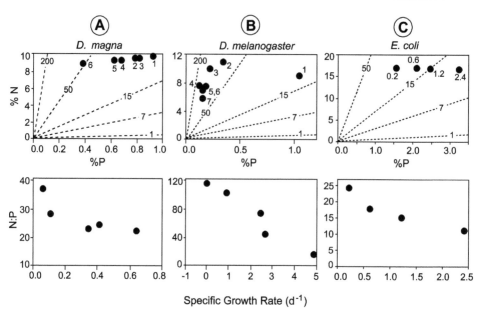

Fig. 4.6. Estimated variation in elemental composition and maximal growth rate in three well-known heterotrophs: A. *Daphnia magna*, B. *Drosophila melanogaster*, and C. *Escherichia coli*. For *Daphnia* and *Drosophila*, growth rates were calculated from changes in body mass. Estimated N:P was plotted against growth rate in the bottom panels. In each case elemental composition was computed from reported data on major biochemical pools. For *Daphnia* and *Drosophila* the numbers next to the symbols on the top panels indicate the sequence of consecutive instars of juvenile animals. For *E. coli*, the numbers indicate the specific growth rate of the culture. In each case, estimated changes in P content during development or in different cultures are large compared to estimated changes in N content. In each case, a negative relationship between estimated N:P and growth rate is seen, largely due to disproportionate changes in estimated % P. Panels A and B based on Elser et al. (1996) with original data from McKee and Knowles (1987) (panel A), Church and Robertson (1966) (panel B). Data from Maaløe and Kjeldgaard (1966) in panel C.

and growth, but only under P limitation. When limited by other inorganic substances, autotrophs perform luxury consumption of P and have high P content even at low growth rate. However, the GRH is really about maximal growth rates, not growth under resource limitation, and we also saw that a fast-growing grass species had consistently lower N:P ratios than a slow-growing species (Fig. 3.3) and that fast-growing algae had lower opti-

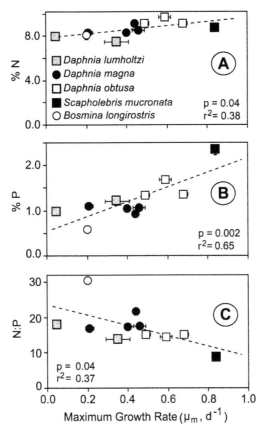

Fig. 4.7. Correlation between C:N:P stoichiometry and growth rate for various size classes of several cladoceran species. Animals were fed high concentrations of high-quality food for determination of maximum growth capacity (μ_m) at a given size. Parallel samples of the animals were analyzed for elemental composition. Based on Main et al. (1996).

mal N:P (Elrifi and Turpin 1985). Whether the GRH can successfully be adapted for autotrophs is unclear at this point.

Getting back to animals, the specific shapes of the relationships between growth and stoichiometry still need investigation. In fact, we saw different shapes in our different examples. In an upcoming section we will present a biochemical allocation model to develop some expectations based on basic principles of cell biology about what the shape of such relationships should be under those assumptions. However, first we take the dangerous step of seeing how much of the mechanistic basis of these relationships might be understood with a rudimentary grasp of cellular-molecular biology.

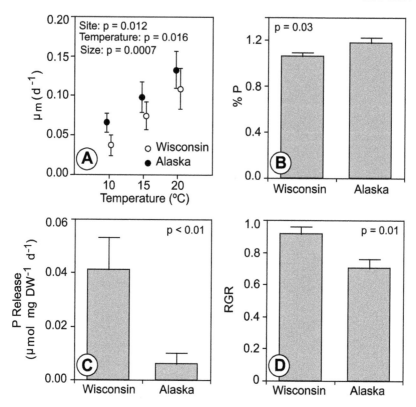

Fig. 4.8. A test of the growth rate hypothesis across a latitudinal gradient using members of the *D. pulex* species complex (western *D. pulicaria* and panarctic *D. pulex*). Juveniles were fed standardized foods to assess their growth rates (A). In addition, body elemental composition (B), nutrient recycling rates on standard food (C), and response to food quality (D) were also measured. The response to food quality is expressed as RGR, with animal growth rate on low-P food divided by animal growth rate on high-P food, and expressed as a percentage (see also Fig. 5.12). A low value of this parameter indicates an animal whose growth is more sensitive to food P content. Based on Elser et al. (2000a).

MOLECULAR BIOLOGY AND THE C:N:P STOICHIOMETRY OF GROWTH, OR ECOSYSTEM SCIENTISTS GO ASTRAY

Now that we have related growth to ribosomes, this suggests the next question, "What are the mechanisms and processes involved in maintaining different concentrations of ribosomes in cells?" Perhaps we can extend our analysis of growth-stoichiometry linkages one layer deeper in the hierarchy of levels of biological organization. Genes encoding rRNA are called rDNA. Mechanisms regulating expression of rDNA genes have been extensively examined, beginning with the classic studies of Maaløe and Kjeldgaard (1966) on *Escherichia coli*. Upon addition of suitable organic substrates and nutrients to nutrient-limited cells, these investigators found that *E. coli* first begins expressing rRNA genes, which then result in the assembly of large numbers of ribosomes that can support high growth rates during the exponential growth phase (μ_m). Various studies have shown that during rapid exponential growth more than 30% of cell dry mass in this organism is comprised of RNA. This allocation value is especially astonishing when one considers that bacterial ribosomes are two parts RNA and one part protein by weight (Chapter 2). Thus, if all this RNA were in ribosomes, at maximal growth rate *E. coli* cells would be approximately 45% ribosomes by dry weight! However, when resources are depleted, the cells enter stationary phase, rRNA genes are no longer expressed, and cellular RNA levels plummet. This up- and down-regulation of RNA contents with growth suggests that, just as was observed for the ontogenetic trajectory of μ_m in *D. magna* and *D. melanogaster* (Figs. 4.6A,B), there should also be an association between growth rate and C:N:P in *E. coli*.

To test for a relationship between growth and elemental composition in these classic studies, we used the biochemical data of Maaløe and colleagues to calculate expected changes in the C:N:P stoichiometry of *E. coli*, and we compared these to growth rate. We did this by using their reported values for protein, DNA, tRNA, and rRNA content along with the information on stoichiometry of these molecules presented in Chapter 2 to estimate cellular % N, % P, and N:P. They found that increased *E. coli* growth rates were associated with a 20% decline in protein content (from 830 to 670 mg g^{-1}) and a 25% decline in DNA (from 40 to 30 mg g^{-1}) but a 158% increase in RNA (from 120 to 310 mg g^{-1}). The increase in RNA was entirely due to huge increases in rRNA (600+%, from 35 to 250 mg g^{-1}) with little change in tRNA (an 18% decrease from 85 to 60 mg g^{-1}). Such molecular shifts indicate little change in total cellular N content associated with growth (Fig. 4.6C, calculated N content declined by less than 1% from 17.2 to 16.8%). Nucleic acids have a slightly lower N content relative to proteins (Fig. 2.2); thus, this small decline reflects the replacement of proteins by RNA as a percentage of biomass. In contrast to these small changes in N, calculated P content increased by over 100% (from 1.5 to 3.2%) with increased growth. Cellular N:P declined from 25 at slow growth to 11.6 at rapid growth. It is thought-provoking to note the qualitative resemblance between the relationships among % N, % P, and N:P and growth rate for *E. coli* (Fig. 4.6C) and for *Drosophila* and *Daphnia* (Figs. 4.6A,B). In the eucaryotes, the relationship of stoichiometry is with μ_m

while in the procaryote, with its very plastic RNA content, it is with μ. Our analysis of the work of Maaløe and colleagues supports the idea that there is a fundamental syndrome linking elemental composition and growth rate in bacteria, as there is in metazoans. In all cases, this syndrome flows directly from the fundamental mixture of biomolecules and elements needed to build the ribosomal hardware of cellular growth.

Since we are seeing patterns that seem to work across such a wide gap of biological diversity from bacteria to insects, it is worth asking whether the genetic mechanisms linking growth, biochemistry, and elemental composition in eucaryotic organisms should operate similarly to those in procaryotes like *E. coli*. This must be considered because several important aspects of the cell biology of ribosomal gene expression and protein synthesis differ for eucaryotes relative to procaryotes. In procaryotes, where the genetic material mingles throughout the cell, transcription of rDNA genes occurs in intimate association with ribosomal protein synthesis. Thus we might expect an intimate and immediate connection between production of rRNA and rates of protein synthesis. In eucaryotes, in contrast, rDNA transcription occurs in the nucleus while most ribosomal synthetic activity occurs in the cytoplasm. Considerable progress in understanding the cell biology of ribosomal gene expression and protein synthesis in eucaryotes has been made over the past 20 years. We will touch on a few highlights here. While our focus is on the genetics of RNA in animals, the genetic mechanisms are general and thus our discussion in places will also touch on studies of plants and Protozoa as appropriate.

There are several connections between cellular growth and the nucleolus, the most prominent feature of the eucaryotic nucleus. The nucleolus is a large, diffuse structure where rDNA sequences are actively transcribed and newly synthesized rRNA is immediately packaged with ribosomal proteins to produce ribosomes. First, the size of the nucleolus differs greatly in different cell types and with different growth conditions. The nucleolus can occupy as much as 25% of nuclear volume in cells characterized by high rates of protein synthesis (Alberts et al. 1994). Second, organisms or life stages characterized by high growth rates are often associated with amplified rRNA gene sequences (extra copies of the rRNA gene duplicated outside the chromosomes). For example, in young oocytes of the frog *Xenopus*, rRNA genes are amplified more than 1000-fold, generating large quantities of free extrachromosomal rDNA that clusters together to form ~1000 "extra" nucleoli (Alberts et al. 1994). Third, a variety of studies have documented a close connection between rDNA transcription and rates of protein synthesis and cell growth in eucaryotes. Sollner-Webb and Tower (1986) summarize studies showing that rRNA production increases when cells are stimulated to reproduce by any of several factors, including partial removal of an organ, feeding after starvation, treatment with various chemical agents, or infection with the tumor-promoting virus SV40. Conversely, rRNA production decreases when cells are treated with protein synthesis inhibitors, the organism or cell is starved for essential nutrients, the cell is infected with certain viruses, or the cell enters the stationary phase. From the preceding evidence we believe that there is a strong theoretical basis at the cellular and biochemical level for a P-rich elemental syndrome associated with the RNA demands of active growth in eucaryotic cells just as there is for procaryotes such as heterotrophic bacteria. Both eucaryotes and procaryotes can regulate rDNA transcription (the manufacture of rRNA), although the genetic mechanisms involved are somewhat different. We will now look at those genetic mechanisms, including the structure and organization of the rDNA genes themselves, in more detail.

If increased allocation to rRNA is indeed associated with increased growth rates of organisms and tissues, then what genetic mechanisms generate this sort of phenotypic variation? In other words, what is the mechanism at the level of the genome itself that might lead to whole-organism nutrient signals which, as we will see later in the book, impinge on a host of ecological phenomena from the growth of single species to global nutrient cycling? To find some answers, once again we must wander into niches in the library infrequently occupied by ecosystem ecologists, in this case the literature related to rDNA coding for rRNA. Much is known about variation in the structure of the rDNA, but it seems that surprisingly little is known about the functional significance at the scale of the whole organism.

Because their transcription products are essential and are needed in high quantities, eucaryotic rRNA genes are clustered as tandem repeats (see Fig. 4.9) in high copy number at one or more chromosomal sites termed nucleolus organizer regions (discussed above, and see the reviews by Long and Dawid 1980; Cortadas and Pavon 1982; Sollner-Webb and Tower 1986). Each rDNA array consists of a transcription unit, including the genes for the 18S, 5.8S, and 28S subunits of the ribosome (the "*rrn*" genes), the internal transcribed spacers (ITS) separating the ribosomal subunit genes within a unit, and an intergenic spacer (IGS) separating adjoining units. (Interestingly, the coding region for 5S rRNA is found elsewhere in the genome, separated from these tandem repeat units.) The IGS contains various transcription-enhancing sequences that can occur in multiple copies that differ both intra- and interspecifically (Reeder et al. 1983; Paule and Lofquist 1996). Thus, the total size of the complete rDNA array in the genome is the copy number times the gene plus the IGS lengths.

Various studies have shown that the organization of the transcription unit as a whole is highly conserved across a wide range of species, exhibit-

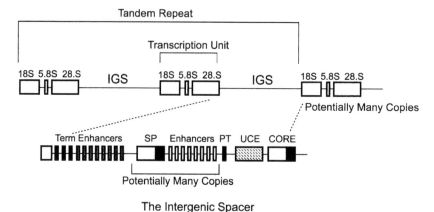

The Intergenic Spacer

Fig. 4.9. The structure of the eucaryotic ribosomal RNA gene (rDNA). One tandem repeat plus the end of another are shown. Potentially, many copies of such tandem repeats are found in the genome of a particular organism. The units labeled 18S, 5.8S, and 28S represent the DNA regions coding for these RNA components of the ribosome. Transcription occurs from right to left. Of particular interest to us in understanding variation in organism RNA content and thus C:N:P stoichiometry is variation in the number of SP-enhancer units (manifested in the length of the IGS) and in the number of copies of the rDNA in the genome. Based on Paule (1994).

ing high sequence similarity among diverse taxa (Goldman et al. 1983; Elwood et al. 1985). In contrast, however, the subrepetitive elements within the IGS appear to evolve rapidly (Tautz et al. 1987). This rapid rate of evolution occurs because a variety of mechanisms can produce structural mutations (rather than point-source mutations such as when an "A" becomes a "T") in these genes, resulting in variations in the number and arrangement of repeat elements within the IGS. These mechanisms include unequal crossing over in the IGS region during both sexual and asexual reproduction or even in somatic cell lineages (Cullis and Charlton 1981; Schlötterer and Tautz 1994; Ganley and Scott 1998). As a result, IGS regions can vary not only between different species but also between populations, individuals, and, because there are many copies of the rDNA tandem repeat in the genome, even within a single cell (Boseley et al. 1979; Paule and Lofquist 1996; Ganley and Scott 1998). Variation in both restriction sites (the particular sequence where transcription enzymes bind) and, more commonly, in length occur (Rogers and Bendich 1987). Differences in length are often associated with differences in the number and length of the variable repeat units that enhance transcription rate ("enhancers," Grimaldi and Di Nocera 1988). One would expect, then, that there should be a positive association between the rRNA demands of an organism's developmental program and the length of the IGS in its rDNA since the IGS is crucial in regulating rDNA activity (Reeder et al. 1983) and, potentially, ultimately growth rate. Given these regulatory functions, variation in sequence and organization of the IGS is likely to be functionally significant, resulting in rapid "molecular coevolution" (Dover 1982) of the transcription apparatus (Evers and Grummt 1995).

To date, rDNA heterogeneity has largely been studied for establishing phylogenetic relationships and for quantifying gene flow between populations (Gray and Schnare 1996). While various studies have documented connections between rDNA heterogeneity and growth and development, performance of different organismal functions, and adaptive response (Table 4.2), much remains uncertain about the functional significance of rDNA variations. Needless to say, the consequences of this genetic variation at the level of ecosystem dynamics have not yet been widely contemplated! Existing evidence indicates that variation in the intergenic spacers has important functional consequences due to its effects on cellular RNA levels and organism growth rate (Table 4.2). For example, when selected for rapid development (high μ_m), IGS length in *Drosophila melanogaster* increased (Cluster et al. 1987), presumably due to increased numbers of enhancer repeats (Paule and Lofquist 1996). An *in vitro* study has genetically engineered *D. melanogaster* cells to show that transcriptional production of pre-rRNA (immature rRNA molecules not quite ready for assembly into ribosomes) is directly proportional to the number of rDNA enhancers present (Table 4.2; Grimaldi and Di Nocera 1988). These examples indicate that there is a fundamental association among the number of enhancers in the IGS (and hence its length), RNA content, and μ_m, as increased numbers of rDNA enhancers should increase the rate of rRNA production by the transcription enzyme RNA polymerase I (Pol I).

However, rates of transcription per rDNA gene must have an upper limit. One might then expect that under strong selective pressures for rapid growth, individuals with increased numbers of copies of the rRNA genes should be favored, as they will be able to generate the required high rRNA production rate for rapid growth. Indeed, it appears that organisms have adopted a variety of strategies to assure sufficient rRNA production according to the demands of their life history (Table 4.3). The tendency to increase rDNA copy number is rather obvious and widespread. While ribosomal genes can be amplified extrachromosomally in certain organisms and tissue types (e.g., *Xenopus* oocytes, mentioned above), at the interspecific level most variation occurs in chromosomal rDNA copy number. According to Long and Dawid (1980), rDNA gene copy number within the class Insecta alone varies at least sevenfold (and at least 3.5-fold within

TABLE 4.2
Observations related to variation in the structure of rDNA genes (as *rrn* copy
number, IGS length) and growth- or rRNA-related parameters in different species

Organism	Observations	Reference
	Ribosomal DNA Copy Number	
Flax	Levels of rDNA at the apex (fast-growing cells) changed in response to the growth rate and nutrient regime, whereas rDNA at the base of the stem (slow-growing cells) remained constant	(Cullis and Charlton 1981)
Soybean cell cultures	Reduction in rDNA (~30%) was concomitant with reduction in growth rate caused by poor nutrient supply	(Jackson 1980)
Fruit fly (*Drosophila melanogaster, D. hydei, D. mercatorum*)	Strains containing 30 to 100 copies have slower development compared to those with 100 copies or more; mutant individuals carrying a reduced number of *rrn* genes exhibited a bobbed phenotype, characterized by a retarded development, reduced growth, thinner cuticle, smaller bristles, etc.	(Shermoen and Kiefer 1975; Ritossa 1976; Franz and Kunz 1981; Templeton et al. 1985; Desalle et al. 1986)
Amphibian oocytes	Temporary amplification of rRNA genes occurred as an adaptation to large cell volume and increased need for rRNA synthesis	(Gall 1968)
Chicken	Increased rDNA gene dosage and larger nucleoli in chickens selected for rapid growth	(Delany et al. 1994a)
Chick embryo	Lower cellular RNA levels and reduced development occurred in embryos with deficiency for *rrn* gene copy number	(Delany et al. 1994b)
	IGS Length Variation	
Maize (*Zea mays*)	Frequency of long-spacer variant was higher and frequency of short-spacer variant lower in maize strains selected for increased yield	(Rocheford et al. 1990)

TABLE 4.2 (*Continued*)

Organism	Observations	Reference
Wheat (*Triticum dicoccoides*)	Spacer diversities in natural populations were significantly correlated and predictable in terms of climatic variables	(Flavell et al. 1986)
Oats (*Avena sativa*)	Bred cultivars had significantly longer IGS regions than ancestral forms	(Polanco and Perez de la Vega 1997)
Fruit fly (*D. melanogaster*)	Under selection for development time, spacer length increased in concert with shortened development time; rate of transcriptional production of pre-rRNA was directly proportional to the number of enhancers located in the IGS	(Cluster et al. 1987; Grimaldi and Di Nocera 1988)
Xenopus oocytes	In an *in vitro* transcription assay, longer IGS's were more competitive than shorter IGS's in binding Pol 1	(Reeder et al. 1983)
Chicken	Selection for reproductive (egg layers) and somatic (broilers) growth resulted in specific variability in two different regions of IGS	(Delany and Krupkin 1999)

Drosophila). In plants at least, it seems that many of the additional copies of rDNA genes are superfluous (Rogers and Bendich 1987). Thus, to evaluate these ideas it will be necessary to distinguish between active and inactive rDNA genes in the genome.

The number of rRNA genes *relative to* total genome size also appears to be important. For example, in a study of copepod development rates, taxa with the highest number of copies of rDNA genes relative to total genome DNA developed most rapidly (White and McLaren 2000); taxa with the highest absolute numbers of copies did not necessarily have the fastest development. In any case, while various studies have associated ecogeographic factors with genetic variation in rDNA in particular taxa, it seems that few studies have considered the functional significance and evolutionary ecological implications of rDNA copy number and spacer length variation across a wide range of taxa and selective environments. If we are correct that major features of the organization of rDNA genes underpin

TABLE 4.3
Genetic mechanisms to satisfy a large demand for rRNA in different organisms

Way to Increase rRNA Synthesis	Organism	Reference
Transcription rate per gene varies to match the growth rate	Bacteria	(Nomura et al. 1984)
Extra-chromosomal amplification: one to several copies of rDNA are used as templates to produce hundreds to thousands of extra-chromosomal copies. Usually this event occurs only at specific stages; however, in some species rRNA genes may be entirely extrachromosomal	Protozoans Slime molds Animal oocytes	(Kafatos et al. 1985) (Vogt and Braun 1976, 1977; Lewin 1980; Long and Dawid 1980; Welker et al. 1985; Williams 1986) (Thiebaud 1979; Nomura et al. 1984; John and Miklos 1988)
Replication of the entire genome: polyploids are created through rounds of DNA replication without cell division	Many plants	(Rogers and Bendich 1987)
Large quantities of highly repetitive rDNA are present in nucleoli of oocytes and then reduced, leaving somatic cells with low number of copies but oocytes rich in rRNA	Some cyclopoid copepods	(Standiford 1988)
Each nucleus contains a large number of rRNA genes, 100 to 1000 copies per diploid cell in animals and 500 to 40,000 copies in plants	Most eucaryotes	(Rogers and Bendich 1987; John and Miklos 1988)

phenotypic variation in rRNA synthesis, ribosomal production, and thus also μ_m, RNA content, and C:N:P, then variation in rDNA has major implications for ecological dynamics due to the ramifications that follow from its effects on C:N:P stoichiometry. We perhaps have a glimpse now of the genetic architecture of a major life history trait: biomass growth rate. Ultimately, the manifold ecological aspects of C:N:P stoichiometry discussed in this book may originate here in the rDNA genome. Work in this area will bring new meaning to the term "functional genomics," bringing genomics together with completely new facets of evolutionary biology and ecology.

In the preceding discussion we have emphasized the genes coding for rRNA. However, ribosomes are made of both rRNA and ribosomal proteins in a closely specified proportion (Chapter 2). How are genes coding for ribosomal proteins regulated so that just the right amount of the right ribosomal proteins are produced to keep up with production of rRNA? The regulatory processes associated with rDNA fall into a class of genetic mechanisms referred to as "transcriptional regulation" and indeed the machinations of rDNA transcription are the prime example of this type of regulation in biology. Ribosomal proteins are regulated instead by a process referred to as "translational regulation" and, again, this is the main example of this type of gene regulation in living things (Alberts et al. 1994). Translational regulation assures that expression of ribosomal protein genes is tightly coupled to changes in rDNA transcription; the ribosomal protein genes are "downstream." Ribosomal proteins readily bind rRNA; this indeed is what self-assembles the ribosome. It also seems that a given ribosomal protein will also bind its own mRNA, thus preventing translation of more of that ribosomal protein. So, as long as there is free rRNA, ribosomal protein will be "soaked up" and translational production of ribosomal protein will continue. However, as rDNA expression decreases and free rRNA declines, ribosomal proteins accumulate, bind to their own mRNA, and production is turned off. If this regulatory dance seems somewhat confusing, don't worry; the take-home points are relatively simple. First, there are fundamental genetic mechanisms that ensure that, during growth, ribosomal RNA and proteins are formed in strict proportions, thus establishing a molecular genetic basis for a tightly constrained C:N:P stoichiometry of ribosome production. Second, the rDNA genes are "in charge" and evolutionary changes impinging on growth and C:N:P stoichiometry are likely to be connected to changes in the rDNA and not to changes in ribosomal protein genes.

With this molecular genetic information we are attempting to build a deeper understanding of the biological stoichiometry of animal growth. Under conditions of sufficient nutritious food, an animal will grow roughly at its genetically determined maximum rate (μ_m). As we will see in the next chapter, when food of severely deficient nutrient content is presented to animals, animal growth (μ) is decreased. The linkages between reduced growth rates and RNA content within species are complex and not yet all worked out. Undoubtedly, they have a lot to do with the precise patterns of stoichiometric homeostasis in any given species, and as we saw in Chapter 1, these patterns may be complex. Key trade-offs in this area might include limits on the ability to down-regulate the growth apparatus to avoid investment in this expensive hardware at inappropriate times. For now we suggest that it is best to think of relationships between body biochemical and elemental composition and growth rate in evolutionary contexts as relationships between composition and *potential* growth rate (μ_m). An animal's biochemical allocation establishes the maximum rate at which it *can* grow (μ_m) but the specific nutritional situation (food quantity, food quality) establishes the rate at which it *actually* grows (μ). Furthermore, it seems justifiable on these grounds to consider that μ_m and biochemical or elemental composition are features characteristic of that species or life stage. We propose that some species have high genetically programmed μ_m values and thus low minimum C:P and N:P. Other taxa have lower genetically programmed ceilings on their growth rate (low μ_m) and thus will generally have high characteristic body C:P and N:P. The suggestions just made hint that it might be possible to use simple principles of cellular allocation and growth kinetics to make explicit predictions about the nature of growth rate–C:N:P relationships. We describe a model such as this in the following section.

A SIMPLE MOLECULAR-KINETIC MODEL OF THE GROWTH RATE–C:N:P CONNECTION

We have proposed and begun to establish a link between animal body elemental composition and growth rate due to P-rich nucleic acids in the ribosome. To investigate this relationship quantitatively, Dobberfuhl and Elser built a stoichiometric growth model (hereafter, the "DE model") based on theoretical biochemical animals that grow at rates dependent on their mix of biochemicals (Dobberfuhl 1999). This effort has shed some light on the possible shape of growth versus elemental composition relationships and provides further insight into how biochemical allocation influences growth. While the model was constructed primarily for heuristic purposes to clarify the consequences of protein-RNA allocation for growth and C:N:P stoichiometry, as we will show below, the DE model shows remarkable correspondence between the characteristics of these theoretical animals and real ones.

The approach and an example calculation of the DE model are presented in Table 4.4. Each theoretical animal (where "animal" means a specific mixture of biochemicals) was assigned a fixed, background allocation of biochemicals known to be relatively constant among taxa or with growth rate (DNA, phospholipid, ATP, chitin; Chapter 2). Since these investigators were interested primarily in crustacean zooplankton, values characteristic of crustaceans were used, but the same principles could be applied to other groups. These fixed components added up to about 12% of body

TABLE 4.4
An example illustrating how the DE model calculates body elemental composition
and growth rate for a given allocation to protein and RNA. In this case, there is
an allocation of 40% protein and 12% RNA. The end products are estimates of
body % N, % P, N:P, and specific growth rate made directly from biochemical
allocation and the biology of protein synthesis

A. Baseline composition (nonvarying); Total ≈ 12%.

Constituent	% of Body Mass	% N	% P
DNA	1	15.5	9.2
ATP	0.75	14.7	15.5
Phospholipids	5	1.6	4.2
Chitin	5	7	0

B. Variable composition for hypothetical mixture of 40% protein and 12% RNA.
The total mass is assumed to sum to 100 units. Thus, in this animal, 36 mass units
are contributed by biochemicals lacking N or P. This mixture has an N:P ratio of
13.7 (molar).

	% N	% P	Mass in 100-Unit Animal	Mass of N in 100-Unit Biomass	Mass of P in 100-Unit Biomass
Baseline	5.9	3.5	11.75	0.69	0.41
Protein	17.2	0	40	6.88	0
RNA	15.5	9.2	12	1.86	1.10
Whole animal				9.43	1.51

C. Determination of growth rate.

Quantity of RNA	12	From table above
Quantity of rRNA	9.6	Assumes 80% of RNA is in ribosomes
Number of ribosomes	2.28×10^{12}	= rRNA divided by mass of RNA per single ribosome
Protein synthesis rate	2.5×10^{17}	= number of ribosomes times protein synthesis rate per ribosome
Protein accumulation rate	29.4	= protein biomass production per day, assuming 60% protein retention
Specific growth rate (d^{-1})	0.55	= [ln(protein biomass after 1 d/starting protein biomass)]/1 d, i.e., [ln(29.4 + 40)]/40

composition, leaving about 88% to account for. Above the baseline alloca-tion, animals were constructed with different amounts of protein and RNA. Protein and RNA as a percentage of body mass were allowed to vary in combinations ranging from 30–75% protein and 1–20% RNA. Any alloca-tion remaining above the fixed allocation plus the particular combination of RNA and protein allocations was given over to biochemicals lacking N and P (e.g., neutral lipids, carbohydrates).

For each protein-RNA combination, two sets of calculations were per-formed. First, the elemental composition (% N, % P, N:P) was calculated based on the contribution of different biochemicals and their elemental composition. This procedure is the same as used earlier in assessing the elemental composition of various biological entities based on direct bio-chemical measurements (e.g., Fig. 2.4). Second, the growth rate of the animal in terms of specific increase in protein biomass was calculated from information on the protein synthesis kinetics of ribosomes. Here the as-sumption is that the animal is exhibiting balanced growth, and thus the growth of the entire body must be proportional to the growth rate of body protein. The goal was not to calculate actual growth rates of an organism as a function of various food or environmental conditions but rather to esti-mate how fast an animal could grow given its particular RNA and protein allocations and to see how body C:N:P at those allocations relates to that growth potential.

To calculate protein synthesis rate based on RNA allocation, the animal's total RNA level was first converted to rRNA, assuming that 80% of total RNA was rRNA (Schwartz and Lazar 1981). Next, ribosome number was estimated using the molecular weight of RNA in a ribosome (4.2×10^{-12} μg RNA ribosome^{-1}, Schwartz and Lazar 1981). Protein synthesis rates were then estimated by multiplying the number of ribosomes by a rate of amino acid polymerization per ribosome typical of eucaryotic ribosome machinery at 20°C (2.5×10^{-12} μg protein ribosome^{-1} d^{-1}) (Lewin 1980; Sadava 1993). Since not all synthesized protein is retained as new biomass, this rate was corrected by a factor of 0.60 (Mathers et al. 1993). Use of a fixed value for protein synthesis kinetics is consistent with observations at fixed temperature that protein synthesis rate per ribosome is the same across species as long as growth is near maximal (Maaløe and Kjeldgaard 1966; Marr 1991). Later, in Chapter 8, we will see that growth-stoichiome-try couplings such as these are to be expected whenever the rate of pro-duction of one substance is established by the concentrations of another.

In the DE model, maximum specific growth rate is calculated as

$$\mu_m = \frac{\ln\left(\dfrac{(Q'_{\text{protein,initial}} + Q'_{\text{protein,new}})}{Q'_{\text{protein,initial}}}\right)}{t}, \tag{4.1}$$

where t is a time interval and Q'_{protein} is a mass of protein within an individual (the prime refers to the use of mass units; see the Appendix). The subscripts "initial" and "new" refer to protein biomass present at the beginning of the interval and produced during the interval, respectively. The increment to protein over the interval t is calculated from ribosomal RNA × net rate of protein biomass produced per ribosomal RNA × t (Table 4.4).

Thus, this approach uses a framework built on very simple parameters of cell biology to explore the nature of the relationship between elemental composition and μ_m as the balance of major biochemicals in the body varies. Writing $Q'_{\text{protein,initial}}$ and $Q'_{\text{protein,new}}$ as separate terms in Equation (4.1) shows that growth rate can be thought of as being a function of two biochemical parameters. The first is RNA allocation, which determines the protein production rate and thus new proteins, or $Q'_{\text{protein,new}}$. The second is protein allocation, which determines the initial quantity of protein biomass that must be compounded to achieve a given growth rate. That is, at a given protein allocation, animals of increased RNA allocation will grow at increased rates due to higher protein synthesis rates. What is less obvious is that animals will grow at different rates at a given RNA content if they differ in protein content: high-protein animals will grow more slowly, low-protein animals will grow more rapidly. One way of thinking about this is to consider an organism's rRNA to be the protein-output machinery driving growth while the organism's protein biomass is the "overhead" that must be replicated in order for the organism to grow at some balanced rate. In such a scheme, increased allocation to protein overhead will tend to slow the organism's growth, because it represents a larger overhead that must be reproduced by the organism's rRNA catalytic capacity. Thus, alternative allocation strategies might be used to achieve a given μ_m. These strategies produce different stoichiometric signatures and suggest strong fundamental trade-offs. That is, increased allocation to protein directly causes lower growth (all else being equal) both by decreasing allocation to biosynthetic apparatus, and by imposing a greater burden of overhead. Further, an attempt to increase growth rate seems to bring with it an inevitable increase in biomass nutrient content, especially for P. Such an increase might impose more stringent food quality demands on such an organism, if maintaining homeostasis at high nutrient levels imposes growth penalties for unbalanced diets. We'll see evidence in Chapter 5 that this indeed appears to be the case.

The relationships between μ_m and % N, % P, and N:P predicted by the DE model are shown in Figure 4.10. In this figure, the calculated value of elemental composition at a given combination of protein and RNA content is plotted against the predicted μ_m. Since % C differed little, % N and % P are good surrogates for N:C and P:C. This figure allows us to examine how the model predicts elemental composition should vary with growth rate at a fixed RNA level (i.e., as protein % changes), or with growth rate at a fixed protein level (i.e., as RNA % changes).

The theory predicts that, at a given protein content, % N increases slightly with μ_m as allocation to RNA increases (Fig. 4.10A). However, these changes in N are small: % N at a fixed protein allocation increases at most only 1.2-fold across the full range of growth rate. Across all combinations of RNA and protein, % N varies only a little more than twofold. These modest trends reflect the relatively high N content of nucleic acids relative to the suite of other biochemicals being displaced as RNA allocation increases. However, % N does decline strongly with increasing μ_m if protein allocation is lowered at a fixed RNA content (Fig. 4.10A, dashed lines). Thus, under the assumption of balanced growth and with growth keyed to protein synthesis rate, increased specific growth rate due to lower protein allocation lowers % N while increased growth rate due to increased RNA content raises % N somewhat. This leads to an overall expectation that changes in organismal N content with differences in growth rate will be relatively modest. In addition, the model predicts that the nature of the change (i.e., whether % N increases or decreases with increasing μ_m) will depend on the particular mechanism by which growth rate is altered (i.e., by modulation of protein allocation or of RNA allocation).

Predicted changes in P content with growth are proportionately much larger. The model predicts fivefold increases in % P with growth rate (Fig. 4.10B). These increases are nonlinear, with % P increasing more steeply with μ_m at high growth rates than at low growth rates. This nonlinearity reflects the fact that at low μ_m most P is in the invariant background biochemical allocation (DNA, phospholipids, ATP) but as growth rate increases due to increased RNA the importance of that background P diminishes. The model also suggests how RNA and protein content might interact in affecting growth rate. At low protein allocation, a massive increase in RNA allocation from 1 to 20% increases μ_m from 0.1 to 1.17 d^{-1} (an increase of 1.07 d^{-1}) but at high protein allocation (75%) the same RNA increase only raises μ_m from 0.04 to 0.53 d^{-1}, an increase of only 0.49 d^{-1}. Being a high-protein animal may inherently constrain achievable growth rates.

The DE model also predicts that N:P should decline strongly with increasing growth rate (Fig. 4.10C), changing fivefold across the full range of allocation patterns. At low μ_m, predicted N:P is generally high but variable. Relatively low N:P in slowly growing animals occurs for combinations involving low RNA content with simultaneously low protein content. Various combinations of protein and RNA investment result in variation in N:P for a given growth rate within the overall decreasing pattern. Note that N:P at fixed protein allocation declines nonlinearly with increasing growth rate. This shape of the predicted relationship bears some resemblance to the observations for N:P and growth rate in Figure 4.6 for *D. magna* and *D. melanogaster*. Finally, the DE model predicts a convergence of N:P at high μ_m, asymptotically approaching an N:P of approximately 9. Thus, at high growth, N:P stoichiometry approaches the stoichiometry of the ribosome itself (N:P ~7; Chapter 2)! The zooplankton species with the lowest N:P so far measured is *Scapholeberis mucronata*, N:P $\approx$ 7.5. Is this organism approximately a swimming ribosome? Its extremely high μ_m (Fig. 4.7) makes one think so. In any case, since animals obviously must devote protein to other, nonribosomal uses (for example, to constructing mitochondria to release energy to support high growth rates), actual somatic μ_m will generally be lower and N:P higher than these values. Differences in amino acid polymerization rate could also generate variability among μ_m-C:N:P relationships for different groups of organisms. For example, *Drosophila* larvae seem to achieve higher μ_m than *Daphnia* at similar values of % P and N:P (Fig. 4.6; Elser et al. 1996).

These analyses illustrate theoretical expectations for a connection between growth rate and C:N:P. They show that, under biologically reasonable values of protein allocation, there are a limited number of ways of being a high-growth-rate animal (Fig. 4.10). The rapid-growth strategy requires high RNA investment, possibly in combination with reduced protein allocation. This inevitably results in an animal with high nutrient content (high % N and especially high % P) and low N:P. In contrast, the broad range of N:P at low growth rate (Fig. 4.10) indicates that there are many ways of growing slowly. On this basis we might speculate that there should be a greater degree of morphological and perhaps phylogenetic diversity for taxa with slow growth rates than for those with fast growth rates. As far as we know, this prediction has not been tested.

Thus, the DE model uses simple biochemical assumptions to predict patterns relating C:N:P stoichiometry and growth. However, do these predictions bear any relation to actual relationships between elemental composition and growth rate in real animals? Since this version of the ribosome kinetics model was developed using parameters relevant to crustacean zooplankton, the DE model was tested using the data presented earlier on elemental composition and μ_m for cladoceran zooplankton (Fig. 4.7; Main et al. 1997). Model predictions were generated using a range of values for protein percentage (30–50%) characteristic of cladocerans (Berberovic 1990).

The correspondence between theory and observation is good (Fig. 4.11). Note that the lines on the figures are not regression lines fitted *a posteriori* to the data. They are mechanistic *a priori* predictions of growth rate and elemental composition from the molecular-kinetic model. The model correctly predicts qualitative aspects of the relationships between elemental

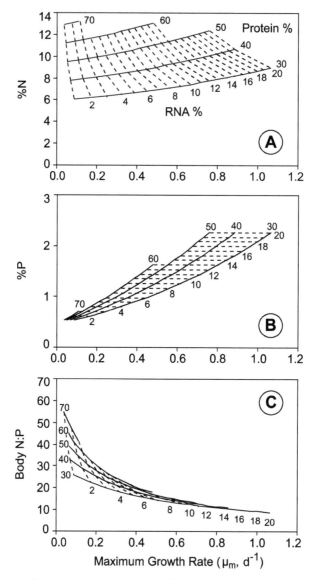

Fig. 4.10. Predicted relationships between body elemental composition and maximal growth rate as a function of allocation to protein and RNA according to the DE ribosome kinetics model. Solid lines connect identical protein allocations but variable RNA allocation. Dashed lines connect identical RNA but variable protein allocations. A. % N vs. maximum growth rate. B. % P vs. maximum growth rate. C. N:P vs. maximum growth rate.

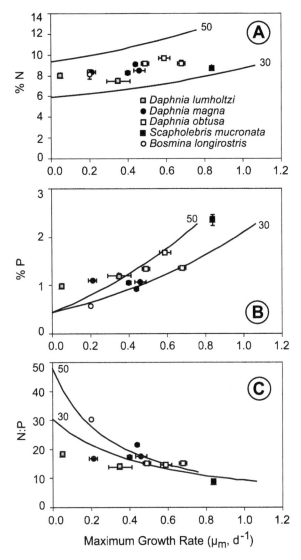

Fig. 4.11. Test of the DE ribosome kinetics model against the data of Figure 4.7. A, B, and C indicate % N, % P, and N:P, respectively. The lines indicate the predicted elemental composition–growth rate relationships for either 30 or 50% allocation to protein, a range characteristic of cladoceran zooplankton. The model predicts growth-stoichiometry relationships that encompass the data over most of the range of growth rates. Therefore, the DE model makes biologically realistic predictions about the growth potential and C:N:P stoichiometry of these animals.

composition and μ_m. It accurately predicts the slight increase in cladoc-eran % N with growth rate (Fig. 4.11A). Likewise, it predicts a strong and nonlinear increase in % P with growth rate (Fig. 4.11B). The model there-fore captures aspects of the N:P versus μ_m relationship in the data (Fig. 4.11C). More specifically, the observations suggest an accelerating increase in N:P as growth rate declines, higher variability in N:P at low growth rates, and convergence of N:P on ~9.5 at high growth rate. All of these are features of the theoretical predictions. We also note that the range of values for N:P predicted by the model matches quite well with the range of values for zooplankton N:P compiled by the NCEAS working group (Fig. 4.3) and P and N content also seems reasonable relative to C:N and C:P values summarized there. Thus, while more data—especially at low growth rates—are needed, the DE model successfully links elemental composition and maximal growth rates over most of the range.

Combining what we have seen in Chapters 2, 3, and this one, we begin to see a fundamental biochemical signature of rapid organism growth that is linked to biochemical investment and cellular machinery. In this chapter, we have seen that rapid animal maximal growth is accomplished by pos-sessing particular proportions of C, N, and P in biomass. This association is not necessarily confined to animals; recall from Chapter 3 that the optimal N:P of a green alga declined with increasing growth rate in chemostats (Elrifi and Turpin 1985; Terry et al. 1985; Turpin 1986) and also recall the tendencies for rapidly growing autotroph biomass to have low N:P (Figs. 3.3 and 3.8). As Redfield surmised (Chapter 1) and as we will describe in later pages, the C:N:P stoichiometry of organisms has an important influ-ence at the level of the global ecosystem. From all this, a thread of causal-ity emerges that runs from the ribosome to the biosphere. As we saw in the logical flow chart of Reiners earlier (Fig. 1.12), with only a few steps, one can bridge molecular- and ecosystem-level scales of organization. Per-haps we can see the beginnings of a unifying fabric of biological study that is built up from the threads of physical-chemical principles.

STRUCTURAL INVESTMENT AND THE
STOICHIOMETRY OF VERTEBRATES

Our discussion of the stoichiometry of invertebrates hinged on the connec-tion between RNA and P. We saw that different biochemical investments produced variation in whole-organism N but especially P content. In turn-ing our attention now to vertebrates, we must take into account the fact that bone—with its very high P content—is an important constituent of body mass. We will see that the same overall pattern of strong variation in total organism P content holds for vertebrates as for invertebrates, but for somewhat different reasons.

The N:P of bone (0.8) is the lowest of any biological material we have seen so far, a reflection of the high amounts of P within the mineral hydroxyapatite (Chapter 2). Bone N:P is in fact much lower than the N:P of other organs or organ systems (Fig. 4.12A). The mineral content of bone and bony materials (teeth, antlers) is also known to vary with hardness and stiffness (Ascenzi and Bell 1972). Thus, a vertebrate requiring bones, antlers, or teeth of high mechanical strength must invest increased mineral resources (P, Ca) in those structures. Bone can contribute substantially to animal biomass and thus significantly influence whole-organism elemental composition. For example, in a 1 kg (dry mass) vertebrate, the skeleton comprises ~7% of body mass (Fig. 4.12B). Since bone averages ~12% P (Fig. 4.12A), the skeleton in such an animal will contribute as much as ~0.8% P to overall animal P content. As Figure 4.12B shows and as we will consider in more detail below, the skeleton comprises an even higher percentage of body biomass in larger vertebrates. In humans, for example, bones contain 85% of total body P (with 14% in soft tissue and 1% in blood; Berner and Shike 1988) and bones contain more than 75% of total body P in some ruminant livestock (Council 1980). Bones are clearly an important pool that may differ significantly among vertebrate species according to important aspects of their biology.

Figure 4.13 shows stoichiometric variation among fish species, where we see some familiar patterns. First, the proportional variation in P content is much greater than for N. Second, much of the stoichiometric variability is taxon based. Nitrogen content ranged from about 8 to about 12%, P content ranged from about 1.5 to about 4.5%, and N:P ranged from about 5 to about 15. In fish, the families Percidae and Centrarchidae have particularly high P content. In another study, members of the Loricariidae, a family containing the tropical armored catfish, have also been found to be very high in P (Vanni et al. 2002); we will encounter these fish again in Chapter 6. High-P fish are among the "bonier" groups, suggesting that mechanical stiffness (due to both bones and scales) is the largest factor accounting for interspecific variation in N and P stoichiometry. Not all studies have seen such large variation in P content, however. Tanner et al. (2000) noted relatively modest variation among 20 taxa of fish from Lake Superior. In their study, the means (ranges) of C:N, C:P, and N:P were 4.7 (4.1–5.9), 48.2 (40.6–64.6), and 10.3 (8.4–12.8), respectively. We do not know why some fish assemblages, such as those of Lake Superior, are similar in their stoichiometry while others are different. Perhaps this will be a new dimension for work in fish ecology. Earlier in this chapter, we discussed the taxonomic variability in zooplankton P content; this was related to particular genera or even species within genera. Now we see that a high-P fish has even

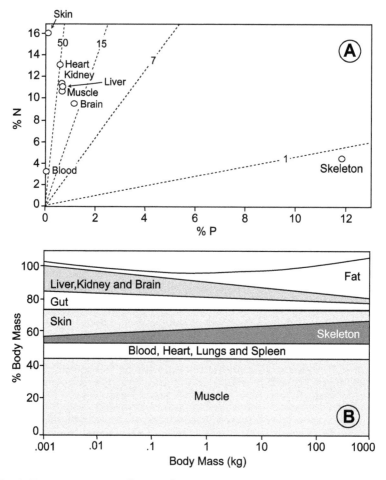

Fig. 4.12. Factors potentially contributing to variation in C:N:P stoichiometry of vertebrates. A. Stoichiometric diagram of mammalian organs. B. Allometric variation in organ % contribution to body mass. The net result of increased allocation to skeleton in large vertebrates would be increased body P content and lowered body N:P. Panel A based on Elser et al. (1996) with original data from Bowen (1979) and panel B based on Calder (1984).

higher P content than a high-P zooplankton, and its N:P is lower. Much more work needs to be done on understanding the causes and consequences of this stoichiometric variation in fish.

Now we've seen similar patterns of taxon-based variation in P content in invertebrates and vertebrates. In the next section, we will explore the al-

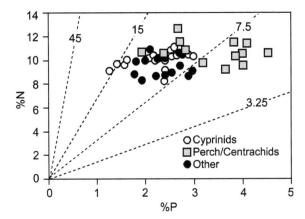

Fig. 4.13. A stoichiometric diagram of fish taxa. Note that the variation in P content exceeds the variation in N content. This variation in P content likely reflects interspecific differences in relative allocation to bone or bony scales. Based on Sterner and George (2000).

lometry of stoichiometry, extending our consideration to include not only animals but also other heterotrophs such as protozoa and bacteria.

ELEMENTAL COMPOSITION AND BODY SIZE

Heterotrophs from bacteria to large vertebrates range over some 20 orders of magnitude in body size. We have seen two primary biological determinants of heterotroph P content: growth rate, with its demands for rRNA-P, and structural investment, with its demands for hydroxy apatite-P. Fundamental scaling laws say that growth rate and structural investment will vary with body size. Given these two primary factors, perhaps we can make some generalizations about how body elemental composition should differ among heterotroph species differing in body size.

Body size is central to life history theory because of its many fundamental physiological and ecological consequences (Peters 1983; Calder 1984; Schmidt-Nielsen 1984). Peters (1983) compiled a considerable array of allometric relationships for metabolic parameters, behavioral characteristics, and population features and discussed a number of ecological consequences. Additionally, a number of theoretical and experimental studies in ecology (reviewed by Stein et al. 1988) have been performed using "size-structured" approaches in communities. For example, in many aquatic systems, the size relationships between predators and prey strongly influence predator-prey interactions and thus can affect the nature of food-web dynamics

(Werner and Gilliam 1984). Indeed, recent theoretical breakthroughs involving fractal scaling relationships have injected new life into the study of allometric relationships (Brown and West 2000) and suggestions for linking the new scaling framework with stoichiometric theory have begun to appear (Enquist and Niklas 2001).

Given the importance of organism size in physiology, ecology, and evolution, we might wonder whether there are any patterns for body C:N:P stoichiometry as a function of body size. Although this is a conceptually simple question, unfortunately we are not aware of a sufficient quantity of data in suitable form to address this issue directly (i.e., there is as yet no good data table containing the stoichiometry of different heterotrophs across a wide range of body sizes). Elser et al. (1996) presented an allometric hypothesis based on much of the biology of body elemental composition that we have presented in this chapter. Here the focus is on N:P as an integrating parameter.

If body elemental composition is linked to organismal growth rate and if there is significant variation in growth rate with body size, then it follows that there should also be variation in elemental composition with body size. Indeed, it is well known that characteristic growth rate (μ_m) declines with increasing organism size (Peters 1983). The RNA content in both crustaceans (Båmstedt 1986) and mammal "soft tissue" (liver, thyroid; Munro 1969) decreases significantly with increasing body size. Thus, if growth rate is a primary determinant of body N:P via its association with RNA allocation, one would predict that, as μ declines with increasing body size, body N:P should increase (Fig. 4.14A). However, we also recognize that such a broad trend would also have considerable scatter. For example, we saw earlier that *Daphnia* and calanoid copepods have strongly divergent N:P despite roughly similar body size. As in the *Daphnia* versus copepod example, strong divergence in N:P at a given body size is likely associated with interesting and important ecological and evolutionary factors.

The hypothesized increase in N:P with increasing body size (Fig. 4.14A) due to decreasing specific growth rate probably does not continue indefinitely. Vertebrates, with their bone investments, enter the body size continuum at a size of ~100 mg and they become the dominant group of metazoans with body size greater than ~100 g (dry mass). Indeed, it is tempting to speculate that one factor facilitating the evolution of bony vertebrates was a lowered demand for P for RNA at the slow growth rates that characterize large animals. Dietary P would then be available for use for other purposes, such as construction of stiff support structures.

We can predict the allometry of N:P for vertebrates from two basic sets of information: data on the elemental composition of major vertebrate organs and organ systems and information regarding the body-size dependence of biomass allocation to those organs (Figs. 4.12A, B). The most

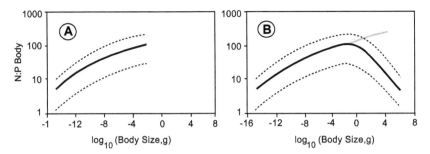

Fig. 4.14. Hypothesized variation in heterotroph N:P as a function of size (g dry weight of the individual). Dotted lines above and below the main trend line (black line) indicate ecologically relevant variation in elemental composition at given body size. A. The hypothesized allometry for small, nonvertebrate heterotrophs (bacteria, Protozoa, invertebrate metazoa). B. The hypothesized allometry for large organisms, including vertebrates (main trend line) and large invertebrates (gray line; giant squid?). Somatic N:P is predicted to increase with increasing body size in invertebrates due to the known negative allometric scaling of growth rate. However, in vertebrates, N:P is predicted to decline with body size because relative skeletal allocation is known to increase with body size (Fig. 4.12B) and bone is extremely P rich (Fig. 4.12A). Based on Elser et al. (1996).

striking allometric pattern for organ contribution from such studies is the disproportionate increase in skeletal mass with body mass (Anderson et al. 1979). For mammals, skeletal mass increases with body mass with an allometric coefficient of 1.07 (Prange et al. 1979). The value for teleost fish (1.09) is similar (Cassadevall et al. 1990). Thus, the skeleton of a shrew contributes ~5% of total body mass, compared to the ~27% contributed to total body mass in an elephant. We can use this information on organ elemental composition and allometric scaling to estimate the elemental composition of two "theoretical" mammals of contrasting body size: 1 kg (a guinea pig perhaps) versus 1000 kg (perhaps a cow).

In Table 4.5 we present the major organ contributions of these theoretical mammals and the predicted elemental composition that results from using those biomass contributions to calculate a weighted average of organ % N and % P values. In this example, the large animal has a somewhat lower N content relative to the small animal (7% N vs. 10.8% N), a reduction that primarily reflects a decrease in the contribution of skin to the larger animal. The large animal, however, has a substantially higher P content than the small animal (1.6% vs. 0.84% P). This increase is nearly entirely a result of increased skeletal allocation in the large animal. As a result, body N:P is predicted to be ~28 for a 1-g mammal and ~9.6 for a 1000-kg mammal. So we suggest that body N:P should decline with in-

TABLE 4.5
Effects of variation in allocation to major tissues, organs, and organ systems on
body elemental composition for small (1 g) and large (1000 kg) mammals. All
values in this table are percentage of body mass except for the N:P ratio, which is
a molar ratio. Note that the whole-body difference in estimated % P (0.76%) is
largely the result of increased allocation to skeleton in the large vertebrate
(11.4–3.2% = 8.2% difference in skeletal allocation; since skeleton is 12% P, this
is equivalent to ~0.98% P). In making these estimates we used data for
allometric variation in allocation from Calder (1984) and for elemental
composition of various organs from Bowen (1979)

Body Mass	Muscle	Skin	Brain	Liver and Kidney	Skeleton	N	P	N:P
1 g	45.9	23.2	5.7	10.1	3.2	10.8	0.84	28.4
1000 kg	45.9	7.7	0.2	1.5	11.4	7.2	1.6	9.6

creasing body size for large animals (Fig. 4.14B), again with potentially
meaningful variation in N:P at a given body size. Note, however, that the
difference in N:P that we estimate for large versus small vertebrates is
substantial but is likely to be somewhat of an overestimate. As mentioned
above, the P content of vertebrate "soft tissue" probably declines in large
animals due to decreasing % RNA (Munro 1969), an effect that would
tend to offset somewhat the predicted decline in N:P with increased body
size.

Skeletal allocation implies that elephant P content can be no lower than
~3.1%, a value exceeding even the highest known % P value for crustacean zooplankton (~2.5% for *Scapholeberis mucronata*; Fig. 4.2). Phosphorus content as high as 5.5% has been recorded in fish (data of Vanni et
al. 2002, discussed in Chapter 6), which should have lower skeletal demands than terrestrial vertebrates (Fig. 4.13). With body P levels this high,
and knowing the generally high C:P ratios of terrestrial vegetation (Chapter 3), it is easy to imagine that a large vertebrate foraging on relatively
nutrient-poor terrestrial vegetation must take extraordinary measures to
meet the mineral requirements for building and maintaining a Ca- and
P-rich skeleton that makes up so much of its body mass. Considering the
allometry of P stoichiometry in vertebrates, one shudders to think of the
mineral hunger experienced by the largest herbivorous dinosaurs! We will
consider questions of mineral limitation of growth in detail in Chapter 5.

Body-size differences in C:N:P stoichiometry have been examined in
some detail for various fish taxa. For example, body P content increases
with body size in centrarchids (bluegills and largemouth bass) while N
content decreases with increasing body size (Davis and Boyd 1978). Since

% N and % P trend in opposite directions Vanni (1996) suggested that there should be a general allometric trend of decreasing N:P with increasing fish size. However, Sterner and George (2000) observed no, or weakly negative, associations between body mass and % P and % N in four cyprinid species. Nevertheless, across a wide range of data, the allometric scaling exponent for skeletal mass versus fish mass is 1.09 (Cassadevall et al. 1990), indicating that, in spite of living in aqueous medium, evolutionary scaling pressures affect fish as they do terrestrial vertebrates. Within more restricted samples though (either taxonomically or in terms of body size), allometric patterns may differ from this overall trend (Cassadevall et al. 1990).

Body-size trends in bone mineral pools may then have important implications for understanding both the evolutionary ecology of large and small vertebrates and the role of such animals in an ecosystem context. As we will see in the next chapter, differences in body P content relate to an animal species' ability to grow on P-deficient foods. Further, grazing and nutrient recycling by vertebrate herbivores are thought be to key processes in many terrestrial ecosystems (Ruess and McNaughton 1987; Pastor and Naiman 1992; Pastor and Cohen 1997). Thus, vertebrate elemental composition may be a key but underappreciated factor affecting trophic dynamics and consumer-driven nutrient cycling in such systems. These points will be considered further in Chapters 5 and 6. For now, our consideration of potential allometric variation in elemental composition suggests to us that stoichiometric analysis may be useful in understanding ecological and evolutionary factors operating on animals of contrasting body size and, in turn, in evaluating how size-structured ecological interactions play out in nature.

CATALYSTS FOR ECOLOGICAL STOICHIOMETRY

In this chapter we examined how animal C:N:P stoichiometry is associated with growth rate and allocation to major structures. Many of the principles we have discussed should hold for many types of heterotrophs. How stoichiometry maps onto different nutritional modes, including such newfound exotic forms as microbial photoheterotrophy (Beja et al. 2001), is a major open question. Our catalysts revolve around the main unresolved questions concerning the biological stoichiometry of animals, and they are organized by habitat. First, what do we need to know for planktonic animals?

- In the stoichiometric models we presented here, as elsewhere, we often rely on the utility of laboratory model organisms as accurate analogs of what we should also find in the field. In microbiological growth studies of RNA, for instance, we are currently forced to use mainly values for ribosome function derived from *E. coli* measured in laboratory culture. Better bridges between laboratory studies of model organisms and the field will improve stoichiometric studies, as they will other kinds of studies.

- Evidence points to a major contribution of P-rich RNA in establishing the C:N:P stoichiometry of zooplankton and thus suggests a connection between selection on biomass growth rate and body elemental composition via the cellular and biochemical machinery required for growth. This seems an exciting and plausible hypothesis but we still need many more data on growth rate, biochemical (RNA) allocation, and elemental composition for a variety of animal taxa to be more confident in building a theoretical structure on this foundation.

- Even in groups such as the zooplankton where considerable data on C:N:P ratios have been collected, we still wonder about the full extent of variation in the entire functional group. What are the C:N:P ratios of species representative of the full suite of freshwater zooplankton? For example, we know very little about noncrustacean zooplankton (e.g., rotifers in freshwater and, in marine plankton, a variety of other major groups, such as coelenterates, larvaceans, and other gelatinous zooplankton, and immature forms of groups such as echinoderms) and almost nothing about protozoans. For example, we wonder if there are species-specific differences in C:N:P among protozoans and how tightly those ratios are regulated. Are they more animal-like or plantlike in their homeostatic patterns?

- With information about a greater suite of zooplankton taxa available, it might be possible to test the patterns predicted by the DE model and depicted in Figure 4.10. That is, are there more ways of being a slow-growth-rate animal than a fast-growth-rate animal? Are there many zooplankton species capable of slow growth rate but only a few rapid-growth specialty species?

While considerable information and the beginnings of understanding seem to be in hand for pelagic invertebrates, little is known about consumer taxa that dominate other food webs, such as streams, forest canopies, or soils. Here are some questions that will help to determine if ecological stoichiometry has a role to play in the study of consumers in lotic and terrestrial food webs.

- What differences are there in the elemental composition of important stream consumers, including herbivores ("scrapers") but also detritivores ("collectors")?

- How does body C:N:P vary among important terrestrial herbivore and detritivore species?

Questions general to all animal taxa include the following.

- Are there systematic differences in C:N:P stoichiometry associated with various levels of phylogenetic association (e.g., phylum, class, order, family, genus)?
- Is variation in body C:N:P associated with patterns of allocation to RNA and protein? Is this variation correlated with growth rates, reproductive investment, or other major aspects of a species' life history strategy?
- What genetic mechanisms (e.g., polymorphisms in rDNA genes) are responsible for variation in biochemical allocation and thus C:N:P stoichiometry? How are these genes regulated in generating phenotypic plasticity in both growth and C:N:P stoichiometry? Does possessing the genetic machinery for maintaining massive rRNA allocation during rapid growth make it difficult to keep these genes down-regulated when times are hard? That is, is there a molecular genetic trade-off that means that high-growth-rate organisms have inherently high minimal P content (Q_{min} in the parlance of Chapter 3)? If so, then we may begin to understand the fundamental basis of food quality requirements of stoichiometrically contrasting consumers, as will be discussed in Chapter 5.

Finally, we need to know more about the storage of nutrient elements, especially P, in the hard structures of animals.

- How does the sequestration of P in hard mineral structures differ among *invertebrate* taxa? Our emphasis in this chapter has been on the role of biochemical P (RNA, phospholipids, etc.) in the stoichiometry of invertebrate growth. However, ecologically significant amounts of P are likely stored in the obvious hard mineral deposits of bivalves (for example, see data for zebra mussels by Arnott and Vanni [1996]) but may also be found in less obvious forms. Sterner and Schulz (1998) showed that *Daphnia* molts contained 0.80 %P and Vrede et al. (1999) showed that P complexed with Ca within the carapace of *Daphnia* contributes up to ~10% of whole-body P.
- We need to know more about differences or similarities in the allometry of N and P stoichiometry in aquatic versus terrestrial vertebrates. It seems intuitively obvious that mechanical support against the force of gravity should be less of a factor for the former than the latter. Fish bones are likely thinner and weaker than mammalian bones. However, the P content of fish integument, including its scales, is a significant factor and varies with body size in a different way from that of the internal skeleton. Groups such as sharks, whales, and giant squid would help us resolve these questions of evolutionary and allometric

pressures on N:P stoichiometry and how mechanical structural forces influence those. Needless to say, getting the data for such beasts will be challenging.

- Finally, given a wider documentation and synthesis of body C:N:P in diverse heterotroph taxa, we eventually will be able to replace the speculation in Figure 4.14 with real data.

SUMMARY AND SYNTHESIS

In this chapter we have seen the basic mechanics of how the stoichio-metric program connects organismal patterns to biological mechanisms in animals. This approach consists of three major steps. First, identify a set of chemical elements of interest. We focus on C, N, and P. In other contexts, one might wish to focus on other potentially important elements, such as Fe or Si or Ca, coupled with major elements such as C or N. For example, the first studies applying stoichiometric analysis to iron dynamics in pelagic ecosystems have recently appeared (Chase and Price 1997; Schmidt et al. 1999). Second, determine which biomolecules, cellular structures, or cellu-lar processes differ strongly in their demands with respect to the elements of interest. Doing so may identify specific biological functions associated with different contents of those elements. In this chapter, key associations were between high P content (low C:P and N:P) and either increased ribosome or increased bone allocation. Third, apply that insight in evolu-tionary or ecological contexts where biological functions associated with differential elemental composition are likely to be important. In our case, we contrasted zooplankton taxa with differing life history traits and saw that differences in body elemental composition were linked to those traits. Indeed, these analyses showed that there appears to be a characteristic biogeochemical signature of body growth in invertebrate animals: rapid growth seems to be fundamentally a P-intensive process. Further, we saw how the special elemental demands required for structural support in ver-tebrates introduce a further mechanism by which the evolved features of an animal are connected by biological mechanisms to its elemental compo-sition and, perhaps, ultimately to food-web dynamics and biogeochemical cycling in ecosystems. In the upcoming pages we will more fully explore these implications. To do so we will need to combine our insights into the factors associated with consumer elemental composition from this chapter with insights from Chapter 3 regarding the mineral nutrition and elemen-tal composition of the things that herbivorous consumers consume: auto-trophs. The effects of the sometimes striking contrasts between autotroph and animal stoichiometry will become apparent in the next chapter.

5

Imbalanced Resources and Animal Growth

A child said "What is the grass?" fetching it to me with full hands;
How could I answer the child? I do not know what it is any more than he.
–Walt Whitman (1819–1892)

Our book now shifts focus from the chemical composition of individuals and their component parts and we start to ask questions about how ecological stoichiometry regulates or constrains ecological processes. This chapter considers how consumer growth is affected by the chemical composition of resources. As we will see, there are many ways that consumers react to foods of differing composition, including changes in feeding rate, food selection, production efficiency, biomass growth rate, and population growth rate. Living consumers and prey are complex biological systems constantly undergoing ontogenetic and physiological adjustments that may influence their stoichiometry. One of the most important conclusions of this chapter will be that something very simple, namely, chemical composition, exerts control over things that are potentially far more complex. Our approaches in this chapter will be both theoretical and empirical.

We have already described variation and patterns in the chemical composition of different species across broad taxonomic groups and across major feeding and nutritional modes. We have seen that the chemical composition of biomolecules, organelles, and whole organisms varies in consistent and interdependent ways (Chapter 2). The reasons why organisms have the element composition they do range from evolutionary pressure on life histories (Chapter 4) to the need for support structure (Chapter 3). Variation is further based on the distribution of availability of energy and materials in ecosystems (Chapter 3). In Chapter 3 we saw that autotrophs exhibit great plasticity in their chemical composition. In fact, a single autotroph species can come close to expressing the full range of variation of nutrient content seen in all autotrophs put together. In contrast, we saw in Chapters 1 and 4 that the chemical composition of individual metazoan species, although not perfectly constant, was less variable than that across the entire suite of animal species, a consequence of physiological homeostasis. From the above information alone, it is easy to deduce that sometimes food will have similar composition to consumers and at other times the match will be poor. This information also suggests that we should expect differences in the stoichiometry of trophic interactions depending upon whether we are interested in animals consuming plants, animals consuming animals, or some other trophic strategy.

Now it is time to put these pieces together and start exploring ecological interactions of various kinds. This chapter considers what happens when a homeostatic consumer processes strongly imbalanced food. First, we will study consumer growth using a series of simple mass balance models that incorporate different assumptions and realities. Using these models, we show how stoichiometric theory pares consumer growth and food processing to its most fundamental basics to delineate formally how consumers turn food into new consumer biomass of fixed chemical composition by adjusting their own mass balance. We describe various perspectives on growth and yield, which permit us to specify characteristic stoichiometric ratios where control of growth shifts from one element to another (these ratios will be called threshold element ratios, or TER's). We then develop a new, minimal model of secondary production that takes us a few steps beyond existing theory. After we have finished developing these theoretical ideas, some real world examples of the effects of stoichiometric constraints on consumer growth and assimilation efficiency will be considered, touching on organisms from zooplankton to fungi to protozoa to moose. By considering this great variety of consumer taxa from detritivores to herbivores to carnivores and from unicellular to multicellular we will see how stoichiometric imbalance is ubiquitous and how consumers maintain homeostasis on imbalanced resources. At the end, these examples come together in a test of the general predictions of the minimal stoichiometric model of consumer production and assimilation. The goal will be to derive a general picture of how growth efficiency varies with stoichiometric balance. The new minimal model, we will see, does quite well in predicting consumer performance from simple chemical parameters.

MASS BALANCE IN GROWTH PROCESSES

The law of conservation of matter says that a general representation of a consumer-resource interaction is

$$X \text{ consumer biomass} + Y \text{ resources} \rightarrow X \text{ consumer biomass} +$$
$$aY \text{ new consumer biomass} + (1 - a)Y \text{ waste products}. \quad (5.1)$$

Stoichiometric relationships must be preserved in any interaction such as this because all mass in resources must also appear in new consumer biomass (in the broad sense, including reproductive tissue, exoskeletons, support structures, and all other forms of biomass) or wastes. The power of stoichiometric analysis comes from its recognition of constraints on system behavior. Constraints can be associated with one or more of these terms. If the chemical composition of consumer biomass new and old is constrained to be equal to some species-specific value, we should be able to predict something about wastes given only information on resources and the species of the consumer (see Chapter 6). The yield of this chemical reaction (production of new consumer biomass) should also have a strong stoichiometric component. Limiting substances from a nutritional or Liebigian standpoint may be similar in principle to limiting reactants of this chemical reaction, so this perspective should be helpful in sorting out which factors limit product formation, i.e., consumer growth.

Discussing the stoichiometry of a trophic relationship makes sense only if we consider more than one currency at a time. For simplicity we will usually examine two. From the standpoint of understanding consumer nutrition, we usually want one of the currencies to be the one that limits production. Furthermore, most of the time one of these two currencies is going to be carbon. In most biological materials, carbon is 40–50% of dry weight (Chapter 2). Hence, any relationship dealing with carbon is easily transposed to one for total mass. This chapter makes great use of the concept of stoichiometric "imbalance" defined for consumer-resource pairs in Chapter 1. Recall that if a consumer and its resource have similar chemical composition, we refer to this as a "balanced interaction." Differences in chemical composition make consumers and their resources stoichiometrically imbalanced. According to this definition the most balanced consumer-resource pair is cannibalism. Indeed, the widespread nature of cannibalism in organisms routinely considered to be herbivores (Polis 1981) might be related to the highly nutritious nature of balanced (i.e., same animal species) compared to imbalanced (plants) food. Actually, in the case of carbon, where some percentage is necessarily respired for metabolic needs, the ideal resource ratio for a consumer may not be precisely identical to its own composition (some extra carbon is needed). Nevertheless, we will simplify things and just refer to balance as the match between the actual chemical compositions of consumer and resource. An example of a highly imbalanced consumer-resource pair is petroleum-consuming microbes. Adding limiting nutrients in balanced proportions greatly speeds up how quickly these consumers can make use of this otherwise highly imbalanced resource (Smith et al. 1998). This interaction illustrates a general principle of this chapter, namely, that the growth and activity of consumers are greatly affected by stoichiometric imbalance The threshold element ratios that we derive in this chapter are stoichiometric ratios in the food that result in equal limitation by two substances (e.g., C and P). In some models, TER's are true optimal foods.

There are three ways that a consumer can maintain its stoichiometry when confronted with imbalanced resources. First, it may select different or supplemental food items. We will see some likely examples of that in this chapter when we discuss such behaviors as geophagy (soil eating). Second, it may alter assimilation patterns, regulating the passage of materials across the gut wall so that they match consumer needs. Thinking about such adjustments takes us into the physiology of the digestive system. If a consumer maintained its stoichiometry by altering assimilation processes we would expect to see digestion and assimilation of an element up-regulated when the element was rare in the food (a greater percentage assimilated when the substance is rare). A contrary pattern, the antithesis of homeostatic regulation of nutrient content, would occur if, for example, the presence of high quantities of protein in the diet caused the secretion of additional peptidases, which resulted in enhanced digestion efficiency of N. We'll see some patterns in fish that look like this contrary pattern in Chapter 6. Third and finally, homeostatic regulation of stoichiometry may occur primarily in metabolism, such as increased protein degradation and subsequent increased release of N wastes when the diet is high in N. Or, under nutrient scarcity, biochemical pathways that conserve nutrients may be brought into play. We will see an example of this when we talk about termite N conservation in this chapter. There is no single way that consumers maintain their stoichiometry in the face of imbalanced resources.

Consumers in fact come in a wondrous variety of shapes, sizes, and forms. Really, any living thing can be considered a consumer. As noted in Chapter 3, plants consume water, nutrients, and sunlight. Animals may consume a variety of different kinds of resources. However, heterotrophic consumers share a variety of stoichiometric features, and so that is the group we will deal with in this chapter as we move from autotrophs to other parts of the food web. Recall that heterotrophs are dependent on the uptake of organic matter for carbon for growth and energy. Heterotrophic consumers include such diverse creatures as herring gulls, bacteria, fungi, wolves, and beetles. Heterotrophic nutritional strategies include detritivory, herbivory, carnivory, and omnivory. In addition, we will consider some less famous nutritional strategies. Osmotrophs are organisms like fungi or bacteria that absorb organic matter across their cell membranes. Microphages and macrophages ingest solid food and may process it through some sort of alimentary canal. Macrophages select relatively large individual food items one at a time while microphages handle collections of small bits of food in bulk. Microphages include deposit and suspension feeders and regularly ingest "unsuitable" items (items of low quality) in their diets. These nutritional strategies (defined in terms of mode of carbon acquisition) do more for us than provide a convenient scheme for organizing information. As we will see, the manner in which a consumer

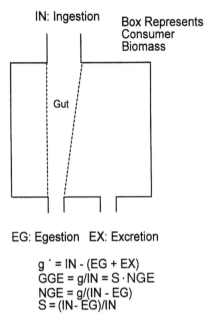

IN: Ingestion   Box Represents
                 Consumer
                 Biomass

Gut

EG: Egestion   EX: Excretion

$$g\,' = IN - (EG + EX)$$
$$GGE = g/IN = S \cdot NGE$$
$$NGE = g/(IN - EG)$$
$$S = (IN - EG)/IN$$

Fig. 5.1. Generalized consumer mass flow, showing ingestion (IN), egestion (EG), and excretion (EX). Formulas for growth rate (g'), gross growth efficiency (GGE), net growth efficiency (NGE), and assimilation efficiency (S) also are given. As most frequently used, these terms relate to energy or carbon flux; however, they may be just as meaningful for nutrients such as N or P.

makes a living—whether it eats plants or animals, or whether it ingests particles or absorbs dissolved material—has stoichiometric implications.

Our major focus for this chapter is the stoichiometry of secondary production in animals. Secondary production is the rate of creation of new mass (or, in other contexts, energy) in heterotrophic organisms. The term may refer to all heterotrophs in a single ecosystem, or to subsets, such as all the herbivores. It may also refer to a single species. A simple mass flow diagram for a generalized animal with an alimentary canal is shown in Figure 5.1. The box represents the animal consumer. Resources are ingested at rate IN. Some resources fail to be absorbed in the gut (indicated by dotted lines) and are egested at rate EG, and the rest are taken into the consumer's body. Of those assimilated, some will be rereleased to the environment as a waste product of metabolism, which occurs at rate EX. This simplified mass flux model fails to account explicitly for some kinds of mass loss from consumers, such as milk from lactating mammals, exuvia from arthropods, etc., but incorporating these would be counterproductive to keeping things general for the time being. These miscellaneous, species-specific fluxes could be included in either EX or g', or could be represented by a new term as the situation warrants. Also note that the scheme presented here does not distinguish between production of somatic tissue and reproductive tissue: both of these are included in the general term for growth.

Given these rates of intake and loss, the consumer grows at rate g',

$$g' = IN - (EG + EX). \tag{5.2}$$

The rate g' is important to us (and to the consumers!)—it is the rate of growth or secondary production. Note that the two terms "growth" and "secondary production" are used in different contexts, as sometimes "secondary production" refers to the summed growth of all members of a trophic level or of all animal consumers in an ecosystem. Even so, Equation (5.2) is clearly of key importance in both contexts. An efficiency of growth can be calculated as growth as a fraction of ingestion,

$$GGE = g' / IN, \tag{5.3}$$

which is commonly called the "gross growth efficiency." In general, GGE_X is a measure of the relative amount of ingested substance X that goes into biosynthesis or anabolic processes as opposed to catabolic processes such as respiration. To those more familiar with population dynamics models than bioenergetics models, you may wish to think of GGE as analogous to a predator's numerical response in that it relates the units of predator biomass gained per unit prey ingested. The GGE actually is a composite measure made up of two component efficiencies. The ratio of assimilation to ingestion, or the assimilation efficiency (S), is one of these. The other is the ratio of growth to assimilation, or the net growth efficiency (NGE). So,

$$GGE = S \cdot NGE. \tag{5.4}$$

The framework shown in Figure 5.1 deals with mass flux through living biomass but does not consider rates of growth compared to rates of death. Population and community dynamics are governed by how growth is offset by losses to predation and other forms of mortality.

Note that we have not specified exactly what type of "mass" Figure 5.1 refers to. Total mass itself may well be of interest. However, this diagram can be used for virtually any specific flux through the consumer, including carbon, particular nutrients, or energy. One should be aware that metabolic release products may be referred to as excretion (hence the EX), but in other situations they may be referred to as mineralization, as for nitrogen release by microbes, or they may be called respiration in the case of carbon. From our vantage point, these are just nuisances of language. In ecological stoichiometry, we must account for multiple substances; thus, we will adopt the use of subscripts to refer to different forms of mass. For example, GGE_C will refer to the gross growth efficiency for carbon.

The animal ecologist clearly is interested in factors determining the rate of growth of individual organisms or populations. The ecosystem ecologist must often know growth efficiencies and rates of many individuals or functional groups in order to estimate rates of flux through food webs. These are fundamental parameters linking individual, population, and ecosystem ecology. Now we will see what more stoichiometry has to say about them.

MAXIMIZING YIELD IN CHEMISTRY AND IN ECOLOGY

The goal of a chemical engineer is to make a chemical product from a reaction as quickly, cheaply, and efficiently as possible, while at the same time minimizing waste products. In Chapter 1 we introduced the concept of stoichiometric yield. One way of calculating the yield of a chemical reaction is to measure the rate of formation of a desired product as a function of the rate of consumption of reactants (cf. Fig. 1.10). The stoichiometry of a reaction is one of the key aspects of yield maximization (Williams and Johnson 1958). An understanding of stoichiometric constraints helps the chemical engineer determine what reactants, and in what proportions, will maximize yield. But what, you might ask, does all that have to do with animal growth in ecological contexts?

The answer is that evolution has performed a similar optimization process for millions of species of consumers over billions of years. Although life history evolution teaches us that there is much more to fitness than simply growing as quickly as possible (Roff 1992; Stearns 1992; Arendt 1997), either rapid or efficient biomass gain clearly is an important aspect of fitness. As intimated in Chapter 4 and considered further in Chapter 8, there may in fact be a fundamental trade-off for rapid versus nutrient-efficient growth. Even where rapid growth is not strongly selected for, a consumer that could attain a given growth rate on a lower intake of food would be favored by not expending energy to gather that food or by not having to expose itself to predators while foraging, etc. At the risk of over-simplifying the study of life history evolution, we will assume that marginal increases (meaning that everything else is held constant) in either growth or GGE improve a consumer's fitness (Wittenberger 1970; Calow 1977b). Negative fitness consequences of reduced or inefficient biomass growth rate include the need to go into diapause (Hunter and McNeil 1997) and an increased exposure to predators due to trade-offs between foraging and predator-safe microhabitats (Benrey and Denno 1997; Torres-Contreras and Bozinovic 1997). Thus, it seems safe to assume that under conditions of food limitation, yield maximization is generally favored by selection. In making this assertion, we are aware of the many just criticisms that have been leveled against strict adherence to an "adaptationist" paradigm, and that inexorable evolutionary movement toward some perceived optimum is often a risky assumption. In fact, some of the most interesting applications of stoichiometric analysis in ecology are when apparently optimum solutions are prohibited because of chemical constraints. Here we only wish to assert a general tendency even if it may not be achievable.

The second law of thermodynamics imposes an upper limit for carbon conversion for a maximally efficient heterotroph. That limit is thought to be about 70–90% for unicellular organisms and 35–50% for metazoans (Calow 1977a; Schroeder 1981). However, many if not most consumers fail to achieve this high level of efficiency. Valiela (1984) compiled measurements of carbon efficiencies for a variety of consumers and grouped them into the three nutritional strategies detritivore, herbivore, and carnivore (Fig. 5.2). Differences among these three groups were large and—interesting to us—consistent with the stoichiometric imbalance between consumer and resource. Carnivores consume the most nutritionally balanced foods (animals eating animals), and they showed consistently high efficiencies, particularly for assimilation. At the other end of the scale, detritivores consume the least nutritionally balanced foods (animals eating nonliving foods), and they had consistently low assimilation and net growth efficiencies. Herbivores are intermediate (animals eating living plants) and they had a wide range of efficiencies. Thus, our first general stoichiometric principle of consumer growth under homeostasis is that when the composition of the resource differs from the composition of the consumer, the GGE for individual elements must vary, whether due to differences in assimilation or in production or both. Recognition that many consumers have stricter homeostasis than their resources means that any search for a single general GGE for such consumers is doomed to failure. This is amply illustrated by the variations in GGE_C for animals with different trophic strategies that we have just seen.

Digging now a little deeper into patterns of gross growth efficiency, Straile (1997) presented an analysis of a large number of field measurements of GGE_C for a wide variety of herbivorous or bacterivorous planktonic invertebrates to see if he could find any taxonomic patterns or other kinds of trend. These diverse taxa (flagellates, ciliates, cladocerans, dinoflagellates, copepods, and rotifers) exhibited a similar mean GGE_C, of about 30% (Fig. 5.3). However, each of these groups also exhibited considerable variation, with 95% confidence limits in GGE_C ranging from below 5% to above 60%. It is also interesting to note that unicellular and multicellular consumers in the field had similar mean GGE_C values, in spite of theoretical arguments for differences (Calow 1977a; Schroeder 1981). Taxonomic groupings such as those examined by Straile seem to be poor pre-

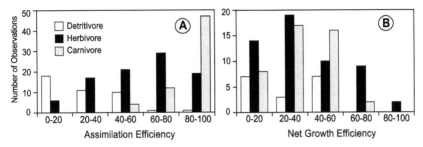

Fig. 5.2. Frequency histograms of (A) assimilation and (B) net growth efficiencies for different nutritional strategies in aquatic consumers. Based on Valiela (1984).

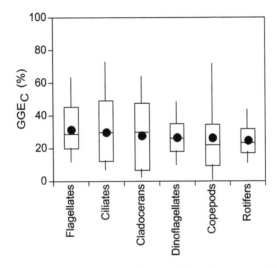

Fig. 5.3. Means (dots), medians (central lines of boxes), and variability (boxes, 25 and 75% quartiles; whiskers, 5 and 95% percentages) in GGE_C in different aquatic consumers. Note the large variability within taxonomic groups. Based on Straile (1997).

dictors of growth efficiency in the field. How can we account for the large variation in gross growth efficiency within taxa? Straile found that none of the commonly measured parameters (e.g., temperature or consumer body size) explained much of this wide variation. He hypothesized that stoichiometric balance might be an important source of variation in growth efficiency within taxa, but he lacked the data to test this hypothesis. As we will see by the end of this chapter, individual studies have indeed found that stoichiometric balance can account for a wide range of GGE_C for individual taxa, supporting Straile's hypothesis.

The figures discussed in the above paragraph raise an important issue. Trophic pyramids, one of the central concepts in ecology, are often taught as being primarily the result of the second law of thermodynamics. However, this paints a misleading picture of how food webs work. Not all of the losses in carbon or energy in going from one trophic level to another are due to unavoidable thermodynamic inefficiencies described by the second law. Mass balance and other issues may in fact be more important (see also Hessen 1997). The energetics and food limitation studies we examine in this chapter suggest that organisms are often far from their energetic optima because of stoichiometric constraints. Other simple realities of natural systems also play a role. For example, much primary production is never encountered by herbivores, and even if encountered it is not ingested because it is defended, toxic, or otherwise unpalatable. While the second law of thermodynamics imposes a calculable upper limit for gross growth efficiency for all heterotrophic consumers, Straile's data show that in nature something usually reduces GGE_C even further. Although a few species may come close to their potential energetic upper limit, most don't. Valiela's study indicates that those that are far from their upper limit are those that are most imbalanced compared to their resources. For imbalanced consumer-resource pairs, a stoichiometric limit is far more important than a purely thermodynamic limit based on the inexorable losses of energy and increase in entropy. The trophic pyramids described in today's textbooks are in need of some of the plumbing referred to by Deevey (1970) in the quotation in Chapter 1.

A more direct comparison between the concept of yield maximization by a chemical engineer and GGE for a living organism is in biotechnology. There, rapid and efficient growth of particular microorganisms (Xie and Wang 1997) or synthesis of a metabolic product (Stephanopoulos and Vallino 1991) is desired. Stoichiometric constraints have been studied in these systems in staggering detail. For example, Varma et al. (1993) examined the numerous anabolic pathways responsible for the biosynthesis of 20 amino acids and 4 nucleotides in *E. coli*. Their analysis consists of precise specifications of approximately 75 linked biochemical reactions. They determined the stoichiometry of the entire network of these reactions, linking growth substrates absorbed from the environment to the biosynthesis of each of the amino acids and nucleotides, as well as total biomass. From this analysis, the authors were able to calculate a stoichiometrically bounded upper limit on gross growth efficiency for *E. coli* growing on a substrate of acetate, glycerol, or glucose. Efficiency of growth [measured as g dry weight/mmol substrate] on these three organic molecules was 0.018, 0.054, and 0.097. Although all these numbers are small, they indicate a relative difference of more than fivefold. The authors concluded from this detailed analysis that biomass production had combined energetic and stoichiometric constraints.

Biotechnology is one example of practical application of ecological stoichiometry. Another application is in waste removal systems; for example, ideal C:N:P ratios for maximum removal of dissolved N and P from wastewaters have been described (Delgenes et al. 1998). Another is livestock production, which may often be constrained by food quality (e.g., Breman and de Wit 1983). There exist many sophisticated tools for determining the optimal forage or combination of forages for maximal growth of individual animals at particular life stages or body sizes (see, e.g., Demment and Van Soest 1983). In the next section we review the development of one such approach from zooplankton ecology, threshold element ratio (TER) theory.

LIMITING FACTORS FOR HETEROTROPH GROWTH: DEVELOPMENT OF THRESHOLD ELEMENT RATIO THEORY

Heterotrophic organisms have a fascinatingly complex nutritional ecology. They ingest or absorb a wide diversity of molecules from their food resources and extract organic and inorganic building blocks for their own biosynthesis reactions as well as energy for their growth and maintenance. Chemical elements of course cannot be biosynthesized, so any of the approximately 20 elements that heterotrophic organisms need for their structure or catalytic functions (Fig. 1.1) must be obtained from their food or otherwise obtained from their surroundings. Likewise, heterotrophs have requirements for a host of biomolecules that they either cannot synthesize at all, or have such limited abilities to synthesize that they need to be ingested or otherwise obtained from the environment.

This nutritional complexity opens up interesting questions about the nature of food limitation in heterotrophic organisms. Heterotrophs might have access to large quantities of bulk "food" but if that food is nutritionally imbalanced, they might be growth limited in the presence of a seeming abundance of food. Clinical problems in humans associated with single-factor dietary deficiencies are well known and include scurvy (deficiency of vitamin C), anemia (iron), and rickets (vitamin D). Mineral limitation of complex heterotrophs is also known in agriculture and aquaculture (McDowell 1992).

Even when food is not abundant, we still need to question precisely what factor or factors of an animal's bulk food limit growth and success. Early ecosystem theory and a long tradition of the study of bioenergetics have used the single currency of energy to capture the full multidimensionality of heterotroph growth (Morowitz 1968). Bioenergetic modeling has become quite sophisticated and can be predictive (e.g., Kooijman 1993; Nisbet et al. 1996). There are two reasons why single-currency models might succeed. The first is that the mechanistic assumption that one factor (energy) truly is a predominant limiting resource and all other resources (such as nutrient elements) are present in relative surplus might be correct (although this seems unlikely). The second is that single-currency energy models may be a sufficient means of summarizing a multidimensional set ("food" composed of many elements and biochemicals) using a single dimension. In the latter case, energy or any other single substance might not be truly limiting, but it may provide an adequate statistical description of a complex set of parameters.

Threshold Element Ratios

We now consider this issue of limiting factors for heterotroph growth in some detail. Our specific questions will include the following. When a heterotroph is "food" limited, what precisely is the rate-limiting quantity? Is it energy? Is it carbon? One or more biomolecules? A single element? Some combination of all of these? As the child asked, "What is the grass?" One very basic question in considering the role of elements for heterotrophic consumer growth is how well Liebig's law of the minimum (see last chapter) will work. Liebig's law states that an organism's growth will be limited by whichever single resource is in lowest abundance in its environment relative to the organism's needs, and it is normally thought of in terms of growth of plants. We can address these questions both theoretically (building up from mechanistic principles) and empirically. It will turn out that stoichiometry can be an extremely valuable approach for answering questions such as these. Many of the basic concepts in consumer-resource stoichiometry have been around for a long time (see, e.g., Russell-Hunter 1970). The questions we pose here are quite general to many kinds of consumers. However, in examining those questions in this section we focus mainly on the freshwater crustacean zooplankton, a well-known group (Chapter 4). These organisms provide an opportunity to discuss many nuances in this subject. A body of literature has explored the stoichiometry of limiting substances in animal growth, calculating threshold element ratios for switches between limiting substances. Conceptually, TER's are similar to the optimal N:P ratios for autotrophs we discussed in Chapter 3.

   Urabe and Watanabe (1992) were among the first to try to understand limiting factors for heterotroph growth with a simple, stoichiometrically explicit model. They calculated C:nutrient TER's for two taxa of freshwater zooplankton. When the C:nutrient ratio in the food equaled the TER, animal growth was limited by both C and the nutrient. Deviations from the TER meant limitation by either carbon or the nutrient. According to this scheme, animal growth is limited either by carbon or by one of the nutrients N or P, i.e., it follows Liebig's law of the minimum. Their model was based on two kinds of information: the C:nutrient ratio in the zooplankton (assumed to exhibit strict homeostasis) and the GGE_C. In their calculations, the latter varied with food concentration. Urabe and Watanabe (1992) calculated several TER's for two taxa of freshwater zooplankton (Fig. 5.4). Their results suggest that limiting factors depended to a certain extent on food quantity. *Daphnia* had higher C:N TER's at all food abundances, meaning *Bosmina* should be more easily N limited than *Daphnia* (Figs. 5.4A,C). At low to moderate food abundance, *Daphnia* should be more frequently limited by P than *Bosmina* (Fig. 5.4B). At very high food abundance, *Daphnia*'s GGE_C dropped to very low values, making the C:P TER rather high (Fig. 5.4D). Over most naturally occurring food abundances, however, this analysis indicates that *Daphnia* is be more frequently P limited than *Bosmina*. This interspecific difference is mainly due to differences in the C:N:P stoichiometry in these two consumers, *Daphnia* being more P and less N rich (Fig. 4.2). We will soon see that consumer nutrient content does indeed account for patterns in growth limitation on low-nutrient foods. This model, like many others, indicates that when the composition of the resource differs from the composition of the consumer, the GGE for individual elements must vary, whether because of differences in absorption or in production or both.

Whether heterotroph growth can really be understood as such a simple stoichiometric process has been questioned by a number of authors on both theoretical and empirical grounds (Brett 1993; Müller-Navarra 1995a; Tang and Dam 1999). One argument has been that elements in organisms are in the form of biochemicals consisting of more than one element; hence elements may lack the mutual "independence" needed to make such models correct. Anderson and Hessen (1995) investigated these issues in developing a simple stoichiometric model that links C and N by considering two biochemical pools, one with carbon alone and the other representing nitrogenous compounds with both C and N. Assimilation efficiency of nitrogenous compounds was modeled to be higher than for compounds lacking N. Differences in assimilation of nitrogenous versus non-nitrogenous compounds mean that TER's calculated with this kind of stoichiometric model are different from those when C and N are considered to be independent. Specifically, nitrogen limitation of the consumer's growth is less likely in the model where C and N are linked because of the improved assimilation efficiency. This model shows that interdependence of elements (as parts of compounds) can be modeled with a stoichiometric framework when the extra complication is justified. It seems likely that an assumption of independence of elements will generally be better for some elements than others. Phosphorus is known to be subject to action by a variety of extracellular phosphatase enzymes, which readily cleave the PO_4^{3-} group off of biochemicals, allowing this anion to be easily taken up independent of most of the C- and N-containing molecules with which it is associated (we won't worry about the dependence on O). Also, elements may be present in a range of different kinds of compounds, such that the use of single parameters for things like growth efficiency can be open to question. For example, Anderson (1994) showed that carbon in phytoplankton could be usefully divided into three fractions, carbohydrate, lipids, and protein, each with different assimilation efficiencies. Average carbon assimilation efficiency would be expected to differ between different foods because of differences in the ratios of the biochemical fractions. Anderson and Pond (2000) considered the simultaneous influence of nutrient elements (N and C) and two essential fatty acids. Although they emphasized the need for better data linking elements and biochemicals, their model is perhaps the first to utilize a stoichiometric approach to understanding biochemicals as limiting factors (an issue mentioned in Chapter 1). It will be interesting to incorporate elemental dependencies in the context of biochemicals further, but we believe that most of the work we discuss in this book supports the conclusion that stoichiometric models with independent elements are often surprisingly predictive given their simplicity and we keep our focus there; indeed, that is where most of the literature is.

Another argument against stoichiometric element models of consumer growth is empirical. Several studies have shown responses of zooplankton consumers to dietary additions of biochemicals, particularly fatty acids (DeMott and Müller-Navarra 1997; Weers and Gulati 1997). The correlation between animal growth and selected biochemicals in the food resource in the field can be impressive (Müller-Navarra 1995b; Müller-Navarra and Lampert 1996). From these observations, some have deduced that zooplankton are regulated by key biochemicals, not by elements. However, similar evidence can be cited for individual elements. Additions of P alone have been shown to stimulate growth of *Daphnia* feeding on P-poor but not P-rich algae (Rothhaupt 1995; Urabe et al. 1997; Elser et al. 2001), and good correlations between zooplankton community structure and algal P content have been found (see below). One of the great powers of stoichiometric theory is that it predicts where and when one would expect to see such correlations. Stoichiometry predicts that P-rich consumer taxa should be particularly sensitive to food P content. Threshold element ratios also predict absolute ranges in food chemical content where correlations should be strong. Müller-Navarra (1995b) found an excellent correlation between *Daphnia* growth and algal fatty acid content, but only a weak correlation with algal P content; her work is supportive of stoichiometric theory because the seston food in her study was on the carbon-limited side of the TER. Detailed reviews covering studies with biochemicals and elements are already available (Gulati and DeMott 1997; Sterner and Schulz 1998). Happily, work in this area is moving beyond an unnecessary dialectic between elements versus biochemicals and into a more productive comprehensive view that attempts to synthesize the elemental and biochemical views.

Another criticism of the Urabe and Watanabe model is that it over-emphasizes anabolic processes, failing to account for the importance of catabolic ones (Tang and Dam 1999). For instance, when body growth is zero (e.g., due to strong energy limitation at low food abundance), all assimilate is used for metabolism and the stoichiometry of growth becomes irrelevant (Schindler and Eby 1997). Even when food is abundant but low in quality, some mass loss due to catabolic processes can be anticipated. Urabe and Watanabe lacked good data for N and P loss rates by consumers feeding on N- or P-deficient food, although Olsen et al. (1986) have suggested that P loss by consumers feeding on P-deficient food may be zero. Urabe and Watanabe assumed zero nutrient loss under such conditions and wrote about this assumption and what effect it had on their calculations of TER's. In particular, if consumers are unable to reduce nutrient loss to zero, limitation of consumer growth by that nutrient becomes more likely. This expected contrast has been supported by work on benthic invertebrates, including albalones, ascidians, chiton, and others (Hatcher 1991b,a, 1994). In these organisms, P excretion remains measurable even when the food supply is P poor. Hence, Urabe and Watanabe's assumption was conservative in terms of estimating the regions where P rather than C limitation is to be expected. More generally, nonzero loss is hardly a reason to abandon a stoichiometric modeling approach; these models need not make such an assumption (in the next section we develop such a model).

In presenting their model, Urabe and Watanabe recognized that there should be a dependence of TER on food quantity. As food quantity declines to limiting levels, growth approaches zero and catabolic needs dominate nutrient demands. This suggests that consumers lack a single TER; instead, TER's come in sets determined by factors such as food concentration. A model explicitly considering such interactions between quantity and quality has been analyzed (Sterner 1997) and predicts how TER's should depend upon food quantity. Figure 5.5 shows how the TER expressed as C:nutrient increases as food quantity decreases. The TER shows two kinds of asymptotic behavior at high and low food. At high food quantity, it approaches the threshold calculated by Urabe and Watanabe (1992). At a finite low food quantity, it approaches infinity. This value for food quantity can be shown to be identical to the bioenergetics parameter referred to as the "individual threshold for growth" (Lampert and Schober 1980). The

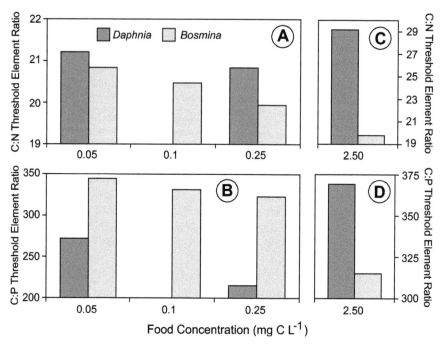

Fig. 5.4. TER's for C:N (A,C) and C:P (B,D) for two herbivorous freshwater zoo-plankton (*Daphnia* and *Bosmina*). When food C:N or C:P is less than the TER, carbon limitation of zooplankton growth is predicted, and when food C:N or C:P is greater than the TER limitation by either N or P is predicted. A,B. Under low to moderate food abundance, *Daphnia* have higher C:N and lower C:P TER's than *Bosmina*. C,D. At very high food, TER's are quite different than at lower food. Values for *Daphnia* in one food level were not given in the original paper. Based on Urabe and Watanabe (1992).

individual threshold for growth is defined as the food abundance where assimilation balances metabolic requirements, and consumer growth is zero. Hence, the stoichiometric model of food quantity and quality (Sterner 1997) provides a general framework to unite certain fundamental aspects of bioenergetics with principles of stoichiometry. By introducing a maintenance respiration term, Anderson (1992) also showed that food quantity as well as quality could be included in stoichiometric models of copepod growth.

Relaxing the Assumption of Strict Homeostasis

A key assumption used in all stoichiometric models discussed so far is that the stoichiometry of consumer biomass is fixed. As we saw earlier (Chapter

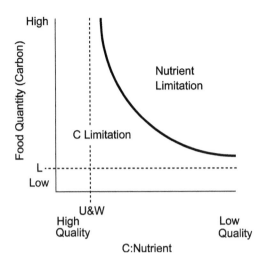

Fig. 5.5. TER set for C or nutrient limitation of a consumer with varying proportion of its carbon economy going toward maintenance. Above and to the right of the curve the consumer is nutrient limited. Below and to the left of the curve the consumer is carbon limited. At a finite low food quantity (marked by "L"), growth approaches zero, and the TER approaches infinity. Lampert and Schober (1980) defined this quantity of food as the "individual food threshold," and it represents a food concentration where the consumer has zero mass change. As food quantity approaches infinity, the TER approaches the expression defined by Urabe and Watanabe (1992) (U&W) as "the" threshold element ratio. Based on Sterner (1997).

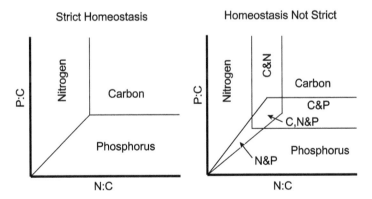

Fig. 5.6. TER sets for C, N, or P limitation in bacteria with strict vs. variable C:N:P stoichiometry. Axes indicate the chemical content of the substrate. Left: bacteria under strict stoichiometry. Right: bacteria under variable stoichiometry. "Mixed" limitation as defined by Thingstad (1987) is indicated by regions with multiple elements. Under mixed limitation, a single element limits cell number growth, but biomass elemental composition varies. Based on Thingstad (1987).

1), this is often an approximation. Stoichiometric variability is taxon and element dependent: for some elements in some organisms nutrient content is tightly fixed, in others (including consumers) it is less tightly fixed. Because the assumption of strict homeostasis influences most of the results of the growth models discussed here it is important to see what happens if we relax that assumption.

Thingstad (1987) analyzed a model of bacterial growth where regions of C, N, or P limitation were calculated for varying substrate P:C and N:C ratios. His model shares many of the assumptions of previously discussed stoichiometric models, but he also explored the consequence of a variable stoichiometry in the consumer. Thingstad's model says that for certain ratios of P:C and N:C in the substrate, two or even all three of these substrates are completely depleted from the medium simultaneously (Fig. 5.6). He termed these regions of "mixed" limitation. Within those regions, bacterial growth is uniquely determined by a single substrate, but up to three substrates are controlling biomass composition. In the mixed limitation region, if the concentration of two or more elements in the resource pools are increased, the bacteria respond by enhanced uptake and therefore total mass increases. However, cell division is uniquely determined by one element at a time. As he stated, "Whether or not this should be regarded as a violation of Liebig's law seems to be a matter of definition" (Thingstad 1987). Another interesting feature of this model is that as growth approaches zero, regions of mixed limitation increase in area. At very low growth, areas of pure N or P limitation disappear altogether. Mixed limitation should therefore be particularly important at low growth.

A general point we can take from the Thingstad model is that variable stoichiometry creates qualitatively different solutions from those we see under strict stoichiometry. In the regions of mixed limitation, we may be approaching the complex realities of real nutritional effects in real animals and other heterotrophs. Although a Liebigian frame of reference is not wrong in this context, it requires some additional layers of complexity.

Whether such a model works for micro- and macrophages as well as osmotrophs (which can deplete the surrounding medium of more than one limiting substance by more or less independent uptake mechanisms) has not been explored yet. Thingstad's model points the way to constructing other models with flexible but constrained stoichiometry due to a nonstrict homeostasis.

We have seen that stoichiometric models of consumer growth can be extremely simple or they can be embellished with more complexity as the situation warrants. The studies described here provide a single conceptual framework. Their major contribution is to indicate how consumer growth is related to resource composition under several very basic assumptions.

These include the assumption that consumer chemical composition is more fixed than composition in resources. They also assume that food can be adequately described by a manageably small set of essential resources. Later in this chapter, we will relate some real world patterns to these simple models. Before doing that, though, we will introduce a new stoichiometric model for secondary production.

A NEW MINIMAL MODEL OF THE STOICHIOMETRY
OF SECONDARY PRODUCTION

The models described above share a number of features. For one, the GGE (growth as a fraction of ingestion) for the most deficient resource is constant and high while the GGE's of other resources decline as the most deficient resource becomes rarer in the food. These are commonsense predictions of what a stoichiometric consumer must do: retain deficient substances efficiently and dispose of those found in excess. However, a constant maximal GGE value for the most limiting element is not the only commonsense way to do this. Another approach is to assume that homeostatic consumers are able to reduce loss rates of a scarce element to some small but finite level. Physiologically, it appears to be more mechanistically accurate to relate nutrient losses to some constant proportion of existing body amounts than to a fraction of what the consumers ingest. Nutrient losses enter into the composite efficiency of GGE, but focusing on the loss itself allows the model to approach more closely the animal's physiology, which should be responding to the current state of the animal, not to the composition of the food. In other words, rather than controlling their homeostasis by increasing growth efficiency to some maximum, consumers minimize "leakiness" of previously assimilated nutrients to some finite low value. Loss dynamics might well be the controlling factor of animal homeostasis.

Here, we present a new and highly general stoichiometrically explicit model of homeostatic consumers to explore stoichiometric nutrient balance. The model is based on just two assumptions. The first is our familiar one of strict homeostasis in the consumer. The second is new: that the rate of loss of a limiting nutrient is reduced to a constant specific rate. This model is another member in the family of TER models just discussed. Although there is some interest in seeing what happens when we model homeostatic mechanisms in this new way, our chief reason for presenting this new model in detail here is that from these two very simple assumptions, the entire mass balance of the organism can be derived. Therefore, this model makes the implications of these two assumptions very apparent. Hence, we consider this a "minimal model." This model demonstrates in a highly transparent way how stoichiometric homeostasis inherently influences consumer growth, nutrient release, and production dynamics. The model has been parameterized for a three-dimensional liquid world, but the insights derived should hold for two-dimensional worlds too. We will see a wide range of interesting predictions that can be made.

Strict homeostasis (the first assumption) means:

$$\overbrace{\frac{g_{X_1}}{g_{X_2}}}^{A} = \overbrace{\frac{G_{X_1}}{G_{X_2}}}^{B}, \tag{5.5}$$

where g_X stands for growth based on element X (moles per time) and G_X stands for the environmental concentration of element X in consumer biomass (moles L^{-1}, see the Appendix). Equation (5.5) states that the ratio of two elements X_1 and X_2 entering consumer biomass as growth (or, synonymously, production) (expression marked A) equals that same ratio already in the consumer's biomass (expression marked B). We can expand this equation by recognizing that production is the difference of incorporation and loss:

$$\frac{\overbrace{eA_{X_1}S_{X_1}}^{A} - \overbrace{l_{X_1}G_{X_1}}^{B}}{\underbrace{eA_{X_2}S_{X_2} - l_{X_2}G_{X_2}}_{C}} = \frac{G_{X_1}}{G_{X_2}}. \tag{5.6}$$

Here, e represents the volumetric feeding rate, A_X is the environmental concentration of element X bound in prey biomass, S_X is the assimilation efficiency of element X, and l_X is the specific loss rate of element X from the consumer. The term marked A in the numerator says that incorporation of an element is the product of the volumetric feeding rate, abundance of the element in the resource pool, and assimilation efficiency. The term marked B in the numerator says that loss is the product of the specific rate of loss of the element and the abundance of the element in the consumer. The term marked C in the denominator gives the ingestion rate for element X_2.

We will hold volumetric feeding rate (e) constant for all our analyses. Some herbivores exhibit compensatory feeding, or an increase in ingestion of bulk food when food quality decreases (Cruz-Rivera and Hay 2000). Compensatory feeding has recently been observed in our model organism *Daphnia* (Plath and Boersma 2001). Other herbivores do not alter feeding rates with changes in food quality (Rothhaupt 1995; Cruz-Rivera and Hay

2000). Compensatory feeding needs to be examined more thoroughly in future modeling studies. Now, from our second assumption, let us find out what happens if the specific loss rate of one of the nutrients, l_X, is minimal and constant, such as might happen if element X_1 is scarce in the consumer's food. Consider first the case where l_{X_1} is constant $(l_{X_1,\min})$. What must the loss of the second element (l_{X_2}) be then for the consumer to maintain its stoichiometry? We substitute into the last equation, and after a little algebra we find,

$$\overbrace{l_{X_2}}^{A} = \overbrace{l_{X_1,\min}}^{B} + \overbrace{\frac{eA_{X_2}S_{X_2}}{G_{X_2}}}^{C} - \overbrace{\frac{eA_{X_1}S_{X_1}}{G_{X_1}}}^{D}, \tag{5.7a}$$

where the parameter marked A is the specific loss rate of element X_2, the parameter marked B is the minimal specific loss rate of element X_1, the term marked C is the consumer element-specific rate of intake of element X_2, and the term marked D is the consumer element-specific rate of intake of element X_1. This equation shows that at stoichiometric homeostasis, element-specific loss rates of all elements must be equal, and that loss rates of elements ingested in surplus (the difference between terms D and C) are given by the minimal loss rate of the limiting element plus the specific amounts of other elements ingested in surplus. By analogy we can write the equation for l_{X_1} where l_{X_2} is at a minimum,

$$l_{X_1} = l_{X_2,\min} + \frac{eA_{X_1}S_{X_1}}{G_{X_1}} - \frac{eA_{X_2}S_{X_2}}{G_{X_2}}, \tag{5.7b}$$

which has the same functional form as Equation (5.7a).

Predicted shapes of loss rates versus resource stoichiometry for a specific set of parameters are given in Figure 5.7. Note that on each side of the breakpoint between consumer growth limitation by X_1 and by X_2, the release of one of the nutrients as a percent of consumer nutrient pools is very low (and nonzero, although it cannot be discerned visually in the plot). So now we have calculated the means by which a strictly homeostatic consumer can maintain its body composition despite chemically varying food under two very general assumptions (strict homeostasis and first-order loss of whichever nutrient is relatively scarce).

The location of the breakpoint separating deficiency of the two elements is a particularly interesting feature of this model. It provides an alternative and possibly more general derivation of threshold element ratios. The TER occurs where both loss rates are at their minimum values because both elements are equally limiting to the consumer. We therefore find A_{X_1} and A_{X_2} under those conditions [using Eq. (5.7a) and letting $l_{X_2} = l_{X_2,\min}$],

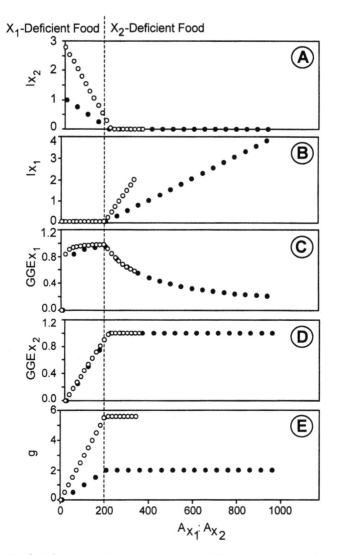

Fig. 5.7. Predicted nutrient flux of a consumer following the minimal stoichio-metric model. These functions were calculated using the following parameter values: $l_{X_1,\min} = 0.05$, $l_{X_2,\min} = 0.01$, $G_{X_1} = 2$, $G_{X_2} = 0.01$, $S_{X_1} = S_{X_2} = 1$, and $e = 0.1$. Resource nutrient concentrations were $1 < A_{X_1} < 96$ and $A_{X_2} = 0.10$ (closed circles) or $A_{X_2} = 0.28$ (open circles). These values approximate what would be expected for the elements C (X_1) and P (X_2) for zooplankton grazing on algae. A. Loss rate of element X_2, l_{X_2}. B. Loss rate of element X_1, l_{X_1}. C. Gross growth efficiency of element X_1, GGE_{X_1}. D. Gross growth efficiency of element X_2, GGE_{X_2}. E. Growth rate of consumer expressed relative to element X_1, g_{X_1}. Dashed line indicates the transition between X_1 and X_2 limitation. To the left of this line, the food is X_1 deficient; to the right, it is X_2 deficient.

and take their ratio. The TER (the ratio of A_{X_1} to A_{X_2} where the consumer has minimal loss of both elements) is given by,

$$
\underset{A}{\overbrace{\frac{A_{X_1}}{A_{X_2}}}} = \underset{B}{\overbrace{\frac{G_{X_1}S_{X_2}}{G_{X_2}S_{X_1}}}} \cdot \frac{l_{X_1,\min} - l_{X_2,\min} + \overbrace{\dfrac{eA_{X_2}S_{X_2}}{G_{X_2}}}^{C}}{l_{X_2,\min} - l_{X_1,\min} + \dfrac{eA_{X_1}S_{X_1}}{G_{X_1}}}. \tag{5.8}
$$

Equation (5.8) for a TER is a more complex expression than we saw in other models. It is actually rather ugly and not easily interpreted in words. The entire term marked B bears a resemblance to the TER models previously described, such as that of Urabe and Watanabe, discussed above, except that the term B is based on assimilation efficiency, not gross growth efficiency. However, it shows that under the simple assumptions we made, the TER is a function of many parameters. Consumer nutrient pools (G_X) as well as their stoichiometric ratio $(G_{X_1}{:}G_{X_2})$ enter into the formula. Also, the minimum loss rates and the resource nutrient levels play a role.

Under these assumptions, there is not a single TER for a consumer but rather a function describing a set of TER's. Unfortunately, the TER cannot be solved as a function of resource ratios in closed form, which means that this TER is not just a function of the ratio, but also of the absolute concentrations of the food elements. Although consumer and resource nutrients appeared solely in their respective ratios in previous models, in this model the absolute amounts of each of these pools also occur. As in previous models, the TER is predicted to vary with consumer nutrient content: consumers with high concentration of a given nutrient are more susceptible to limitation by that nutrient (for example, if G_{X_1} increases disproportionately to G_{X_2}, the TER also increases). However, unlike previous models, this model indicates that the TER is also a function of minimal loss rates of both elements as well as the absolute abundances of the food as represented by the elements in the food. Nevertheless, although these factors enter into the calculation of the TER, inspection of the term marked C shows that these adjustments, at least for this parameter set, are small. For the parameters used to calculate Figure 5.7, for instance, the TER is so close to the value of $G_{X_1}{:}G_{X_2} = 200$ that the difference would not be discernable in the graph.

As we have discussed, nutrient losses and consumer production are intimately related to each other. We can now use this new minimal model to understand some things about secondary production in homeostatic consumers. One major thing we are interested in knowing is how the gross growth efficiency of the consumer varies as the elemental composition of its food varies. To do this, we need to derive four different quantities: the

GGE's for each of the two elements where each of the two loss rates are minimized. Together, these will give us a complete description of the GGE for both elements throughout all food types.

In general, the GGE for element X can be expressed as

$$\text{GGE}_X = \overbrace{\underbrace{\frac{\overbrace{g_{X_1}}^{A}}{eA_{X_1}}}_{B}} = \overbrace{S_{X_1}}^{C} - \frac{\overbrace{l_{X_1}G_{X_1}}^{D}}{eA_{X_1}}. \tag{5.9}$$

Here, the parameter in the numerator marked A is the growth (production) rate of the consumer in terms of element X_1, the term in the denominator marked B is the ingestion rate of element X_1, which is a product of feeding rate e and element X_1's abundance in the resources, A_{X_1}. The parameter marked C is the assimilation efficiency of element X_1, and the entire expression marked D is the fraction of ingested element that is lost by metabolism. Determining the gross growth efficiency for the element that is deficient in the resource is easy. For example, when resource 1 is scarce (l_{X_1} is minimized),

$$\text{GGE}_{X_1} = S_{X_1} - \frac{l_{X_1,\min}G_{X_1}}{eA_{X_1}}. \tag{5.10}$$

Increased "leakiness" as given by higher values of the parameter $l_{X_1,\min}$ would decrease growth efficiency. To find how the consumer balances its stoichiometry by adjusting the efficiency of use of surplus resources, we solve for the gross growth efficiency for the surplus element (e.g., GGE_{X_1} for $l_{X_2} = l_{X_2,\min}$). To do this, substitute the relationship we derived above for l_{X_1} under these circumstances [Eq. (5.7b)] into Equation (5.9) and simplify,

$$\text{GGE}_{X_1} = \frac{G_{X_1}}{A_{X_1}} \left(\frac{A_{X_2}S_{X_2}}{G_{X_2}} - \frac{l_{X_2,\min}}{e} \right). \tag{5.11}$$

Equation (5.11) also is difficult to interpret in words. The above two results, along with analogous ones for growth efficiency for the other resource, are summarized in Table 5.1. Specific calculations of these functions are presented in Figures 5.7C,D using parameters chosen to approximate carbon and phosphorus stoichiometry for zooplankton consuming algae. For this figure, resource ratios in the food (x axis) were produced by holding A_{X_2} constant while varying A_{X_1}. GGE_{X_1} varies with A_{X_1} over the whole range, while GGE_{X_2} varies with A_{X_1} over only part of the range. Hence, the plots for GGE_{X_1} and GGE_{X_2} have some differences. For GGE_{X_1}, we can see that the homeostatic consumer has a maximal gross growth efficiency at an intermediate value of the $A_{X_1}:A_{X_2}$ ratio (GGE_{X_1} is

TABLE 5.1
Gross growth efficiencies for homeostatic consumers in the minimal model

	GGE_{X_1}	GGE_{X_2}
l_{X_1} minimized (X_1-deficient resources)	$S_{X_1} - \dfrac{l_{X_1,\min} G_{X_1}}{eA_{X_1}}$	$\dfrac{G_{X_2}}{A_{X_2}}\left(\dfrac{A_{X_1} S_{X_1}}{G_{X_1}} - \dfrac{l_{X_1,\mathrm{mim}}}{e}\right)$
l_{X_2} minimized (X_2-deficient resources)	$\dfrac{G_{X_1}}{A_{X_1}}\left(\dfrac{A_{X_2} S_{X_2}}{G_{X_2}} - \dfrac{l_{X_2,\mathrm{mim}}}{e}\right)$	$S_{X_2} - \dfrac{l_{X_2,\min} G_{X_2}}{eA_{X_2}}$

very low at the lowest value of $A_{X_1}{:}A_{X_2}$) . This aspect of the behavior of this TER model is quite different from previous ones, which have been based on constant GGE_{X_1} under X_1 deficiency. We can think of this intermediate ratio as a point of optimal balance between the chemical composition of the consumer and that of its food resource. It can be shown that the resource ratio where the GGE reaches a maximum occurs at an identical breakpoint to that already determined for nutrient loss rates (l_X, Figs. 5.7A,B). In this TER model, ratios of nutrients in consumer foods equal to the TER are optimal foods, and imbalance in either direction decreases the quality of the food. Such behavior is consistent with some recent observations of *Daphnia* and algae (Plath and Boersma 2001). In contrast, a plot of GGE_{X_2} versus $A_{X_2}{:}A_{X_2}$ is superficially similar to predictions from previous models, namely, a high constant growth efficiency under X_2 deficiency. Understanding these differences in behavior of the two growth efficiencies is rather subtle, but it is made simpler with a good understanding of Table 5.1 as well as the way that the two growth efficiencies vary with the separate resource concentrations and resource ratios. If we had calculated $A_{X_1}{:}A_{X_2}$ ratios in a different way, such as by holding A_{X_1} constant and varying A_{X_2}, the graphs would have appeared different. We leave it to the interested reader to explore this topic in more depth.

As previous models have shown, foods very rich in a resource will show reduced GGE for that resource because the consumer must elevate loss of that substance as a waste product to maintain its homeostasis. Elevated loss reduces growth efficiency. Such is the case far to the right in Figure 5.7C and near the y axis in Figure 5.7D. Another interesting feature of this model is that these functions have no roots (do not cross the x axis) in the meaningful regions of the graphs, which means that there occurs no food too "rich" in a particular substance for the homeostatic consumer to be able to maintain its homeostasis. We have modeled this consumer as being able to increase its loss of particular nutrients without limit (Figs. 5.7A,B), which is probably an unrealistic assumption.

Previous models have not explored in detail how the GGE of an element changes in the region where it is deficient. Previously, the assumption has been made that the homeostatic consumer will remain at some maximal GGE in this region. However, this model predicts that GGE_{X_1} crosses the x axis at a positive nonzero value, sending GGE_{X_1} into negative values for foods very deficient in resource X_1 (close to the y axis in Fig. 5.7C). A negative GGE can occur only when a consumer has lower production than ingestion; such a deficit obviously cannot be sustained indefinitely. The model thus predicts that there is some food composition too poor to allow for positive growth of the consumer at all. Let us refer to these values as "hard TER's." It is commonly held in bioenergetics studies that with insufficient carbon or energy growth cannot be sustained (the animal goes on a diet and loses energy or mass). However, in most stoichiometrically explicit models so far, the same is not true for elements: growth, although perhaps small, is sustained no matter what the element abundance is in the food. This new stoichiometrically explicit model is more like the standard bioenergetic results in that foods can be too poor to sustain growth.

Hard TER's are found by setting equations for GGE's equal to zero and solving. For GGE_{X_1}, the hard TER occurs at a value $l_{x,\min} \cdot G_{X_1} = eS_{X_1}A_{X_1}$, and for GGE_{X_2} it is at the comparable value of $l_{x,\min} \cdot G_{X_2} = eS_{X_2}A_{X_2}$. Only the absolute abundance of the limiting element enters into the expressions in this case, not a ratio. In both of these equations, it is clear that the hard TER occurs when resource concentration in the food falls below a level that can at least make up for finite positive losses. This threshold for positive growth using the parameters of C and P balance modeled in Figure 5.7 is probably not too important since it happens at $A_{X_1}:A_{X_2} = 7$, an unnaturally low C:P. However, hard TER's might be important under different scenarios, particularly when minimal nutrient loss rates are large. Hence, this model may be particularly appropriate for identifying foods too nutritionally dilute to support positive growth of the homeostatic consumer.

We can summarize the major features of our new minimal model of secondary production in organisms with strict stoichiometry:

- Gross growth efficiency is maximized in stoichiometrically balanced foods with elemental ratios defined by the body composition, metabolic loss rates of the consumer, and other factors.
- In progressing from foods extremely deficient in a resource toward the point of balanced composition, specific resource losses by the consumer fall to minimal and constant levels. Meanwhile, growth efficiency for that resource should increase from potentially negative values toward a maximal value.
- In progressing from foods with balanced composition toward greater and greater excess of a particular element, elevated specific loss of

that element is coupled to progressively decreasing growth efficiency for that element.

So which of all these stoichiometric models of consumer growth is best, including this new one and the several presented in the last section? The answer to this question depends on the assumptions that one is willing to live with. Our goal in presenting several different, but related, models was partly to emphasize their common points (the assumption of homeostasis has large and similar effects in all these models) as well as to discuss their differences. As a group, these models provide a heuristic set of studies that help us understand the probable consequences of one of the major patterns discussed earlier in the book—that major classes of consumers (detritivores, herbivores, osmotrophs) tend to have more tightly fixed chemical composition than the resources they eat. This difference between consumer and resource is a consequence of one of the fundamental features of biological systems, homeostasis, and it has manifold ecological consequences.

SOME REAL WORLD PROBLEMS
IN STOICHIOMETRIC BALANCE

Now that we have seen some theoretical predictions about consumer nutrient balance, growth, and growth efficiency, we will look at some real world problems in stoichiometric balance. The biodiverse world has created numerous stoichiometric challenges and solutions to imbalance in ecological and evolutionary time. This section provides an overview of this great diversity. We have organized this section primarily by feeding modes (saprotrophs, bacterivores, herbivores, etc.). Although we sought examples and illustrations for this chapter far and wide, we purposely avoided a large literature on nutrient balance and nutrition in agricultural (livestock) systems. There must be useful information there, but we worry about the effect of strong selection for biomass production, particularly given the connections between selection on growth rate and stoichiometry that we outlined in Chapter 4. When we look only at "wild" consumers, the available data poorly represent the great diversity in nature. The groups best studied so far are zooplankton and insects, and possibly bacteria. In the section after this one we will bring data and theory together and look for empirical support for some of the main theoretical predictions discussed above.

Saprotrophs and Other Osmotrophs

A saprotroph is an organism that feeds on dead or decaying organic matter. This feeding mode includes heterotrophic bacteria, fungi, and some Protozoa. An osmotroph is a heterotrophic organism that obtains its nutrition by absorbing organic matter in solution from its surroundings. Saprotrophs are osmotrophs; they do not engulf food in discrete packets. Understanding the stoichiometry of saprotrophs and other osmotrophs is important for a number of reasons, not the least of which is that they are the largest catabolizers of organic matter in the biosphere. Decomposition by saprotrophs is probably the second largest flux in the global cycling of organic matter, next to primary production itself. Most saprotrophs are also detritivores, and because most detritus generally has low concentrations of elements like N and P, most saprotrophs must subsist on low-nutrient resources. This mode of feeding allows the organism, within limits, to choose the elements it will take into the cell either by altering extracellular lytic processes or by regulating transmembrane fluxes. For example, extracellular phosphatase enzymes are known to be induced under P limitation and can increase the availability of P to the microorganism by giving it access to organic P (Gauthier et al. 1990). This selectivity would seem to give saprotrophs an inherent advantage from the standpoint of nutrient balance, sparing the organism from dealing with surpluses of any one nutrient. However, nature does not always (or may hardly ever!) provide a good balance of available resources relative to saprotroph demands.

Saprotrophs present a wide range of stoichiometric values and flexibility. For example, fungi have been described to have low levels of N (2–4%) and P (0.1–0.4%) but strikingly high levels of Ca (10%) in specific tissues (Cromack and Caldwell 1992). Variability in the stoichiometry of the C:N ratio in a fungal species was described in some detail in Chapter 1 (Fig. 1.7). Interspecific variation in fungal C:N ratios (5–17) and low carbon assimilation efficiencies (0.2–0.6) have also been reported elsewhere (Moorhead and Reynolds 1992). Bacteria, on the other hand, are N and P rich, with values of C:N < 7 and C:P < 70 being typical (Vadstein 2000; Cotner 2001). We saw a strict stoichiometry of bacterial C:N ratios in Chapter 1 (Fig. 1.6). Bacteria—as should be expected by their general lack of carbonaceous support structures (Chapter 2) and their high potential growth rates (Chapter 4)—are generally the most N- and P-rich of all living things. A comprehensive treatment of the stoichiometry of bacterial growth was presented years ago by Herbert (1961; Herbert et al. 1971, 1976). More recently, Chrzanowski and Kyle (1996) analyzed bacterial C, N, and P content as a function of nutrient supply and growth rate in their own cultures as well as in a collection of studies from the literature (Fig. 5.8). We have replotted some of their data using the logarithmic homeostasis model of Chapter 1. A relatively strict homeostatic pattern is seen. We are unaware of stoichiometric studies on saprotrophic protozoa, but some studies on bacterivorous protozoa are discussed below.

Our focus in this chapter is on consumer growth, but for saprotrophs it

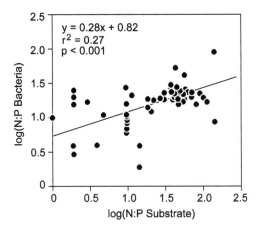

Fig. 5.8. Bacterial cellular N:P as a function of N:P in their substrates. Using the logarithmic model of homeostasis developed in Chapter 1, we see a relatively close homeostasis of the bacterial N:P ratio (slope = 0.28, H = 1/slope = 3.57). Based on Chrzanowski and Kyle (1996).

is often the case that data are taken on rates of consumption and catalysis of major resources rather than the growth of the consumers per se. This is because rates of organic matter breakdown (relatively easily measured) and rates of growth of the heterotrophic consumers (often not often easily measured for these organisms) should be related (although not always with the same stoichiometric yield). The importance of the stoichiometric imbalance between soil microbes with relatively low C:N compared to plant-derived detritus with high C:N has been known for a very long time. In fact, one of the most important factors in making organic compost successfully is getting the substrate C:N properly balanced so that both microbial growth and N mineralization are favored (Minnich 1979). On high C:N litter, soil microbes "immobilize N," i.e., they take up inorganic N and retain it while mineralizing C. Paul and Clark (1989) stated that a C:N of 25 (by mass) divides net mineralization (at lower C:N ratios) and immobilization (at higher C:N). Similarly, carbon fluxes are affected in predictable ways by soil C:N. Aitkenhead and McDowell (2000) recently showed that carbon export in the form of riverine DOC was closely related to soil C:N, accounting for 90% of the variance at the global scale. Dissolved organic carbon export is much larger from soils of high C:N than from low-C:N soils.

Detrital breakdown is known to have a very strong stoichiometric component. C:N:P ratios of detritus are excellent statistical predictors of decomposition rate. Enriquez et al. (1993) found that detritus C:N:P stoichiometry explained 89% of the variability in decomposition rate in a set of plant-derived substrates ranging from unicellular algae to trees. We will discuss this result again in Chapter 6 when our focus is on nutrient recycling. The carbon and nitrogen balance alone was once considered to have nearly completely solved the whole issue of rates of decomposition: Harmsen and van Schreven (1955) wrote, "The study of the general course of mineralization of organic N in soil was practically completed before 1935. It is surprising that many of the modern publications still consider it worthwhile to consider parenthetically observations dealing with these entirely solved problems."

Streams are widely regarded as being largely heterotrophic, with major energy and nutrient inputs from surrounding riparian zones (Hynes 1970). The action of saprophytes breaking down terrestrial material is thus a major component of stream ecosystem metabolism. This breakdown has been divided into three phases: (1) initial rapid leaching; (2) microbial decomposition and conditioning; and (3) mechanical and invertebrate fragmentation (Webster and Benfield 1986). Detrital conditioning consists of the physical and chemical changes in the substrate associated with the early phases of its breakdown. The term "conditioning" implies that the substrate is becoming more suitable for breakdown. Fungi such as aquatic hyphomycetes are often the first important colonizers on fresh detritus in streams (Bärlocher and Kendrick 1974; Arsuffi and Suberkropp 1984). After a period of time, a mixed community of fungi and bacteria or a mainly bacterial community can be found (Suberkropp and Klug 1976; Webster and Benfield 1986). It seems likely that the fact that fungi have a higher C:N ratio than bacteria (Kihlberg 1972), and therefore have a lower imbalance compared to fresh leaves, helps them in colonizing this substrate; however, we are unaware of any direct empirical tests of this hypothesis. Like detrital breakdown in other ecosystems, stream detrital processing is characterized by the input of a resource with very high C:N and C:P ratios compared to the consumers' TER's. As expected from the relative scarcity of N and P in fresh detritus compared to saprophytic consumers, both N and P may be immobilized (retained by microbes) during tree leaf breakdown. At the same time, other elements are lost from the substrate at rates similar to or faster than detrital mass loss (Webster and Benfield 1986). As a result, during microbial decomposition and conditioning, total detrital mass generally decreases, but detrital N either remains nearly constant or declines only marginally such that detrital C:N ratios decline with time (Suberkropp et al. 1976; Arsuffi and Suberkropp 1984). That stream detrital breakdown is affected by nutrient imbalance is consistent with the observation that high dissolved nutrient concentrations in stream water can increase breakdown rates (Suberkropp and Chauvet 1995).

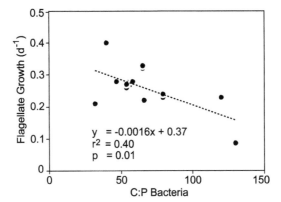

Fig. 5.9. As bacterial C:P and stoichiometric imbalance increase, flagellate growth rate decreases. Based on Nakano (1994a).

Bacterivores

Owing to their ubiquity, bacteria are actually present in the diets of most macrophagous consumers. Some organisms, though, specialize in bacteria consumption. In freshwater and marine water columns, it is generally single-celled Protozoa that are the main bacteria specialists. Because bacteria are generally N and P rich, bacterivores should have a high ingestion rate of N or P compared to C and may more often consume foods near or below their C:nutrient TER's than perhaps any other consumers. Further, a protozoan bacterivore has been shown to exhibit a strict homeostasis of C:P and perhaps also of C:N as nutrient ratios in its bacterial food changed (Nakano 1994a,b,c). Given this stoichiometric scenario with high levels of N and P in their diet, and tight homeostatic control of their own biomass nutrient ratios, it should not be surprising that bacterivores are thought to be major regenerators of nutrients in pelagic environments (Chapter 6). Again, consistent with strict homeostasis, effects of nutrient deficiency on protozoan growth can be observed. Nakano's (1994a) study showed decreased flagellate growth on poorly balanced bacterial food (Fig. 5.9).

Herbivores

The consumption of plant matter by an animal can be a profound stoichiometric and nutritional discontinuity and indeed attracts much of our attention in this book. However, the very large variation in plant nutrient content (Chapter 3) means that the alternative also occurs, namely, at certain places and times plant matter is nutritious and reasonably stoichiometrically balanced. Of all nutritional modes, herbivory is where stoichiometric analysis is the most thoroughly advanced. Much of the literature on the nutritional ecology of plants and herbivores is—broadly speaking—an effort to make sense of this variability. This variability also is why we chose the famous lines from Whitman to begin this chapter; like the child we too wonder, "What is the grass?" The ecology of herbivores feeding on plants varying in nutrient content has been very extensively explored. But there are two ways to affect balance between consumer and resource: variation in resource and variation in consumers. In studies of plant-herbivore interactions, the former is well known but the latter has scarcely been recognized. In this chapter we present a new finding: that the chemical variation in the consumers themselves also contributes to the great variability of nutritional quality of plants for animals.

At the outset, we should recognize that few (if indeed any) consumers can truly be characterized absolutely as herbivores. Even "classic" herbivores like sheep (*Ovis*) and red deer (*Cervus elaphus*) sometimes include animal matter in their diets. Bazely (1989) hypothesized that behaviors such as the eating of tern chicks by sheep, or geophagy (soil eating) by large herbivores, may be a means to alleviate mineral deficiencies, particularly P or Ca. By including small amounts of nutrient-rich matter in their diets, these animals can supplement an otherwise badly balanced diet.

Within mammalian herbivores, Demment and Van Soest (1983) proposed that body size, nutritional strategy (ruminant or nonruminant), and minimal N content in forage were all related. Small herbivores, with short gut passage, lose a large fraction of ingested nutrients in their feces and so we would say that they have higher N:C TER's (are easily N limited) because of low assimilation efficiency for N [see Eq. (5.8)]. Demment and Van Soest further suggested that large herbivores can extract more energy from plant material than small herbivores (due to longer gut passage) but cannot concentrate on the easily digestible foods required by small animals because those foods are rare. In the parlance developed in this book, we would say that their lower N:C TER's allow them to chose a wider range of forages.

There is a vast literature on the interactions of insects and nitrogen, a field that has been comprehensively reviewed more than once (e.g., Mattson 1980; Mattson and Scriber 1987). In his major overview of that literature White (1993) stated, "The evidence for the generality of a relative shortage of nitrogen in the food of herbivorous insects is everywhere apparent." Although it will be apparent to the reader of this book that we do not agree with White's conclusion that it is always N that is the most limiting element in the environment, his thorough review of the role of N in insect ecology makes it unnecessary for us to repeat that exercise here. The potential for other elements, particularly P, to limit insect growth has seldom been considered; one example where P does seem to be closely associated with insect growth is in African armyworms and maize plants (Janssen 1994). But in spite of this wealth of information about terrestrial herbivores and nitrogen, much is still lacking. For one thing, information on the interspecific variation in nutrient content of different insects is only beginning to emerge (see Fig. 4.3). Large interspecific differences in element requirements in insect nutrition are known (Dadd 1970). Markow et al. (1999) found a large range in N and P content in drosophilids, and also showed that insect nutrient content was correlated with the nutrient content of their resources (several species of cacti or various types of rotting fruit). Earlier, we hypothesized that evolutionary pressures associated with body size led to interspecific variation in nutrient content in animals (Chapter 4). Clearly, there is considerable interspecific variation in N and P content and stoichiometric balance within the insects.

Some of the most extreme stoichiometric imbalances are likely experienced by phloem feeders such as aphids. Nutrients in phloem are soluble and easily assimilable such that rates of nutrient intake by phloem feeders may be relatively high (Risebrow and Dixon 1987); however, concentrations of N in this resource are low, meaning that large volumes of phloem must be processed. Leaf sap has been reported to contain only 1.9% N (by mass) (Van Hook et al. 1980), approximately one-tenth the N percentage in typical insect dry mass. Aphids may also have a high P content in their body mass (see Chapter 4). Further, they are obligately constrained to feeding on this single poor-quality resource, which might explain their preference for feeding on phloem with relatively high nutrient content (Blackman and Eastop 1994). One adaptation to this carbon-rich diet has been the evolution of the ability to secrete high-carbon "honeydew," a sure sign that growth of these animals is not energy limited. Stadler and Müller (1996) have shown that production of organic honeydew is a function of the mineral nutritional status of the plant and that changes in honeydew production by aphids can influence biogeochemical cycling in forests. Given this extreme stoichiometric imbalance, it is not surprising that aphids have larger population growth rates when their host plants occur on fertile soils (Stadler et al. 1998). Precise timing mechanisms in aphid feeding and growth, ensuring a good match between peak resource quality and peak resource demands (Dixon and Kundu 1998), also suggest evolutionary responses to stoichiometric imbalance.

Another interesting stoichiometric imbalance occurs in large cervids such as deer or moose. In these vertebrate herbivores, males must replace a large mass of bony (high Ca and P) antler material each year. Antlerogenesis is a process of rapid true bone formation without equal among vertebrates (Grasman and Hellgren 1993). Before each year's mating season, an equivalent to nearly 20% of the animal's skeleton must be produced in a 3–4 month period. This is a very large Ca and P demand. The Ca and P content of antler ash is 36% and 18%, respectively (Brown 1990). It is startling to think that a large moose must place nearly 3 kg of P into its 30 kg antlers each year (Moen and Pastor 1998). At typical foliar P levels, this amount of P would have to be gleaned from 600 kg (dry weight) of leaves! Given that P assimilation is not perfectly efficient and that the animal must also use this resource for things other than antlers, this is clearly a major requirement indeed. It would be impossible for these animals to obtain such large quantities of Ca and P from their forage during these relatively brief times, particularly as peak antler formation is occurring during portions of the growing season when vegetation quality is declining due to nutrient resorption by plants. Instead, Ca and P are translocated from the animal's skeleton each year. The skeleton thus acts as a storage pool for Ca and P. It has been shown that antlerogenesis in deer, which have relatively small antlers, is unlikely to be P limited: antler size was unaffected by forage P content (Grasman and Hellgren 1993). However, P requirements are much larger for moose, and P limitation is thought to be more likely for them (Grasman and Hellgren 1993; Moen et al. 1997). Even larger antlers still were once possessed by the now extinct Irish Elk. It has been argued that it was only the longer growing seasons of their climatic regime in the Younger Dryas of the Late Pleistocene that gave these animals sufficient time to acquire the nutrients to make these huge structures (Guthrie 1984a,b; Moen et al. 1999). By this hypothesis, when the climate changed, stoichiometric imbalance resulted in extinction.

In Chapter 4, we referred to the extensive literature on the stoichiometry of herbivorous zooplankton when we discussed factors contributing to C:N:P ratios in animals. Here, we will consider food quality dynamics in those organisms. Several reviews dealing with stoichiometric food quality in zooplankton-algae interactions have recently appeared and can be consulted for many more details than we will present in this book (Sterner and Hessen 1994; Gulati and DeMott 1997; Sterner and Schulz 1998). Early laboratory work established that the growth of *Daphnia* was sensitive to the growth status of the algae it ate (Mitchell et al. 1992; Sterner et al. 1993). One experiment found that *Daphnia* growth was more closely correlated with P content (Fig. 5.10) than with N, protein, carbohydrate, or several other fractions (Sterner 1993). Whether such relationships indicated bona fide mechanistic relationships between P and animal growth or were artifactual correlations due to other food quality factors was debated (Brett 1993; Hessen 1993; Urabe and Watanabe 1993). Some argued that *Daphnia* were actually responding to individual fatty acids that could be correlated with P content, and that the animals were not responding to phosphorus per se (Müller-Navarra 1995a; Brett and Müller-Navarra 1997). However, laboratory experiments with P supplementation (Urabe et

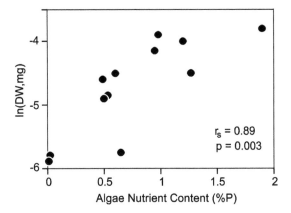

Fig. 5.10. Dry mass of 5-d-old *Daphnia obtusa* (ln transformed) vs. P content in the food alga *Scenedesmus acutus* raised under varying N- and P-limited conditions (Spearman rank order correlation statistics are given). Based on Sterner (1993).

al. 1997) and detailed studies of carbon and phosphorus balance (DeMott et al. 1998) have shown conclusively that the relationship between *Daphnia* growth and algal P content is indeed due to a mechanistic link. Phosphorus supplementation of field-collected seston with high C:P has also been shown to stimulate *Daphnia* growth (Fig. 5.11), indicating that P limitation occurs for natural foods (Elser et al. 2001; Makino et al. 2002). Similar results supporting direct P limitation of zooplankton growth on natural foods were also given by Boersma et al. (2001). The sum of the evidence allows for a confident conclusion that *Daphnia* can indeed be P limited when feeding on algae of low P content. Hence, a paradigm of P limitation in freshwaters can be extended at least partially to upper trophic levels.

Understanding the factors influencing the ecology of *Daphnia* is worthwhile because they are important herbivores in pelagic ecosystems. However, this taxon is not representative of the stoichiometry of all zooplankton taxa. We have already shown that they have relatively high P content, possibly due to life history selection on rapid biomass growth (Fig. 4.2). Stoichiometric principles (see the TER models above) predict that high-P herbivores such as these should be more susceptible to P limitation than species of lower P content [see, e.g., Eq. (5.8)]. In fact, testing this hypothesis that high-P herbivores are more susceptible to P limitation is a powerful test of the importance of nutrient imbalance in consumer growth. Here, we consider the question of whether the nutrient content of different consumers can be used to predict their growth responses on biochemically complex living algal diets. In one study (Schulz and Sterner 1999),

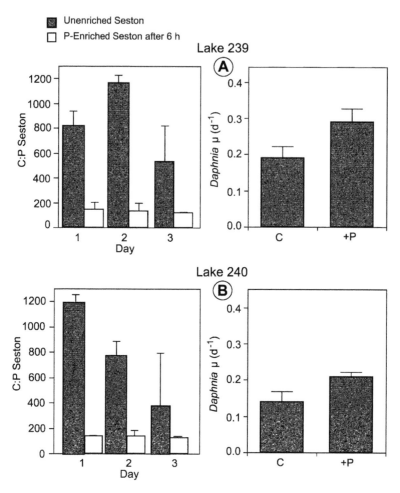

Fig. 5.11. Stimulation of *Daphnia* growth by P supplementation of natural seston from two P-limited lakes having naturally high seston C:P. In this experiment, lake water was collected and spiked with PO_4. Algae and other microplankton rapidly took up the added P, lowering overall seston C:P (left hand panels) but no change in overall seston C concentration occurred (data not shown). In the growth experiments, juvenile *Daphnia* were grown for 5 d on unenriched lake seston (controls, C) or P-supplemented seston for 6 h and unenriched seston for the remaining 18 h each day. Fresh food was re-supplied daily. The P-enrichment period was confined to 6 h in order to minimize potentially confounding effects of changes in digestibility, biochemical quality, or species composition. Lake 239 (top panels) will reappear in Chapter 7, where data will be presented demonstrating that seston C:P, along with zooplankton biomass and species composition, is modulated by light:nutrient balance. Based on Elser et al. (2001).

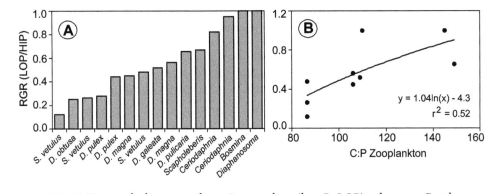

Fig. 5.12. Zooplankton growth on P-poor algae (low P, LOP) relative to P-rich algae (high P, HIP) (RGR = 1.0 means equal growth rates on the two foods, RGR = 0.0 means no growth on P-poor food) as a function of consumer P content. A. RGR for individual taxa ranked from left to right in order of increasing RGR. B. RGR as a function of consumer C:P for the same data as in (A). Spearman rank correlation for data in (B) is highly significant ($p = 0.003$). Data come from a variety of published and unpublished sources. For some taxa, C:P measured in one study had to be paired with growth data from another study.

the small herbivore species *Bosmina*, which has lower P content than *Daphnia* (0.7 vs. 1.5%), had similar growth on food of high and low P content, while low-P food resulted in marked growth reductions for the more P-rich *Daphnia*. A similar trend was also observed during food quality studies of arctic and temperate members of the *D. pulex* species complex (discussed in Chapter 4). Arctic *Daphnia* had higher P content and suffered greater growth declines when fed high C:P food than did their temperate counterparts (Fig. 4.8). As we will see now, empirical tests with larger numbers of species further support these stoichiometric principles.

To further examine the way that consumer P content predicts sensitivity to P-limited foods, we compiled data on zooplankton growth and zooplankton C:P (Fig. 5.12A). Zooplankton response to P-limited food ranged over almost the entire possible range of $0 < RGR < 1$. Some taxa had growth almost completely shut down on low-P foods while other taxa showed no noticeable effect of food P content at all. A scatter plot of RGR versus consumer C:P (Fig. 5.12B) shows that consumers with low C:P (and, by extension, low C:P TER's) are the ones with great growth reductions on P-poor food. Although many previous studies have shown that a single consumer species reacts to the nutrient content of its food, Figure 5.12B indicates that the sensitivity of consumer growth rate to resource nutrient content also relates to consumer nutrient content. Reflecting on these results, we find it astonishing that something so potentially complex as the nutrition of metazoans feeding on a living algal diet can be predicted by something so simple as the relative P content of consumer and resource!

This appreciation for differences in consumer nutrient content and associated interspecific variability in stoichiometric imbalance implies that it is incorrect to think of an entire zooplankton community with diverse species as homogeneously constrained by stoichiometry. Instead, it implies that stoichiometry may be involved in determining zooplankton community structure. If consumer P limitation does have an important influence on zooplankton community structure, we should see patterns between animal community structure and food elemental composition. Such correlational studies must be interpreted with caution because hidden relationships, such as with lake trophic status, fish community structure, or a myriad of other factors, undoubtedly have some influence. However, it stands to reason that if stoichiometric food quality constraints are important, correlations should be observed in nature. One way to look for stoichiometric effects is in direct measures of bulk zooplankton element content. With varying intensity of stoichiometric food quality constraints, one should see positive correlations between bulk zooplankton P content and bulk seston P content as high-P taxa are replaced by low-P taxa with declining food P content. Gulati et al. (1991) found a strong possibility of such a trend in 13 Dutch lakes, if one seemingly unusual observation is considered an outlier (Fig. 5.13). Sterner et al. 2001 found statistically significant, though generally weak, correlations between bulk zooplankton P content and P content of epilimnetic seston in a set of lakes that, unfortunately, did not include any strongly *Daphnia*-dominated systems.

Another way to assess the role of stoichiometry in zooplankton communities is to compare zooplankton community composition, determined through microscope observations, with seston chemistry. Hessen (1992) found that population densities of P-rich zooplankton in 47 Norwegian lakes were better correlated with seston P than with seston N, while N-rich zooplankton were better correlated with N than with P. Hassett et al. (1997) surveyed 31 lakes in temperate North America and similarly found that the contribution of P-rich *Daphnia* to total zooplankton biomass was negatively related to seston C:P ratio. Stoichiometry does not predict that all zooplankton will always respond to seston P content, only that some will respond to seston P content when it is in the correct TER range. DeMott and Gulati (1999) examined *Daphnia* and *Bosmina* abundance from a long-term study of three Dutch lakes subjected to reduced P loading (these three are among the lakes in Fig. 5.13). Over nine years, seston C:P increased in these lakes, while phytoplankton abundance and community composition did not change. Consistent with stoichiometric predictions, the P-rich zooplankter (*Daphnia*) was reduced in abundance at higher seston C:P but the low-P taxon (*Bosmina*) was not (Fig. 5.14).

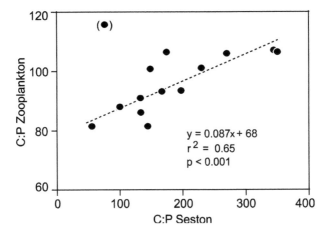

Fig. 5.13. Zooplankton C:P vs. seston C:P in 13 Dutch lakes, showing that the overall stoichiometry of the zooplankton community is related to seston stoichiometry. One biomanipulated lake was sampled in two years; its data points are the two with the lowest seston C:P ratios (one of these marked with parentheses). Clearly, this one lake showed a great deal of variation in the zooplankton between these years. Zooplankton in biomanipulated lakes are known to exhibit rapid and large, but sometimes transient, responses to changes in fish. The regression line is $y = 74.1 + 0.058x$ ($r^2 = 0.24$), including the indicated point. Based on Gulati et al. (1991).

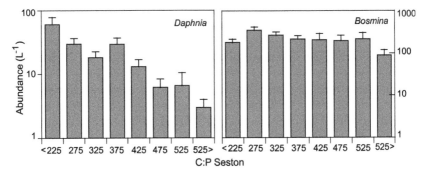

Fig. 5.14. Abundance of P-rich (*Daphnia*) and P-poor (*Bosmina*) zooplankton as a function of seston C:P for three Dutch lakes (note logarithmic scale). Bars show mean abundance ±1 s.e. within the indicated ranges of seston C:P. The negative relationship for *Daphnia* and lack of relationship for *Bosmina* are consistent with stoichiometric predictions. Based on DeMott and Gulati (1999).

Stoichiometry clearly helps us interpret a diverse set of relationships dealing with herbivore dynamics and secondary production in its many guises in zooplankton. This body of literature provides a single self-consistent picture of the role of element balance in growth and community structure in planktonic herbivores. We leave it to future work on other herbivores such as insects to determine if similar patterns exist there.

Termites

Other groups of consumers in this section are categorized by nutritional strategy, but consumption of wood in living and dead trees is not readily pigeonholed in these categories (is it herbivory or detritivory?). Nevertheless, termites are an interesting stoichiometric story in and of themselves. In Chapter 3, we discussed how carbonaceous plant structural material causes large plants like trees to have high C:nutrient ratios. Few consumers would be able to make a living on such an imbalanced diet (Haack and Slansky 1987), which is a good thing for trees and houses. Termites, however, are famous wood consumers. There is nothing particularly special about their stoichiometric biomass ratios. The N content of termites is typical for animals, with a % N of 8–13 and C:N ratios of 4–12 (Matsumoto 1976). Given the extreme stoichiometric imbalance between wood and termites, we might expect some exquisite adaptations to balance out C and N, and that is in fact the case.

In making termite consumers with C:N of about 10 out of wood with C:N of perhaps 1000, we must have means of supplementing or strongly conserving N, of voiding excess C, or both. Examples have been discussed by Higashi et al. (1992, 1997) and White (1993). Many and possibly all termites have symbiotic N fixers in their guts (Tayasu et al. 1994). For N conservation, termites produce uric acid—a metabolic end product that most organisms release to the environment—and excrete it into their hindgut, where resident microbiota convert it to NH_3 or amino acids, available for reuptake (Potrikus and Breznak 1981). Uric acid is also energetically expensive compared to other forms of N release such as ammonia; hence, forming uric acid may also be a means of eliminating surplus carbon. Termites also readily ingest their exuvia as well as dead colony members (White 1993). Under strong N shortage, they are also cannibalistic; it has been found that cannibalism can be reduced or eliminated by supplementing the diet of the colony with nitrogen (White 1993). Termites release considerable quantities of excess C via the methane output from the methanogenic bacteria in their guts (Prestwich and Bentley 1981) at rates significant compared to global greenhouse gas cycles (Schlesinger 1997). Finally, a means of modifying their wood resources to make them more balanced is the practice of "cultivating" specific fungi on their feces

(Higashi and Abe 1997). These fungi modify the high C:N feces by eliminating C as CO_2 by respiration, somewhat reminiscent of what we discussed above in streams. Subsequent reingestion of these modified feces (as well as parts of the fungi themselves) allows the termites a further means of conserving N and relieving their intense N imbalance. Life on such an imbalanced diet has been a great ecological and evolutionary challenge. The relationships between the insects and their different symbionts (bacteria, fungi) were essential to meeting that challenge.

Although there are many ways that termites in general achieve C and N balance, different species use these different strategies to different degrees. Understanding that variability links stoichiometry to foraging ecology. Higashi et al. (1992) hypothesized that the means of achieving C:N balance, whether primarily by loss of C or enhanced use of N, depends on whether the termite colony feeds and nests in a single location (such as a single dead limb) or whether the colony nests in one place with members foraging in other places. Imagine a termite that could not lower its N loss. Its only response would be compensatory feeding, to consume more wood to obtain sufficient N. This would not be a feasible strategy if wood were used both for shelter and for food. Use of excess C would require a correspondingly larger ingestion of C by the consumer, compared to mechanisms that retain or supplement N. Wasting excess C might not be expected to be a favorable strategy for termites that live and forage in the same location, essentially eating themselves out of house and home. According to Higashi et al., these species do, in fact, primarily use N conservation strategies, while species foraging away from their nest use primarily C disposal strategies. It is interesting to note in this example how behavior interacts with stoichiometry.

Detritivores

Long-term changes in chemical content in detritus as it ages are associated with specific stoichiometric shifts. Although it may seem as though fresh detritus, being closer to the living state, would be the best growth substrate for higher consumers, this is not the case.

Recall that heterotrophic breakdown of terrestrially derived organic matter is often a major portion of stream metabolism (Hynes 1970), and that vascular plant breakdown in streams includes a phase of microbial conditioning followed by invertebrate action (Webster and Benfield 1986). Fresh detritus is often a poor substrate for aquatic insects (Lawson et al. 1984). As in herbivores consuming live leaves, the nutrient content of leaf litter can predict the quality of fallen leaves as growth resources for insects. Specifically, the leaves from N-fixing alder supported the highest growth of the detritivorous insect *Sericostoma personatum* (Iversen 1974).

In general, the growth rate and growth efficiency of stream macroinvertebrates such as caddisflies and chironomids is higher after fresh coarse detrital material has been conditioned, which (as discussed above) causes C:N and C:P ratios to decrease. Higher consumers benefit from this modification of the substrate, as well as others such as physical changes (reduced size of particles, reduced physical toughness, etc.), and they also ingest the saprotrophs themselves.

Overall, the succession of detritivores on leaves in streams consists of a system composed of leaves and wood (C:N ratio very high), saprophytes such as fungi (C:N moderate to low) and bacteria (C:N low), and stream insects (C:N low). Even single species of fungi may make good substrates for higher consumers (Bärlocher and Kendrick 1973b,a) although some fungal species are poor-quality food for insect consumers (Arsuffi and Suberkropp 1984, 1986). The factors making one fungal species high-quality food and another species poor-quality food have not yet been sorted out; perhaps there is a stoichiometric component.

The vast majority of autotroph detritus is broken down on land, not in streams. Similar chemical and mechanical processes are known to occur there, and there are many studies of chemical changes of detritus during breakdown. The biotically diverse consumers associated with detrital breakdown in soils, however, seem not to have been studied from a stoichiometric perspective. This looks like intellectually fertile ground to plow!

Carnivores

The etymology of this word refers to "flesh eating." As we have already established, carnivores generally have high assimilation efficiency (Stevens and Hume 1995) and growth efficiencies (Fig. 5.2). Let us start with the obvious question first: What could be a more chemically balanced consumer-resource pair than one animal eating another's flesh? It would seem that stoichiometric imbalance would have little influence on carnivores. However, at least one kind of strong imbalance impinges on carnivores, namely, vertebrates that specialize on consuming nonmolluscan invertebrates. As we saw in Chapter 4, a vertebrate's skeleton, made up largely of Ca and P, can contribute significantly to its total elemental composition. Skeletal investment is a higher proportion of body mass in larger animals. Thus, stoichiometry tells us that large vertebrates may have a difficult time meeting their Ca and P demands (as we indicated earlier when discussing Irish Elk). In contrast, Ca in invertebrates is highly variable, and is especially low in many arthropods, including insects (Sang 1978). Mattson and Scribner (1987) compiled data on elemental content of various species of insects, and found the range in Ca to be from 0.03% to 9.5%. The data are scanty; however, they suggest that at least the folivorous insects are generally lower in P content (0.9%) than mammals (4.3%) or birds (1.9%). Thus, one group of carnivores that might have severe problems with stoichiometric balance is insectivorous vertebrates. The dramatic imbalance in Ca can be illustrated by a specific example. Ovarzun and coworkers (1996) measured the chemical composition of termites and the stomach contents of free-ranging anteaters that specialize on them. The Ca content of arboreal termites was only 0.26% Ca, and the stomach contents of the anteaters was even lower in Ca, 0.11%. These are extremely low numbers for a vertebrate this large.

Birds require Ca both for their skeleton and for their eggshells. Whether Ca might limit some aspect of egg production in birds has been repeatedly examined. Graveland and Vangijzen (1994) measured a Ca budget for the great tit, *Parus major*. According to their findings, there is no significant Ca storage in these birds, meaning that all Ca for egg formation must be obtained during the laying period. Even on Ca-rich soils, Ca intake from arthropods (or seeds) covered only 5–10% of Ca requirements, indicating that there must have been another dietary source. Graveland and Vangijzen indicated that these results likely apply as well to many other passerines. Interestingly, birds that include either woodlice (Isopoda) or millipedes (Diplopoda) in their diet may have less of a problem: these particular arthropods contain 10–13% Ca, a hundred times the value observed typically in insects (Graveland and Vangijzen 1994). Thus, birds may generally avoid Ca deficiency by consuming Ca-rich resources, such as mollusks (Pierotti and Annett 1991; Nisbet 1997), especially during the laying season. Overt Ca deficiency in the wild may therefore be rare because of food choice in these large, highly mobile animals (Johnson and Barclay 1996; Mahony et al. 1997). We can view some aspects of diet choice in some birds as their means of overcoming stoichiometric Ca imbalances.

The Ca needed by a young mammal is provided for it in its mother's milk. Lactation is a period of great Ca demand for the mother; her Ca:C TER is especially high during this time. Successful reproduction in insectivorous mammals must either overcome or be limited by the natural stoichiometric imbalance in Ca between consumer and resource. For the insectivorous little brown bat (*Myotis lucifagus*), it has been calculated that a mother obtains about three times more energy than is needed from a diet containing just enough Ca to meet the demands of midpregnancy (Barclay 1994). To do this, she must consume two to four times her own mass per day. In fact, the timing of reproduction and clutch size in bats have both been hypothesized to be constrained by Ca availability (Barclay 1994; Bernard and Davison 1996). Although throughout this book we have emphasized the stoichiometry of C, N, and P, principles learned from a study of these elements might easily carry over to special cases of stoichiometric imbalance of other substances, such as Ca in some vertebrates.

Homeotherms versus Poikilotherms

We have not dealt much with the important role of carbon substrates in providing energy until now, but clearly this must be considered in the carbon or energy balance. Carbon for growth or reproduction is available only after basal metabolism is satisfied. Metabolisms with large energy requirements increase the need for carbon compared to nutrients. The TER for carbon compared to nutrients should shift in predictable ways.

Our final consideration of real world problems in stoichiometric balance deals not with feeding mode, but rather with this other major way of categorizing animals. It is well known that homeothermy is a physiological lifestyle much more demanding of energy than poikilothermy. Hence, we should expect predictable shifts in TER's for homeotherms versus poikilotherms. Just such a pattern has been suggested by nutritional studies of some economically important animals, including humans (Table 5.2). The optimal ratio of glucose:protein for growth in several different insects ranges between 1 (mass ratio) and 3, while the ratio for homeothermic vertebrates ranges from 4 to 15. This suggests that homeothermic metabolism has a large effect on consumer TER's.

GROWTH EFFICIENCY

Given a physiology of tight homeostasis coupled with variable nutrient content in food resources, it is a logical necessity that growth efficiencies of elements must vary. In bioenergetics, growth efficiency typically is a parameter. In stoichiometrically explicit models, it varies dynamically. This is one fundamental difference in perspective between the multiple-currency approach of ecological stoichiometry and a purely bioenergetic one. On high C:N food, a homeostatic consumer must retain relatively more N and dispose of relatively more C than on low C:N foods. Whether they achieve mass balance by regulating C, regulating N, or both, is likely a function of the circumstances (the organism, the food, etc.). We examined these changes in this chapter primarily at the individual level, but a similar effect is observed at other levels, such as where respiration rates of entire suites of bacteria (disposal of C) are higher with greater C:N and C:P imbalance (Cimbleris and Kalff 1998).

To close this chapter, we bring the theoretical and empirical approaches together using data from a variety of consumers. To start with, general features of the stoichiometric C:nutrient balance in herbivores can be highlighted by comparing studies on contrasting animals. One is the production of eggs by the marine copepod *Acartia* (Kiørboe 1989), and the other the growth of the chrysomelid beetle *Paropsis* (Fox and Macauley

TABLE 5.2
Approximate optimal ratios of available carbohydrate plus fat (expressed as g glucose) to protein in the diets of some homeotherms (top entries) and some poikilotherms (bottom entries). From Bernays (1982). Note that poikilotherms require less glucose per unit protein, which is an indication of a large difference in needs for materials for growth compared to energy for maintenance

Animal	Glucose:protein (g:g)
Homeotherms	
Human child	6:1
Human female (pregnant)	7:1
Cattle	7:1
Calves (very young)	4:1
Heifers	7:1
Buffalo	10:1
Goats	15:1
Poikilotherms	
Silkworm	3:1
Silkworm (artificial diet)	3:2
Locust (artificial diet)	1:1
Cabbage butterfly larva	1:1

1977). We chose these two studies because they consider two very different taxa and present all the information necessary to calculate the relevant rates and efficiencies. In each of these studies, herbivores were raised on relatively high quantities of autotroph diets that were easily ingested, but had variable nutrient content. Rates of feeding, growth, and growth efficiencies for both C and N were either presented in the study or were easily calculated from the data presented. For the copepod, the food was the diatom *Thalassiosira* while for the beetle it was a set of species of *Eucalyptus*. Secondary compounds (tannins) were considered unimportant in the C and N dynamics.

The similarity of the trends for the two consumer species is striking (Fig 5.15). Neither herbivore showed any difference in rates of bulk feeding (i.e., feeding expressed per unit dry weight or unit carbon) (Figs. 5.15A,C) in response to food chemical composition. In other words, there was no compensation in feeding rate as a function of food nutrient content. The rate of ingestion of N, however, was strongly correlated with N content of the autotroph in both studies, as it must have been if feeding rate was constant but food N content varied (Figs. 5.15B,D). In both species, herbivore growth rates were positively related to autotroph N content, no matter whether growth was measured in carbon units (Figs. 5.15E,G) or

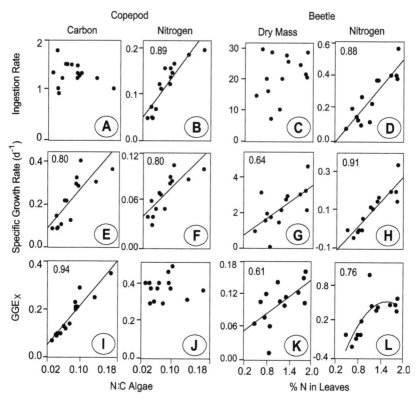

Fig. 5.15. Carbon and nitrogen balance in two herbivores raised on autotrophs of varying N content. The left two columns show results for a marine copepod and the right two columns show results for a beetle. The top row gives information on ingestion rates for carbon (A, C) or nitrogen (B,D). Units are as follows. A. μg C $(\mu g$ C$)^{-1}$ d^{-1}. B. μg N $(\mu g$ C$)^{-1}$ d^{-1}. C. mg C d^{-1}. D. mg N d^{-1}. The middle row gives information on growth rates, again for either carbon (E,G) or nitrogen (F,H). The bottom row gives information on growth efficiencies for carbon (I,K) or nitrogen (K,L). Numbers in panels give correlation coefficients (r) for the trend lines (linear except for L, which is quadratic) shown. Panels A, B, E, F, I, and J based on Kiørboe (1989) and panels C, D, G, H, K, and L based on Fox and Macauley (1977).

nitrogen units (Figs. 5.15F,H). GGE_C increased with autotroph nitrogen content in both species (Figs. 5.15 I,K), and GGE_N increased with auto-troph N content for the beetle (Fig. 5.15L), but not for the copepod (Fig. 5.15J).

The trends observed in Figure 5.15 are almost entirely consistent with the minimal model described earlier in this chapter. The predictions of the minimal model presented in Figure 5.7 correspond to holding food C constant (X_2 in the minimal model), but letting food N vary (corresponding to X_1 in the model), as would be the case with constant food quantity but varying food quality. The model assumes a constant feeding rate so that bulk food ingestion is constant, as observed in both studies (panels A and C). Ingestion of the potentially limiting nutrient (in this case nitrogen) linearly increases with food N:C, as was also observed (panels B and D). The agreement of model and data for feeding rates has more to do with demonstrating that the underlying assumptions of the model are met by the experimental system, rather than that the model was successfully predicting animal behavior or performance.

However, the minimal model does a credible job of describing the nutrient balance in the two herbivores. Compare the trends in Figures 5.7 and 5.15, remember that N corresponds to X_1 and C to X_2 in the model, and assume we are in the left portion of Figure 5.7 because N:C progresses from very low values to higher ones. Both the beetle and the copepod showed increasing growth (expressed in either C or N terms) with increasing N content (Figs. 5.15E–H). The model predicts linear increases, similar to the trends observed. Figure 5.7 shows only growth expressed for one element, but with strict homeostasis, growth for the other element must be similarly linearly increasing. For production efficiency, the model predicts linear increases in terms of C (Fig. 5.7D) but curvilinear increases in terms of N (Fig. 5.7C). The data are somewhat equivocal here, but carbon growth efficiency looks very linear in the copepod (Fig. 5.15I), in agreement with theory, and may be consistent with a linear trend in the beetle (Fig. 5.15K). For nitrogen, where curvilinear changes are expected, the beetle data are consistent (with the exception of one seeming outlier, Fig. 5.15L) and the copepod data are not strongly suggestive of any particular trend (Fig. 5.15J).

These two studies support the broad-scale generalization that growth efficiency expressed per unit carbon is high when the contents of other essential nutrients in the food are also high, and it is low when essential nutrients are scarce. GGE_C values in both studies showed wide ranges, from near zero at low food N content to about 15% in the beetle and about 30% in the copepod. This difference in maxima may also reflect stoichiometry: for the copepod, the algae with highest N:C approached the N:C of the eggs being produced ($\sim$0.2), whereas for the beetle, the eucalypt diet with the highest N content ($\sim$2%) was still considerably lower in N content than the beetles themselves ($\sim$7).

We end this chapter with a figure that shows how stoichiometry as a concept can integrate across numerous diverse species. Increasing GGE_C with increasing food nutrient content is a highly general trend, observable

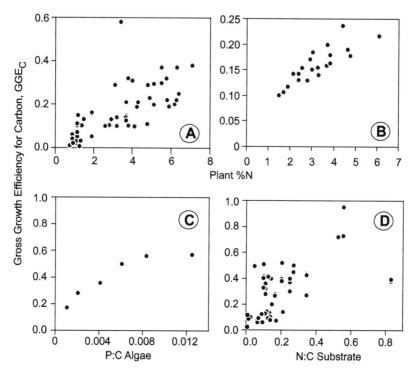

Fig. 5.16. Relationships between carbon gross growth efficiency (GGE$_C$) and re-source nutrient content in different consumers. A. Mixed species of insects. B. Cabbage butterflies. C. *Daphnia*. D. Bacteria in lab culture and in the field. Note the consistent trends in growth efficiency in these diverse consumers. Panel A based on Mattson (1980), Panel B based on Slansky and Feeney (1977), panel B based on DeMott et al. (1998), and panel D based on del Giorgio and Cole (1998).

in many different studies. Examples based on mixed insects, *Daphnia*, a butterfly species, and bacteria are shown in Figure 5.16. With constant food quantity, the minimal model predicts linear increases followed by a horizontal plateau. The *Daphnia* and the cabbage butterfly study had con-stant food quantity, and at least the *Daphnia* data hint at a plateau. Others have noticed similar ties between growth efficiencies and stoichiometry (e.g., Linley and Newell 1984). Given the strong degree of consistency in these trends, it is safe to conclude that observed variability in GGE$_C$ in consumers (Fig. 5.3) does have a very important stoichiometric compo-nent. The yield of secondary production is a function of the C:nutrient balance in the resources consumed relative to the consumer's body de-mands. This relationship will reappear several more times in later chapters.

At the beginning of this chapter, we considered the range of growth efficiencies in diverse aquatic consumers (Fig. 5.3), and we remarked that stoichiometric balance might well explain much of the variation in GGE_C observed by Straile in this work. As we have seen, there are numerous examples of how stoichiometric balance regulates growth efficiencies. Given the failure Straile observed in parameters such as taxonomic group, body size, etc., in explaining growth efficiencies in the field, it seems highly probable that stoichiometric balance will be one of the most important factors to consider.

We will next turn to a consideration of the "inverse" of this chapter. Low gross growth efficiency means that a small fraction of what is ingested is incorporated into consumer growth. It also means that a high fraction of what is ingested is returned to the environment. If consumer growth has a strong stoichiometric component, so too must nutrient cycling, which is our next topic.

CATALYSTS FOR ECOLOGICAL STOICHIOMETRY

The potentially broad importance of issues connected to stoichiometric imbalance across numerous diverse consumer groups unfortunately is not reflected in the distribution of taxa in the scientific literature. Aside from several well-known groups, we are missing far too many basic data on patterns of nutrient balance. Too few consumer groups have been studied across wide stoichiometric gradients of food quality; we often seem to have portions of trends rather than the full dynamic range. Nutrient constraints seem not to have been given the weight they deserve in many consumer taxa. With further study, it will be interesting to see how general are the nutrient balance relationships we saw in a copepod and a beetle. Specific questions concerning various consumers include the following.

- What is the role of the skeleton in element storage in wild verte-brates? As we have seen, it plays a role in antlerogenesis. Does it create a variable stoichiometry? Does it create a unique set of food quality constraints for vertebrate consumers?
- What are the physiological mechanisms of homeostatic regulation in diverse consumers? Do consumers generally regulate the proportions of substances ingested and assimilated from the environment, or do they generally differentially metabolize assimilated substances?
- What is the role of spatial heterogeneity and food choice in consumer-resource balance? Do consumers generally avoid stoichiometric imbal-ances through judicious food choices or habitat preferences?

Our second set of questions focuses on a variable that needs to be mea-sured and reported far more frequently. The literature has many examples of how resource chemical content influences consumers. Until now, the potential importance of interspecific differences in consumer nutrient content seems not to have been understood; perhaps people have wrongly been thinking that most animals have the same stoichiometric nutrient ratios. Chapter 4 showed that this assumption is wrong and in this chapter we saw that zooplankton nutrient content did have a large effect on how growth responded to food nutrient content.

- In general, what is the role of consumer nutrient content in production dynamics, fitness, and community structure?
- How do interspecific differences in insect stoichiometry influence their growth and food choice?
- Does consumer nutrient content relate predictably to consumer niche position?

In Thingstad's model covered in this chapter (Fig. 5.6), we encountered some fundamental issues connected to nonstrict homeostasis, which we defined carefully in Chapter 1. Most models of consumer growth utilize the simplifying assumption of strict homeostasis. However, Thingstad's model shows that variable stoichiometry can be studied in similar ways.

- The regions of "mixed limitation" Thingstad identified need to be explored both mathematically and empirically. What are the linkages between physiology, growth dynamics, and ecological interactions in these regions?

The material considered in this chapter opens up a set of questions about trophic levels. Regulation of consumer N production efficiency has implications for current attempts to characterize trophic position by stable isotope analysis. Based on the observations of Minagawa and Wada (1984), it is commonly supposed that the ^{15}N content of a consumer will be slightly and consistently enriched compared to its food (i.e., the $^{15}N{:}^{14}N$ ratio of a consumer is higher by a fixed amount than that of its food). The basis for this enrichment is a differential loss of lighter ^{14}N during N metabolism (Minagawa and Wada 1984). Most of the examples for consistent isotope enrichment (Minagawa and Wada 1984), however, are from carnivores. Controlled laboratory studies on stable isotope fractionation are clearly needed (Gannes et al. 1997). Stoichiometrically imbalanced consumers and resources might have different degrees of isotope enrichment due to the variable biochemical pathways involved in homeostatic control of N. Indeed, Adams and Sterner (2000) observed a very significant change in isotope fractionation of *Daphnia* consuming algal foods of differing C:N ratio. This is only a single example, however. Thus, we have the following catalyst.

- Is stoichiometric balance an important issue in considerations of stable N and C isotope dynamics in ecosystems?

Additional questions about trophic levels brought up by this chapter are the following.

- What are the general trends of correlations of nutrient content across trophic levels?
- Does high nutrient content in resources generally select for high nutrient content in the consumers?
- If so, what might be the effect on ecological dynamics of selecting for higher-nutrient consumers which likely have higher growth rate (Chapter 4), as the nutrient content at the base of the food web increases?

SUMMARY AND SYNTHESIS

In this chapter we explored numerous diverse challenges and responses to stoichiometric imbalance in consumers. We saw stoichiometric challenges overcome in ways unique to individual taxa (skeletal storage of Ca, retention of N in termite guts, etc.), but we also saw at a very basic level (described by some simple models of nutrient balance) that diverse consumers do similar things. Patterns related to gross growth efficiency seem to be very robust across groups. Specifically, the greater the stoichiometric imbalance between resources and food, the lower the efficiency expressed as a fraction of the element present in surplus. This perhaps should not come as much of a surprise: at a certain level, it is an inevitable consequence of homeostasis and conservation of matter. Given the robustness of these relationships, thoughts should be directed toward understanding the broad and highly general patterns they produce. Single-currency frameworks, whether they be based on energy, carbon, nitrogen, or something else, do not and cannot capture these effects.

This chapter contains a wealth of information, both empirical and theoretical, about some potentially highly complicated things. Nevertheless, this information is self-consistent with regard to the effect of nutrient balance on animal growth. Threshold element ratio models are also consistent, both qualitatively and quantitatively, with empirical results at the single-population and at the community scale. As we will see in Chapter 7, they are also consistent with numerous other experimental results at different scales. Theories that are backed by a wide diversity of observations and analyses are rare in ecology.

This chapter emphasized the stoichiometric diversity of consumers from Protozoa to vertebrates. The most important take-home message of this chapter is that simple stoichiometric models of resource-consumer interaction can be very successful in helping us predict as well as understand consumer production dynamics. We have demonstrated some general trends of growth and production efficiency arising out of stoichiometric imbalance. Stoichiometry helps us understand interspecific and intraspecific patterns in consumer-resource interactions. This chapter underscored the great importance of differences in the nutrient content of consumer biomass.

6

The Stoichiometry of Consumer-Driven
Nutrient Recycling

There is no coming into being, nor any disappearing of ought that perishes; only mingling and separation of what has been mingled.—Empedocles of Acragas (ca. 444 B.C.E.)

Consumers separate what has been mingled in their food and return some of those ingested nutrients to their surroundings; in this chapter we will refer to this process as consumer-driven nutrient recycling (CNR). In Chapter 3 we explored the stoichiometrically flexible lifestyle of autotrophs, whose elemental composition is variable in response to nutrient supply and abiotic factors such as temperature and light intensity. In Chapter 4 we considered major aspects determining the elemental composition of metazoan animals, emphasizing patterns of cellular allocation that determine the characteristic C:N:P of various species. In strong contrast to the situation with autotrophs, our main impression was of a generally tight stoichiometric homeostasis related to an animal's evolved life history and body structure. In Chapter 5 we saw what happens when these constrained requirements of metazoan consumers meet the broad flexibility of autotrophs: disparities between animal and autotroph elemental composition constrain the growth and production of consumers of a wide variety of types. We are now in a position to more explicitly assess how stoichiometric imbalances affect levels of biological organization higher than populations. In Chapter 7 we will consider the impact of elemental imbalance on aggregated secondary production, consumer community structure, and food-web dynamics, and we will examine the even broader ecosystem contexts in Chapter 8. Right now we will consider how elemental imbalances feed back to autotrophs via differential nutrient recycling by consumers. This feedback occurs when consumers convert organically bound nutrients again to inorganic form, thus returning them to the autotroph competitive arena and bringing the autotroph-consumer interaction full circle.

So this chapter will focus on CNR and examine how stoichiometric analysis helps us understand the complex interactions among herbivorous consumers, autotrophs, and nutrient cycling in ecosystems. In this chapter, "consumer" refers primarily to metazoans, although we also consider stoichiometry and mineralization by microorganisms such as Protozoa and bacteria. As before in this book, we will see that two sets of elemental ratios are critical for the analysis: those of the organism or resources being consumed and those of the consumer, because the balance of these ratios directly determines the rates and ratios of limiting nutrients recycled. The material in this chapter and the chapters on animal growth (Chapter 4) and food quality effects (Chapter 5) are intimately linked. Materials not used for growth are recycled. In considering the effects of stoichiometric food quality on animal growth in Chapter 5, we emphasized the benefits of a stoichiometrically balanced resource base to production. Here, we will see what happens to materials in stoichiometric excess. Our examples come from freshwater macrozooplankton and microbes, marine protozoans, soil microbes, and freshwater fish. In presenting them, we hope to show how stoichiometric analysis may help ecologists to understand the role of the food web in regulating nutrient cycling in diverse habitats. Consumer-driven nutrient recycling is one of the easiest places to explore the reciprocal relationships between the biological features of individual species and the ecosystem context in which they are imbedded (Jones and Lawton 1995). Thus, advances in understanding CNR, how CNR is affected by various consumer species, and how CNR affects autotroph growth and community structure will represent considerable progress in understanding fundamental relationships between species dynamics and functional aspects of the ecosystem.

We will first present a brief history of the ideas and findings regarding CNR that suggested that a stoichiometric view of this process would be productive. Second, we will review three primary theoretical constructs that have mathematically formalized CNR in a stoichiometrically explicit framework. Third, we will review studies that have simultaneously considered recycling of both N and P by zooplankton and assess whether these studies support the role of stoichiometry in CNR. We finish by describing a set of findings regarding the potential role of stoichiometry in nutrient cycling by fish, a potentially important component of biogeochemical cycling in aquatic systems (Vanni 1996).

A BRIEF HISTORY OF STUDIES OF CONSUMER-DRIVEN NUTRIENT RECYCLING

Oceanographers and limnologists have long known that herbivorous zooplankton are closely coupled to the phytoplankton not just through their grazing but also via their nutrient cycling (Ketchum 1962; Lehman 1980; Sterner 1986). Work on consumer-driven nutrient recycling in aquatic systems has generally considered only one nutrient (with some notable exceptions; see, for example, the work of Corner discussed below), reflecting classic paradigms of single-nutrient limitation. For example, marine zooplankton ecologists, working under the view that N is a dominant limiting nutrient, have focused on N processing by zooplankton. Meanwhile, in freshwater, limnologists have worked under a paradigm of P as the primary limiting nutrient and thus have centered their efforts on P. In Chapter 8 we will see that thoughts of single limiting nutrients in either marine or freshwaters are being modified. Study of single-nutrient CNR has revolved largely around the question of its relative importance in fueling phytoplankton productivity.

Interestingly, some of the earliest measurements of nutrient recycling by macrozooplankton in the ocean dealt with both N and P. Following in the wake of Redfield and the prescient realization by Ketchum (1962) that zooplankton generally had body N:P exceeding phytoplankton N:P in the ocean and therefore must excrete N:P at a low ratio, E.D.S. Corner and colleagues documented body N:P of various zooplankton species (Corner and Davies 1971), considered the assimilation of both nutrients into new biomass (Corner and Davies 1971; Corner et al. 1972, 1976), and evaluated the resultant rates and ratios of N and P returned to the environment (Corner et al. 1965; Corner and Newell 1967; Corner and Davies 1971). Given this early impetus, one wonders why so much of the more recent marine work has taken such a one-dimensional approach. In freshwaters, Lehman (1984) also noted that zooplankton might not regenerate N and P with equal efficiency and suggested that this difference might affect phytoplankton communities by altering the N:P of the nutrient arena. Olsen et al. (1986) soon showed that the rate of P release by *Daphnia* depends strongly on algal P:C, with P release declining strongly with decreasing algal P content.

Two additional pieces of the puzzle came soon after. Elser et al. (1988) described the results of a whole-lake food-web manipulation that shifted the zooplankton community from *Daphnia* dominance to copepod dominance, resulting in a qualitative change in the identity (P vs. N) of the nutrient limiting phytoplankton growth (Fig. 6.1). When *Daphnia* were the dominant grazer, i.e., under conditions of low planktivorous minnow abundance, phytoplankton growth was P limited, but when copepods dominated in the presence of minnows phytoplankton were N limited. Short-term manipulations of zooplankton biomass or size structure in bags (independent of fish) also qualitatively altered algal N versus P limitation in similar ways. Elser and colleagues were puzzled by these findings and could report no definitive mechanism for this effect. However, they did point to the possibility

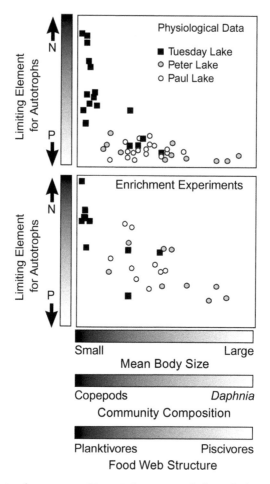

Fig. 6.1. Relationships among N vs. P limitation of phytoplankton growth, zoo-plankton community composition, and food-web structure in three lakes during whole-lake food web manipulations. In this study, predation intensity on zoo-plankton was reduced (by adding minnow-feeding bass; Tuesday Lake) or increased (by removing bass and adding minnows; Peter Lake) while Paul Lake was left unmanipulated to serve as an experimental reference. In the top panel, physiologi-cal assays (alkaline phosphatase activity for P limitation and ammonium enhance-ment of dark C uptake for N limitation) were used to assess algal nutrient limita-tion on approximately weekly intervals in the study lakes. Measures for the N limitation assay were divided by values for the P limitation assay to provide a rela-tive measure of N vs. P limitation on any given day. In the lower panel, data are from biweekly nutrient enrichment bioassays in which lake phytoplankton were enriched with PO_4^{3-} or NH_4^+ and their growth response measured over five days.

that the N:P of zooplankton nutritional requirements might generally be lower than the N:P ratio found in algal biomass, resulting in a general tendency for zooplankton to preferentially recycle N while retaining P. Fortunately, the data from Andersen and Hessen (1991) discussed extensively in Chapter 4 soon clarified the situation. The fact that *Daphnia* has low body N:P (~14) relative to adult calanoid copepods (~30–50) led Sterner et al. (1992) to conclude that the phenomena reported by Elser et al. (1988) were consistent with an altered stoichiometry of CNR. Zooplankton assemblages dominated by *Daphnia* (low N:P) likely retained P in biomass at high efficiency while recycling N at a relatively high rate, and thus shifted phytoplankton growth toward P limitation. In contrast, when calanoid copepods dominated, the zooplankton community retained N while preferentially recycling P; thus, phytoplankton growth was primarily limited by N.

Thus, it seems that taxon-specific differences in elemental composition of different zooplankton can influence the CNR feedback to phytoplankton nutrient status. However, until the mid-1980s and early 1990s, little ecological theory existed that imposed strict mass balance on the algae-grazer interaction and evaluated its effects on grazer nutrient release. Luckily, mass balance constraints on the dynamics of multiple elements (C, N, and P) facilitate the development of such theory.

STOICHIOMETRIC THEORIES OF CONSUMER-DRIVEN NUTRIENT RECYCLING

Ecological stoichiometry is based on the premise that conservation of matter constrains biological dynamics at all levels of organization. As we saw several times already (e.g., the TER models of Chapter 5), the law of conservation of matter also provides a powerful tool for mathematically analyzing the fate of multiple elements in trophic interactions. By constraining system behavior, it simplifies potential dynamical outcomes. Some stoichiometrically explicit models of trophic interactions and nutrient recycling have been developed that help us understand how disparities between food and consumer elemental composition influence CNR.

The models we will consider do not distinguish between assimilated and

---

Again, the index for N limitation (growth response to N) was divided by that for P limitation. In the original study, the *x* axis represented biomass-weighted mean body size of the zooplankton. Variation in this measure was associated primarily with the contribution of *Daphnia* to total zooplankton biomass, which was strongly affected by lake food-web structure. Based on Elser et al. (1988).

then metabolized nutrients released as excreta and nutrients not assimilated and released in solid form as egesta. These two types of loss differ in physical form: excreted nutrients are solutes and egested nutrients are solid or semisolid. The nutrients also likely differ in the time scale over which they become available for reuptake. Thus, the nutrients released may have different fates depending on the form in which they are released: in pelagic systems, excreted nutrients should remain in the system longer while egested nutrients may be lost from surface waters by sinking. Several studies have considered the potential role of zooplankton excretion versus egestion of nutrients in affecting the nutrient balance of surface waters (Mazumder et al. 1989; Elser et al. 1995b; Elser and Foster 1998; Sarnelle 1999). This issue may have particular importance in very deep lakes and in the oceans, where sedimented fecal pellets are a primary way that materials reach deep waters (Knauer et al. 1979; Angel 1989). In contrast, in terrestrial systems nutrients released as excreta may be subject to losses due to volatilization or leaching, whereas egested nutrients might be retained more efficiently, becoming available in a "slow release" mode. Future applications of the models we will discuss to other systems may need to explicitly distinguish excretion and egestion.

Here we describe two stoichiometrically explicit theoretical approaches to instantaneous nutrient release by consumers. The first (Sterner 1990b) considers the N:P of recycled nutrients as a function of food N:P, consumer N:P, and the consumer's ability to sequester a growth-limiting nutrient. This model helps us think about the effects of consumers on the identity of the element limiting phytoplankton (N vs. P), because the N:P requirements of grazer biomass determine the rates and ratios of N and P release that then influence the nutrient regime experienced by phytoplankton. The second approach considers C and P rather than N and P and focuses on how P release is affected by the relationship between grazer P:C and food P:C. These studies originated with Olsen and Østgaard (1985) and Olsen et al. (1986) with subsequent modifications by Hessen and Andersen (1992). Recall from Chapter 5 that the idea underlying C:nutrient-based studies of nutrient cycling is that grazers limited by C (energy) will liberally recycle nutrients but when food is abundant (in terms of C or energy) but has low nutrient content, net release rates of nutrients decline because gross growth efficiencies for the nutrient then are high (Chapter 5).

These models were developed by limnologists for application to pelagic food webs. For example, Sterner's model focuses on zooplankton N:P and algal N:P to predict recycling N:P in the water column. However, there is nothing specific to plankton in these equations. We are also unaware of any differences in animal physiology between consumers in freshwater and those in (for example) soil that would preclude application of these equations in soil systems, provided that suitable information on food and consumer elemental composition were available and that appropriate time lags between consumption, recycling, reuptake, and autotroph growth were incorporated.

An Approach Based on Nitrogen and Phosphorus

Sterner (1990b) modeled zooplankton, phytoplankton, and dissolved pools of both N and P with strictly homeostatic zooplankton consumers. Fluxes between pools are ingestion (from phytoplankton to zooplankton), uptake by algae (from the dissolved pool to phytoplankton), and recycling by zooplankton (from zooplankton to the dissolved pool). Since the focus was on zooplankton CNR, only fluxes into and out of zooplankton biomass needed to be considered. The equations have the following general form:

$$\frac{dG_X}{dt} = \overbrace{eA_X\,G_X}^{A} - \overbrace{l_X\,G_X}^{B} - mG_X, \tag{6.1}$$

where G_X is the environmental concentration of a specific element (X) in zooplankton biomass (moles/L), the expression marked A is the ingestion rate for element X, the expression marked B is the release rate, and m is mortality experienced by the zooplankton (identical for all elements). Here, the volumetric feeding rate e must be in terms of $L\ mol^{-1}\ t^{-1}$ with "mol" referring to the molar quantity of the zooplankton abundance. The elements N and P were considered. The expressions for N and for P can be usefully combined under strict homeostasis, that is, when $d(G_N{:}G_P)/dt = 0$, by adjustment of assimilation efficiencies for N and for P depending on the match between food N:P and grazer N:P, as discussed in Chapter 5. The assumption of elemental homeostasis permits solution of Equation (6.1) to predict the N:P of nutrients released by the animals $(N{:}P)_r$ as a function of the N:P ratio of zooplankton $(N{:}P)_z$ and the N:P ratio of their algal food $(N{:}P)_f$. Two equations are needed, depending on the stoichiometric imbalance between consumer and food. The first equation describes what happens if P is deficient in the food:

$$\overbrace{(N{:}P)_r}^{A} = \underbrace{\dfrac{\overbrace{(N{:}P)_f - \left(GGE_{max} \cdot (N{:}P)_z\right)}^{B}}{1 - GGE_{max}}}_{C}, \qquad \text{when } \overbrace{(N{:}P)_f > (N{:}P)_z}^{D}. \tag{6.2a}$$

The parameter marked A is the N:P of consumer nutrient release. The parameter GGE_{max} is the highest realized gross growth efficiency (Fig. 5.1), which this model assumes occurs for either potentially limiting element. (The original paper described this parameter as "accumulation" efficiency because GGE would not take into account food broken up but not ingested, an important part of the mass balance in some pelagic grazers. We will overlook this bit of trivia here and be satisfied with GGE in this context.) The term marked B in the numerator says that if the growth efficiency for the limiting element is very low (approaches zero), the N:P of nutrient release approaches the N:P of the food. The term marked C in the denominator says that as growth efficiency for the limiting element P approaches 1, the N:P of nutrient release approaches infinity. The inequality marked D states the stoichiometric condition where Equation (6.2a) holds. Equation (6.2a) says that when food N:P is greater than consumer N:P, the N:P of nutrient release changes linearly with changes in food N:P. When the inequality is reversed and N is limiting, the following describes CNR:

$$(N{:}P)_r = \frac{\overbrace{(N{:}P)_f(1 - GGE_{max})}^{A}}{\underbrace{1 - \dfrac{GGE_{max}\,(N{:}P)_f}{(N{:}P)_z}}_{B}}, \qquad \text{when } (N{:}P)_f \leq (N{:}P)_z. \qquad (6.2b)$$

Here, the term marked A in the numerator says that as the gross growth efficiency approaches 1, the N:P of nutrient release approaches zero. From the term marked B it is apparent that the N:P of nutrient release is not linear with food N:P when N is deficient in the food. You can see with a little algebra that under stoichiometric balance (N:P of food and zooplankton are identical), Equations (6.2a) and (6.2b) are identical, as they need to be. Like most TER models we discussed in the last chapter, this model assumes GGE_{max} to be constant and always attained for the element in shortest supply to the consumer. These expressions formally describe the simultaneous dependence of the nutrient recycling ratio on the balance between food and grazer nutrient stoichiometric ratios and on the grazer's ability to extract and retain the limiting resource from its food. They apply for a strictly homeostatic consumer.

The predictions of Equations (6.2a) and (6.2b) regarding how recycling N:P should change with food N:P and grazer N:P are shown in Figure 6.2. To see how grazer N:P affects recycling N:P, consider the model's predictions for the two grazer species consuming the same food (i.e., at a single food N:P). At any given GGE_{max}, recycling N:P is inversely related to N:P of grazer biomass. Thus, when the zooplankton community is dominated by low N:P animals like *Daphnia*, recycling N:P should be high, but when grazers like copepods or *Bosmina* that have high body N:P dominate, re-

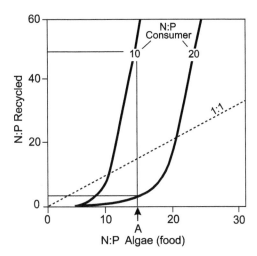

Fig. 6.2. Predicted relationships for the N:P of nutrients released by grazers as a function of algal (food) N:P and grazer N:P based on model of Sterner (1990b). Two curves for grazers with body N:P values of 10 and 20 are shown. Maximum gross growth efficiency for the limiting nutrient (GGE_{max}) was set at 0.75. Note that for any given grazer the predicted relationship between the N:P of released nutrients and food N:P is linear when food N:P is greater than body N:P [Eq. (6.2)] but curvilinear with food N:P when food N:P is less than body N:P [Eq. (6.3)]. Also note that, for any given food (such as point A on the figure), different grazers will recycle nutrients at different ratios. In the case shown, grazers of high N:P (~20, such as *Bosmina*) feeding on algae with N:P ~14 recycle nutrients at ~3 while grazers with low N:P (~10, roughly equivalent to *Daphnia*) recycle at ~50.

cycling N:P should be low. Recycling N:P is also clearly a function of food N:P for a given grazer but in a nonlinear way because $(N:P)_r$ increases curvilinearly with $(N:P)_f$ when N is deficient in the food. High grazer growth efficiency accentuates the deviation from linearity. At GGE_{max} $\ll 1$, the relationship between $(N:P)_r$ and $(N:P)_f$ is relatively straight; indeed, if the grazer accumulates none of its ingested nutrients ($GGE_{max} = 0$), recycled N:P is identical to food N:P, regardless of grazer elemental composition. Such a grazer is a "food processor" and merely converts its prey into inorganic chemical elements. The food processor model would be appropriate if growth were zero. However, the more efficiently grazers can accumulate a scarce nutrient element ($GGE_{max} \rightarrow 1$), the more strongly the $(N:P)_r$ versus $(N:P)_f$ relationship is "bent." This nonlinearity has key implications for the short-term feedbacks between the grazer and its algal food, as it means that grazers skew nutrient supply ratios away from those of the algae themselves. This skewing conceivably accentuates the elemental imbalance between algae and grazers (see Chapter 7). In other words, grazers feeding on P-limited algae will recycle nutrients at even higher N:P, potentially accentuating P limitation in surviving algae and resulting in even higher algal N:P (recall Fig. 1.5), which then feeds back again via further increases in recycling N:P. Sterner et al. (1992) concluded that such mechanisms likely generated the unexpected changes in N- and P-limited algal growth induced by food-web manipulation reported by Elser et al. (1988). Although described and parameterized for zooplankton and algae, the primary assumption in this model is strict homeostasis in the consumer and variable stoichiometry in the resources. For example, Fenchel et al. drew a very similar curve (without an accompanying model) when describing C:N stoichiometry and mineralization by microbes (their Fig. 2.1, Fenchel et al. 1998).

Approaches Based on Carbon and Phosphorus

Some of the first stoichiometrically explicit theoretical approaches to CNR came from Scandinavia via the efforts of Y. Olsen, D. Hessen, and T. Andersen (and co-workers) who used mass balance to develop a series of expressions for the rate of P release per unit biomass C ingested (Olsen and Østgaard 1985; Olsen et al. 1986; Hessen and Andersen 1992; Andersen 1997). Their model used the animal's stoichiometric demands for C and P to calculate the rate of P recycled per unit animal biomass:

$$\lambda' = \overbrace{(P{:}C)'_a f'}^{A} - \overbrace{\mu(P{:}C)'_z}^{B}. \tag{6.3}$$

Here, $(P{:}C)_a$ stands for the P:C ratio of algae and $(P{:}C)_z$ stands for the P:C ratio of zooplankton. This model expresses the rate of P mass release per unit zooplankton carbon mass (λ') as the rate of P ingestion per zooplankton carbon mass (expression marked A) minus the rate of body growth in terms of mass of P per unit zooplankton carbon mass (expression marked B). Here, $(P{:}C)'_A$ is the P:C mass ratio in algae, f' is the specific rate of food ingestion [mass of carbon ingested $\cdot$ (consumer carbon mass)$^{-1}$ $\cdot$ time^{-1}], μ is the specific growth rate of the grazer (time^{-1}), and $(P{:}C)'_z$ is the P:C of grazer biomass (mass ratio). Here again strict homeostasis is assumed in that $(P{:}C)'_z$ is given a constant value. Equation (6.3) shows that specific release of P by the grazer will decline as food P content declines or as grazer P:C increases. This same relationship can be used to define a threshold food elemental content below which animal P release is zero. That value $[(P{:}C)'_a = T')$ is found by setting Equation (6.3) equal to zero and solving.

$$T' = \frac{\mu(P{:}C)'_z}{f'}. \tag{6.4}$$

To test their model, Olsen et al. (1986) measured the rate of P released per unit of food carbon ingested and found a roughly linear relationship with food P:C with an x intercept of around 7 µg P mg^{-1} C (atomic C:P ~370). (Their data are presented along with data from other studies in Fig. 6.4 to be discussed later.) From this result, they concluded that when the food C:P exceeded 370, *Daphnia* no longer recycled P.

Olsen et al. (1986) used this finding to suggest a potential for P limitation of grazer growth under natural conditions, thus becoming perhaps the first to write about such a possibility. However, they also believed that P limitation of herbivore growth would actually be uncommon in nature because they thought a small proportion of bacteria (e.g., Hessen and Andersen 1990) or high-P algae might be sufficient to meet grazer P demand (like birds consuming small amounts of Ca-rich food to make eggshells). As we saw in the last chapter, however, P limitation of grazer growth is likely much more common than Olsen et al. conjectured and has now been demonstrated unequivocally in both the field and the laboratory. In part, this change in view is because, while Olsen et al. were aware of the importance of grazer P:C stoichiometry, they were not aware of *Daphnia*'s high P content compared to other zooplankton. The work of Olsen and colleagues was a key initial advance in understanding the stoichiometry of CNR by zooplankton.

The linearity of Olsen et al.'s model throughout the entire range of P:C in the food bears further consideration. Hessen and Andersen (1992) modified the Olsen approach, addressing whether animals completely stop P release when consuming a P-deficient diet (an aspect we discussed in relation to the assumptions of the new minimal model introduced in Chapter 5). The Olsen et al. model is based on a single animal growth rate (μ). Hessen and Andersen considered the situation where animal growth declines when food P:C decreases below the P:C of the animal's body, food quality effects we've seen happen (see, e.g., Fig. 5.10). The most recent treatment of C- and P-based equations for P release is from Andersen (1997), who noted that the model of Hessen and Andersen (1992) is one case in a continuum of assumptions regarding how animal growth and P economy respond to P-deficient food. He considered a range of potential animal responses from release of a certain percentage of body P even under P-limited growth—such as we described in our new minimal model in the last chapter—to animals that are 100% efficient in reclaiming P from assimilated materials—such as in the Olsen et al. model. Andersen chose the following expressions to describe P release per unit zooplankton biomass as a function of animal P:C and food P:C. We will concentrate on what these equations mean for nutrient release here; the dynamical consequences mediated in part by these expressions will be discussed extensively in Chapter 7. As we saw earlier, the direction of stoichiometric imbalance determines which of the two expressions holds. Let the parameter

μ_m be the specific growth rate of the consumer under good food conditions. When food P content is greater than grazer P content you get

$$\lambda' = f'(P:C)'_a - \mu_m(P:C)'_z, \qquad \text{when } (P:C)'_a > (P:C)'_z. \qquad (6.5a)$$

When the inequality is reversed, you get

$$\lambda' = f'(P:C)'_a - \frac{\overbrace{\mu_m \, (P:C)'_a}^{A}}{(P:C)'_z} (P:C)'_z$$

or, simplifying,

$$\lambda' = f'(P:C)'_a - \mu_m(P:C)'_a, \qquad \text{when } (P:C)'_a < (P:C)'_z. \qquad (6.5b)$$

Equation (6.5a) is the same expression used by Olsen over the whole stoichiometric range (with $\mu = \mu_m$). In the equation leading to Equation (6.5b), the term marked A makes grazer P demand for growth a linearly increasing function of food P content. At extreme imbalance (algal P:C ≈ 0) grazer incorporation of P into new biomass is approximately zero. Equation (6.5b) reflects the assumption that when food is P deficient relative to consumer P content, consumer growth rate declines proportionally to the imbalance in their elemental composition.

The result of these equations in comparison to the Olsen et al. model is shown in Figure 6.3 where P release rate is normalized to food ingestion rate. The Andersen model generates two line segments for a single consumer, shown with gray lines in the figure. The Olsen et al. model is shown with black lines. At high food P:C the two models are identical and predict different nutrient release for two different grazers. For a given food P:C level when P:C is high, the low-P:C consumer has higher C-specific P excretion rates than does the P-rich (high-P:C) consumer. At low food P:C, the Andersen model predicts positive nutrient release all the way down to zero food P:C. However, the Olsen et al. model has the x intercept we have already discussed. The lack of positive x intercept in the Andersen model can be seen by noting that Equation (6.5b) can equal zero only when food P:C is zero (impossible for any essential substance) or when μ_m and f' are identical (i.e., all ingested carbon goes to growth). By comparing the two panels in the figure you can see how the gross growth efficiency in terms of C (GGE_C) affects P release and, similar to the effect of GGE_{max} in the model of Sterner (1990b), accentuates the effect of grazer P content on P release. Animals with low gross growth efficiencies have high P release rates because they simply convert most of what they eat into inorganic materials and release them back to the environment.

Equation (6.5b) says that at food P:C lower than the P:C of the consumers, P release is not a function of grazer P content. Because we have assumed identical μ_m for all consumers, consumer elemental composition

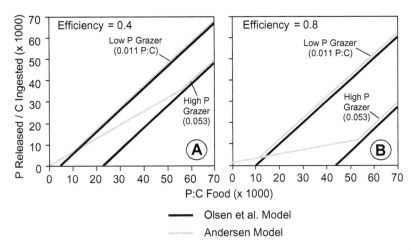

Fig. 6.3. Predicted rate of P release by a consumer vs. the P:C of ingested food from the Olsen et al. and Andersen models. Predictions for a low-P animal (body P:C of 0.011, equivalent to *Bosmina*) and a high-P animal (body P:C of 0.053, equivalent to *Daphnia*) are shown, assuming a gross growth efficiency of the consumer of (A) 0.4 and (B) 0.8. The gray lines with bends indicate the predictions of the Andersen model. The upper line segments are equivalent to the Olsen et al. model at high food P. For low-P food, the models diverge, and the Olsen et al. model is shown as dark lines. Note that this graph shows that the rate of P release per unit of ingested food in the Andersen model is a function of consumer stoichiometry but only when food P:C is greater than the P:C of the consumer with the lowest body P:C. This is because the same μ_m is assumed for the two consumers. Note also that increased grazer growth efficiency increases the predicted differences in P release between consumers.

influences the rate of P release in the Andersen model only when food P:C is higher than the lower consumer P:C. However, species-specific differences in consumer P:C do determine the food P:C for the switch from Equation (6.5a) to Equation (6.5b) and affect P release rates when algal P:C is high [Eq. (6.5a)]. Based on the growth rate hypothesis discussed in Chapter 4, we might expect a positive relationship between $(P:C)_z'$ and μ_m, which would have interesting implications for these patterns. It is important to note that Olsen's expression [Eq. (6.3)] and Andersen's expression [Eq. (6.5b)] make different predictions about CNR at very low nutrient content: Olsen's predicts zero nutrient release while Andersen's predicts reduced but continuous release of nutrient. Distinguishing which of the two expressions, Olsen's or Andersen's, is better, or indeed whether either set is satisfactory, is important because, as we saw in Chapters 3 and 4, seston P:C in lakes is frequently lower than body P:C for most grazing zooplankton in lakes. Likewise, nearly all foliage and detritus in terrestrial systems has lower nutrient:C than consumer bodies.

Overview and Comparison of Models

In this and the last chapters, we explored in detail several different mass balance models for nutrient flux in homeostatic organisms. These included three models in this chapter plus the minimal model of grazer growth in Chapter 5. We also described several other related models in less detail in Chapter 5. Although the models in Chapter 5 were discussed mainly from a standpoint of how much ingested nutrient is retained, and those in Chapter 6 were discussed from a standpoint of how much ingested nutrient is released, these clearly are two sides of the same coin. All these models predict that rates of nutrient loss are functions of nutrient intake as well as nutrient content of the consumer. Although some models were discussed in terms of N versus P balance, and some in terms of C versus P balance, none of them were based on any real differences in physiology among elements (such as the need for C for energy metabolism). In fact, we could just as well have discussed each of these models with generic labels such as X_1 and X_2 for the two nutrients. All models were based on a homeostasis of a nutrient ratio (either N:P or C:P) and adjustments of either growth efficiency or minimal loss rates in the consumer.

So why do the models seem to differ? The Sterner (1990b) model examines ratios of two nutrients released by a homeostatic consumer. No assumptions about growth rate are made (although growth efficiency is represented in the parameter GGE_{max}). In this model, the ratio of nutrient release crosses the axes only at the origin (Fig. 6.2); which means that a consumer could potentially maintain its homeostasis without ever completely turning off the release rate of any one of the elements. The other three models all make additional assumptions, and thereby make somewhat different predictions as well. The model of Olsen et al. (1986) assumes a fixed consumer growth rate under all conditions; this fixed growth rate generates a constant demand for nutrient for growth of the homeostatic consumer. With a constant nutrient drain and a variable nutrient input (ingestion), release becomes zero when drain and input are equal. The model makes nonsensical predictions for conditions where drain exceeds input (very low food P:C where the straight line predicts negative nutrient release). The Hessen and Andersen (1992) model makes a different assumption about growth; in this model, growth is assumed to be zero at zero nutrient intake, and to increase linearly with increased nutrient intake up to some maximum growth. Finally, the minimal model presented in Chapter 5 is built up from the assumption of constant specific element loss rate (≥ 0) for a limiting element. This model makes no assumptions about growth and it predicts similar broken line segment relationships between quota and nutrient release (compare Fig. 5.7B to Fig. 6.3). However, it does not necessarily go through the origin. It predicts nonzero loss at zero nutrient intake (the animal goes on a "nutrient diet" and loses weight). We can see through these differences that the mass balance physiology of consumer growth and consumer nutrient recycling are intimately related.

The different ways that consumers adjust nutrient balance, whether by controlling growth, nutrient loss, or both, create somewhat different predictions, particularly under very poor food conditions. The differences among some of these are very subtle, and may or may not be meaningful in the real world. Homeostasis coupled with different means of adjusting growth and mass loss generates some consistent patterns. Taken as a whole, these models illustrate many different ways that simple stoichiometric relationships generate physiological predictions that can impact on ecosystem dynamics.

EVIDENCE THAT CONSUMERS DIFFERENTIALLY RECYCLE NITROGEN AND PHOSPHORUS

One way to test the validity of a model is to test its assumptions. The most common assumption in stoichiometric models of CNR is that a consumer's body elemental composition is constant for a given consumer species regardless of environmental or dietary conditions. Our view is that this is often a useful assumption as long as the variability in chemical content of a resource species exceeds the variability in the consumer. To date, only Thingstad (1987) has rigorously examined the effect of relaxing this assumption (see Fig. 5.6), and his model was about bacterial growth, not CNR.

Another way to test a model is to compare its predictions to appropriate sets of observations. Elser and Urabe (1999) compiled data from studies that measured both the N and P release rates of zooplankton and the C:N:P of their food. The number of studies was not large, but the observations came from a wide range of environmental conditions, including a small eutrophic pond (Urabe 1993a), a large mesotrophic lake (Lake Biwa) (Urabe et al. 1995), and the tropical Atlantic Ocean (Le Borgne 1982). Although Sterner (1990b) originally used marine data to evaluate his stoichiometrically explicit model, algal elemental ratios in marine systems fall within a limited range around Redfield proportions (Chapters 1 and 3). Thus, values from freshwater where seston N:P is high can more effectively evaluate the effect of food quality on the N:P of nutrient release. In examining trends across all these different environments, Elser and Urabe found that, as predicted, the N:P of nutrient release by grazing zooplankton is not constant but is related to food elemental composition. With increasing food N:P, the N:P of nutrient release increased (Fig. 6.4A). Note too that, with the semilogarithmic plot, the N:P release ratio is an accelerating function of food N:P (i.e., it bends upward from left to right on linear axes). This nonlinearity in the data is due to stoichiometric homeostasis in consumers (e.g., Fig. 6.2). Models lacking homeostasis generate linear functions (Sterner 1990b). A single trend line fitted all sites well: food N:P was a good predictor of the logarithm of N:P release.

To examine the validity of the P:C-based models of Olsen et al. (1986) and Andersen (1997), Elser and Urabe (1999) plotted P released per unit of food C ingested against food P:C (Fig. 6.4B). Again, consistent with the models (cf. Fig. 6.3), the P release rate relative to C ingestion rate increased with increasing food P content. However, the variability in P release relative to C ingestion also tended to increase with increasing food P:C, with a strong suggestion of site- or study-specific variability. One possible cause of this variability is that P release rate may also reflect the contents of other elements in the food, such as N (Urabe 1993a). That is, these models are based on C:P stoichiometry only. In the case that some other aspect of the food were limiting (e.g., N), P release rates relative to C ingestion would be higher than C:P alone would lead you to expect.

Earlier we mentioned the need to evaluate the utilities of Equations (6.3) versus (6.5b) at low food P:C. A correlation line with positive x intercept ($r = 0.86$, $p < 0.001$) fits these compiled data (Fig. 6.4B), but closer scrutiny suggests that there are still questions about P release rate on food of very low P content. The experiments in Lake Biwa where P content was lowest (Urabe et al. 1995) showed detectable (albeit low) P release even when food P:C was considerably lower than grazer P:C (Fig. 6.4B), where Equation (6.3) predicts zero P release. Measurable P release on low P food suggests that the expressions of Hessen and Andersen (1992) and Andersen (1997) may in fact represent P release when food is P deficient. Furthermore, no data at all are available for extremely P-deficient food (P:C $<$ 0.002). Lakes with particulate C:P considerably higher than in Lake Biwa are common (see Fig. 3.13B). More data on P release under low food P:C (say, for less than 0.003 P:C or greater than 300:1 C:P) typical of P-limited lakes are clearly needed.

In addition to effects of food C:N:P ratios, stoichiometric models predict that the grazer's elemental composition also affects nutrient release ratios. Elser and Urabe saw no direct relationship between the N:P of nutrient release and consumer N:P when simply plotting these two variables (Fig. 6.4C). However, since the N:P of food differed considerably among sites and this parameter explained much of the variance, the influence of con-

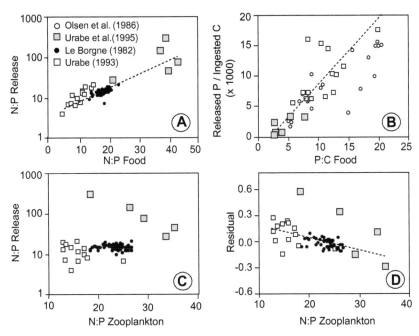

Fig. 6.4. Compilation of measurements from several studies of nutrient release by plankton testing for stoichiometric nutrient processing effects. A. Relationship between N:P release and food N:P. Consistent with stoichiometrically explicit theory (Sterner 1990b), recycling N:P increases with food N:P in a nonlinear way (note semilogarithmic plot), B. Relationship between the rate of P released by zooplankton per unit food C ingested as a function of food P:C. C. Lack of relationship between recycling N:P and zooplankton N:P. The lack of relationship likely reflects the fact that food N:P differed considerably among measurements made for zooplankton of differing N:P. D. Negative correlation between zooplankton N:P and the residuals of the relationship between release N:P and food N:P (from panel A). Thus, for a given food N:P, low-N:P zooplankton communities tend to recycle N and P at a higher ratio than do high-N:P zooplankton, consistent with stoichiometric theory. Based on Elser and Urabe (1999).

sumer elemental composition on nutrient release may have been obscured. The residuals of the relationship between N:P of nutrient release and food N:P (Fig. 6.4A) plotted against consumer N:P (Fig. 6.4D) confirm this expectation and demonstrate that consumer stoichiometry also affects nutrient cycling patterns. Although the correlation is not terribly strong ($r = -0.50$, $p < 0.001$), there is a highly significant negative relationship between the residuals and consumer N:P, indicating that, consistent with the stoichiometric theory of CNR, zooplankton communities of low N:P recy-

TABLE 6.1
Compilation of studies measuring the accumulation of dissolved nutrients or inorganic N:P in treatments in which either *Daphnia* alone or another zooplankton species or assemblage were incubated with ambient lake water. D >, =, < C indicate that accumulation of that nutrient was higher, equal, or lower in *Daphnia* treatments relative to treatments with the comparison species or assemblage

Lake	N accumulation	P accumulation[1]	N:P	Grazer(s)	Citation
Laboratory study	D < C[3]	D < C	D > C	*Eudiaptomus*	Rothaupt (1997)
Castle Lake	D > C[2]	D < C	D > C	*Diaptomus*	Brett et al. (1995)
	D > C[2]	D = C	D > C	*Holopedium*	Brett et al. (1995)
	D > C[3]	D < C	D > C	*Diaptomus*	Elser et al. (1995c)
Lake Mendota	D < C[3]	D < C	D = C	Ambient zooplankton (*Daphnia, Diacyclops*)	Moegenburg and Vanni (1991)
Lake 227	D > C[3]	D = C	D > C	*Epischura*	MacKay and Elser (1998b)

[1]Soluble reactive phosphorus.
[2]Dissolved inorganic nitrogen.
[3]Ammonium nitrogen.

cle nutrients at higher N:P ratios than communities of higher N:P. Stronger correlation of recycling N:P with food N:P than grazer N:P seems reasonable, given that elemental composition is considerably more variable (both intra- and interspecifically) for autotrophs than for Metazoa (compare Figs. 3.5A and 3.13 with Fig. 4.3).

Elser and Urabe (1999) also evaluated the possibility of differential nutrient recycling by consumers with differing body C:N:P by compiling the results of field experiments in which the net change in inorganic nutrient concentrations was assessed in containers with different zooplankton species exposed to similar food (Table 6.1). Residual concentrations of nutrients in containers incubated with living algae and grazers reflect not only nutrient release by consumers but also nutrient uptake by phytoplankton. However, since the N:P of phytoplankton nutrient uptake should be largely independent of zooplankton species, differences in the N:P of accumulated nutrients between treatments should still reflect differences in released

N:P by different zooplankton species being compared. Stoichiometric CNR models predict that inorganic nutrient pools at the end of grazing experiments will have higher N:P when a low N:P animal such as *Daphnia* is the grazer than in treatments with higher N:P species, such as calanoid copepods.

Five of six experiments were consistent with this prediction (Table 6.1) in that the N:P of inorganic nutrients at the end of grazing experiments was higher in treatments containing *Daphnia* alone relative to treatments with other grazer taxa. The exception was the study of Moegenberg and Vanni (1991) in Lake Mendota, which showed a similar ratio of inorganic N and P accumulated in *Daphnia* and non-*Daphnia* treatments. However, this result is not necessarily at odds with stoichiometric expectations because they used an ambient zooplankton assemblage that included *Daphnia* for comparison with *Daphnia*-only treatments while the four other studies used copepods or cladoceran species with high body N:P for comparison with *Daphnia*. Thus, it seems that the overall predictions of stoichiometric models of CNR are generally born out by empirical tests, at least in the planktonic realm.

MICROBIAL MINERALIZATION

Bacteria and other microbes in aquatic and terrestrial ecosystems are often very important mineralizers and we need to consider them separately. Bacteria, like metazoans, are heterotrophic consumers more homeostatic than their resource base, perhaps particularly so given the wide range in C:N:P in nonliving organic matter. Soil bacteria and their consumers, like other organisms, must meet their growth requirements for energy and nutrients while obeying matter conservation. Some excellent studies of stoichiometric balancing in soil microbial food webs (Darbyshire et al. 1994), as well as some multicurrency considerations of effects of macroconsumers in the soil (James 1991), have appeared.

Microbial mineralization is a very large topic, part of which we encountered in Chapter 5 and which we will again touch on in Chapter 8. Here, we shall consider the idea that the release of nutrients in microbial systems is often not done by bacteria (classically regarded as decomposers), but instead is performed mainly by their consumers. Recall from Chapter 5 that stoichiometry says that bacteria will remineralize organically bound nutrients when the concentrations of those nutrients in the substrates are high compared to their growth requirements. Bacteria are nutrient-rich organisms, which means we should expect that their growth would easily be limited by nutrient supply and also that organic matter produced by most other biota, especially terrestrial plants, will commonly be stoichiometrically imbalanced compared to bacterial needs. Hence, stoichiometry says that bacteria should often be nutrient sinks, not sources. Now for the flip side of that coin.

A number of studies have examined the effects of substrate C:N:P stoichiometry on rates and ratios of nutrient remineralization by bacteria (Tezuka 1985; Hessen et al. 1994). One of the best examples is that of Tezuka (1989). He found that when detritus was derived from P-limited algae and thus had high C:P and low C:N ratios, bacterial activity resulted in the release of inorganic N but not in the release of inorganic P. In other words, P-limited bacteria retained P. Conversely, when N-limited algae with high C:N and low C:P were used as a source of detrital material, bacteria remineralized P but retained N. Field studies also support the key role of stoichiometry in determining the role of bacteria in nutrient cycling in pelagic systems. Realizing that bacterial N:P reflects growth limitation patterns (Fig. 5.8) as it does for autotrophs (Chapter 3), rates of nutrient remineralization as a function of bacterial cellular stoichiometry can be instructive. For example, Elser et al. (1995a) showed that when the N:P of bacteria-sized particles (below 1.0 μm) was less than 20, bacteria released inorganic N but when N:P was greater than 20, bacteria immobilized N. Thus, pelagic bacteria are sinks or sources of N depending on their physiological condition. In the same lakes, experiments using naturally derived organic substrates from algae varying in their nutritional status showed that the C, N, and P demands of the bacteria varied considerably during the growing season, closely tracking bacterial population development during growth and stationary phases (Chrzanowski et al. 1997).

Stoichiometrically explicit models should apply to any homeostatic consumers, and as we mentioned earlier, stoichiometric principles may also be usefully applied in understanding the role of microzooplankton in nutrient recycling. Their small size and consequent high mass-specific rates of activity suggest that protists may be very important recyclers of N and P (Dolan 1997). For the stoichiometric aspects, consider the extensive studies of Goldman and colleagues (Caron et al. 1985; Goldman et al. 1985; Andersen et al. 1986; Goldman et al. 1987a), the most thorough to date in examining the stoichiometry of microconsumer nutrient recycling. They studied a system consisting of bacteria, a single species of phytoplankton (either *Phaeodactylum tricornatum* or *Dunaliella tertiolecta*), and the phagotrophic flagellate *Paraphysomonas imperforata*. The alga was grown in chemostats under differing degrees of N or P limitation (so that algal C:N or C:P increased, respectively). Cultures were then inoculated with *P. imperforata* and left in the dark in "batch" mode for 5–6 days (some were left uninoculated with *P. imperforata* to examine nutrient remineralization due to bacterial activity). Continuous monitoring examined changes in biomass of the three populations and in dissolved nutrient concentrations.

In the cultures with bacteria alone there was no appreciable accumulation of inorganic nutrients. Instead, it was growth inefficiency of the flagellate in terms of nutrients that accounted for increases in dissolved nutrients. Stoichiometry says that consumers of resources with high nutrient contents, such as Protozoa consuming bacteria, should be important remineralizers. The experiments showed that the presence of the flagellate was necessary for nutrient remineralization. The timing and efficiency of nutrient release by flagellates depended strongly on the nutrient limitation status of the alga. When the alga was N limited (with high biomass C:N), release of NH_4-N into the medium mediated by the protozoan was delayed and net regeneration efficiencies (% of ingested nutrient released to the medium) decreased from ~25% to ~8%. P release rates were relatively high and independent of N-limitation status of the alga. The converse was true when the flagellate consumed P-limited algae (high biomass C:P): release of dissolved P was nearly nondetectable but N release was high and continuous throughout the flagellate population's growth cycle. These results are clearly in line with stoichiometric theory. In fact, Goldman and colleagues attributed the differential nutrient recycling by the flagellates to their attempt to maintain a relatively constant macromolecular composition (and thus a constant C:N:P) as they grew exponentially, a scenario that conforms well to the assumptions of homeostasis and balanced growth that we have discussed throughout. The studies of Goldman and colleagues indicate that stoichiometry has a clear role in regulating nutrient recycling in the aquatic microbial food web (see also Nakano 1994a).

Thus, in aquatic systems there is a close interdependence of organic matter breakdown by microbes and the balance of multiple elements in the substrate relative to microbial demands. This association is also seen in other ecosystems. In a very broad comparative study, Enriquez et al. (1993) compiled data from 256 studies on the rate constant of litter breakdown (k) relative to the nutrient content of the litter. Litter was derived from diverse autotrophs, ranging from phytoplankton to seagrasses to conifer needles. Within any particular study, k was negatively, positively, or not correlated with litter nutrient content but across all study data combined litter decomposition was faster for nutrient-rich detritus. Decomposition rate was strongly correlated with detritus C:nutrient ratio, for both N and P (Fig. 6.5). In fact, the combined statistical contributions of detritus C:N and C:P explained ~90% of the variance in k across studies, whether the analysis was done via multiple regression or path analysis. Incorporating the lignin content of the detritus had no additional explana-

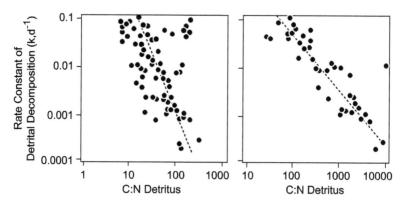

Fig. 6.5. Dependence of the rate constant (k) of detrital decomposition on detritus C:N and C:P ratios. Data points represent separate measurements of detritus breakdown and elemental composition extracted from the published literature and involve detritus derived from diverse autotrophs, from phytoplankton (generally low C:nutrient ratio) to vascular plants (high C:nutrient ratio). Multiple regression and path analysis indicated that C:N and C:P jointly explain ~90% of the variance in the rate constant of detrital breakdown. Based on Enriquez et al. (1993).

tory value. Enriquez et al. noted the importance of this association for ecosystem processes: fast-growing plants have high nutrient content and produce nutrient-rich detritus which decomposes rapidly, maintaining a high nutrient supply. Conversely, slow-growing plants produce low-nutrient litter that breaks down slowly, reinforcing slow nutrient-limited plant growth. We already mentioned such feedbacks in Chapter 3 and we will return them in the next chapter. Figure 6.5 is compelling in part because it relates stoichiometry to a major ecological process across a wide diversity of habitats. This plot illustrates well the power of stoichiometric analysis in integrating diverse ecological studies in a coordinated framework.

THE STOICHIOMETRY OF CONSUMER-DRIVEN NUTRIENT RECYCLING BY VERTEBRATES

The chemical composition of vertebrate bone was reviewed in Chapters 2 and 4. In Chapters 4 and 5 we saw how major allocation of mineral elements (Ca, P) in bone plays a major role in the nutritional physiology of vertebrate animals. This suggests that the mineral requirements of vertebrates might also affect the impact of these consumers on the cycling of nutrients in ecosystems. Here we emphasize how the C:N:P stoichiometry of vertebrates might influence nutrient cycling. Because bone is a major portion of biomass and has a very high P content, it is a major factor in considering CNR in vertebrates. As with previous material in this chapter, our emphasis is aquatic (fish are on the menu).

It's important to recognize that limnologists and fish ecologists have known for some time that fish can play an important, though variable, role in the cycling of nutrients in lakes (Kitchell et al. 1979; Northcote 1988). For example, Kitchell et al. (1975) noted that in some lakes more than 75% of water column phosphorus can be contained in fish biomass! One well-recognized aspect of fish-driven nutrient cycling is the contribution of anadromous fishes to the nutrient budget of lakes when fish return from the sea to spawn, die, and rot in their spawning lakes. Stockner (1987) reviewed studies quantifying such inputs and showed that salmon carcasses contributed between 3 and 42% of annual P loads in sockeye salmon lakes of northwestern North America. Some terrestrial equivalents may exist: for example, aeolian transport of insects to high-elevation areas has been shown to represent a biogeochemically significant source of P to nutrient-deficient alpine habitats (Halfpenny and Heffernan 1991). Fish can also be important in shorter translocations of nutrients, as when they feed in one part of the lake (e.g., the littoral zone) but excrete nutrients elsewhere (e.g., the pelagic zone). Stable isotope data indicate that, in lakes, many fishes derive most of their sustenance from benthic-based food chains (Hecky and Hesslein 1995) and thus their excretion represents a "pump" of nutrients from littoral to pelagic systems. Using a combination of empirical data and modeling, Carpenter and colleagues (Carpenter et al. 1992; Schindler et al. 1993) showed that this is an important pump in many lakes. They estimated that 70% of total consumer P excretion in the pelagic zone of a planktivore-dominated lake was from fish excreting P they obtained by feeding on littoral zone prey. In contrast, in a piscivore-dominated food web, fish excretion of littoral-derived P was only 6% of the total consumer P recycling (reflecting both an absolute decrease in fish excretion and an increase in zooplankton excretion due to higher zooplankton biomass). They concluded that this "external" subsidy amplified the effects of the trophic cascade, such that the effects of top predators (piscivores) on algal standing stocks were greatest when littoral zone feeding by fishes was greatest (Schindler et al. 1996).

These past studies emphasize the overall role of fish in the storage and cycling of particular nutrients but are not stoichiometrically explicit. That is, they do not consider how the balance of energy and multiple elements in the fish and its food affects CNR by fish. Fortunately, several recent studies have made substantive contributions in this area and help in identifying when and where stoichiometric influences will be important or unimportant in this area. We will consider three main aspects: variation among fish in body stoichiometry, effects of predator and prey elemental composition on the stoichiometry of CNR, and how fish maintain C:N:P homeostasis.

First, how much do fish differ in their body C:N:P stoichiometry? Data compiled by Sterner and George (2000) demonstrate a familiar pattern. While % P varied nearly fourfold among taxa (from 1.2 to 4.5% of dry weight), % N was much less variable proportionately, with a range from 8.2 to 12.5% of dry weight (Fig. 4.13). Similar patterns were reported by Vanni et al. (2002), a study discussed in more detail below. In Sterner and George's compilation, fish N:P varied from 13.4 to 34.7 with modest but statistically significant differences in C:N, C:P, and N:P observed among the four cyprinid taxa they studied in detail. Differences among fish taxa in C:N:P stoichiometry may relate to body size. For example, body P content increases with increasing body size in centrarchids (bluegills and large-mouth bass) while N content decreases with increasing body size in the same group (Davis and Boyd 1978). Such an increase in P content with body size might be associated with increased bone allocation due to bio-mechanical considerations (Chapter 4). Since % N and % P trend in opposite directions Vanni (1996) suggested that there should be a general allometric trend of decreasing N:P with increasing fish size. However, Sterner and George (2000) observed no (or weakly negative) correlations between body mass and % P and % N in four cyprinid species. Fish in general may not be subject to the same degree of biomechanical demands for structure as air-dwelling vertebrates, and small species such as these cyprinids may respond in their limited size range to such stresses differently from fish that achieve a larger size. In contrast to the species and size dependencies of body C:N:P in fishes identified by these studies, Tanner et al. (2000) noted only modest stoichiometric variation among 20 fish taxa from Lake Superior. In their study, the means (ranges) of C:N, C:P, and N:P were 4.7 (4.1–5.9), 48.2 (40.6–64.6), and 10.3 (8.4–12.8), respectively. For comparison with the study of Sterner and George, variation in % N was similarly modest but % P ranged only between 1.9 and 2.7%, indicating that the Lake Superior fish fauna has a limited range of P stoichiometry. However, the important overriding pattern in an ecosystem context is that fish biomass is nutrient rich when compared to zooplankton (Figs. 4.2 and 4.3) and especially to freshwater seston (Fig. 3.17). Thus, the overall picture as one ascends the pelagic food web is that trophic groups grow increasingly nutrient and especially P rich (we will consider this trend more explicitly in chapters to come).

Second, how does the elemental balance between food items and fish requirements affect the stoichiometry of CNR by fishes? This question has now been addressed with some thoroughness in a set of similar studies by Kraft (1992), Vanni (1996), and Schindler and Eby (1997). These papers share the strategy of merging existing bioenergetic models of consumption and growth (e.g., Hewitt and Johnson 1987) with information about the elemental composition of predator biomass and of prey. Another feature of these models is the assumption of set digestive assimilation efficiencies for N and P (the fraction of ingested nutrients not lost in egesta). Using this set of tools, the investigators estimated rates and ratios of N and P recycling by fishes consuming various food items. Kraft (1992) modeled yellow perch (*Perca flavescens*) and estimated that the N:P of fish excretion ranged from 29 to 104. His calculations suggest that excretion of P by yellow perch, especially by juvenile fish, is a potentially important source of P for summer phytoplankton production. He also noted that the relative importance of fish in the recycling regime is sensitive to assumptions about the P content of the fish (especially juveniles), about which little information was available.

A similar approach to nutrient excretion by fish, but focusing on the role of detritivorous gizzard shad (*Dorosoma cepedianum*) in reservoirs, involved a combination of modeling (Vanni 1996) and direct measurements (Schaus et al. 1997). The modeling studies show a familiar pattern: rates of P excretion were negatively related to the C:P of ingested detritus (Fig. 6.6B) and the N:P of excreted nutrients was positively but nonlinearly (convex upward) related to ingested N:P. In general, model estimates of nutrient recycling rates and ratios agreed well with measured values for real fish. Further, comparisons of estimated release rates of N and P by gizzard shad with other sources of nutrients in the study reservoir (Acton Lake, Ohio) showed that fish excretion was a potentially substantial source of nutrients. Summer P excretion was of a similar magnitude to P runoff from the watershed as well as to rates of P release from sediments. Finally, both calculated and measured values of the N:P of excretion indicated that gizzard shad recycle at low N:P (<20) and thus may promote the proliferation of cyanobacteria, which tend to dominate at low N:P supply ratios (Smith 1983b).

Schindler and Eby (1997) reported on a taxonomically broader study of CNR involving 18 species of fish. In estimating fish CNR, these investigators explicitly considered the normal diet items of each species as well as the relative nutrient contents of those different prey items. As with similar approaches, predicted nutrient recycling rates and ratios were sensitive to prey nutrient content, predator nutrient content, and predator growth rates and efficiencies. High N:P recycling ratios were obtained for high N:P prey items, for low N:P predators, and for high consumer growth efficiencies. Schindler and Eby's approach also permitted them to calculate a TER for C versus P limitation for each fish species. In comparing calculated TER's to the stoichiometry of dominant prey items, in only 2 of 18 cases was P limitation of fish growth suggested at feasible values of fish

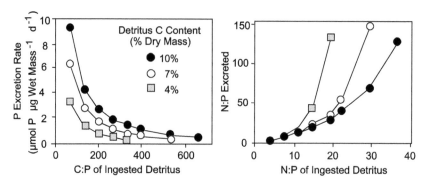

Fig. 6.6. Dependence of nutrient recycling by gizzard shad on the elemental composition of their detrital food base, as estimated by the recycling model of Vanni (1996). Predictions are given for three scenarios in which the C content of detritus differs.

growth rate (the two cases involved sockeye salmon and vendace feeding on copepods, a relatively low-P prey). However, it is important also to note that the species under study were all predatory, consuming zooplankton, insects, and/or fish, all food items with relatively high P content. The possibility of mineral limitation of algivorous fishes feeding on nutrient-limited algae seems a more likely bet in need of further investigation. The low incidence of potential P limitation also reflects the fact that, due to ecological factors generating low prey availability, most fish were growing at rates far below their physiological maxima and thus gross growth efficiencies were low, leaving little opportunity for imbalances in prey and predator N:P to result in a fractionation of N and P release. Indeed, estimates of N:P recycling ratios for piscivorous versus other species yielded only moderate differences (N:P ~14 for piscivores vs. ~29 for other species), a result that primarily reflects differences in the N:P of ingested prey. Further, although these authors recognized different stoichiometries of prey items, they assigned a single value of % N and % P to all the fish species, and the value for N:P they used was high compared to what we now know as the stoichiometric range of fish (Fig. 6.6A and Vanni et al. 2002). More refined analyses taking into account the actual nutrient demands of different fish species are needed. In addition, studies of CNR by strongly algivorous and detritivorous fish would also be of considerable interest.

Such a study has recently been completed by Vanni et al. (2002), who examined nutrient recycling by 28 species of fish and amphibians (generally algivorous and detritivorous) in species-rich tropical streams in Venezuela. Their careful and ambitious study provides striking evidence of strong stoichiometric dependence of fish CNR (Figure 6.7). Among the

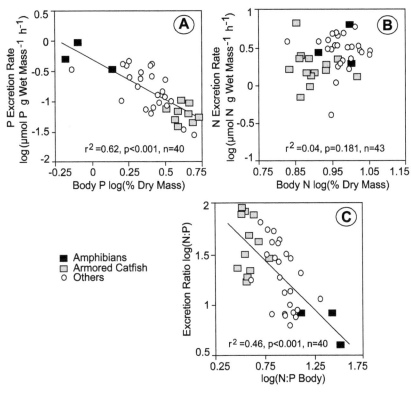

Fig. 6.7. Stoichiometric dependence of consumer-driven nutrient recycling by stream fish and amphibians in Venezuelan streams. Nutrient recycling rates were measured for freshly captured animals held under simulated field conditions. Consistent with stoichiometric theory, strong gradients in body P content are associated with variation in mass-specific P release (A). For example, P-rich armored catfish (gray symbols) exhibit low rates of P release and high N:P of recycled nutrients (C) while low-P amphibian larvae (black symbols) are characterized by high rates of P release (A) at low N:P (C). Little dependence of N release on body N content (B) is seen. The range of P content was from ~0.5% P in the amphibian larvae to ~5.5% in armored catfish while N content varied across a more modest range (7–11% N). Body N:P varied from ~25 (in the amphibians) to less than 5 (in the armored catfish). Based on Vanni et al. (2002).

taxa considered, body P content exhibited tenfold variation, from 0.5% in tadpoles to more than 5% in armored catfish, while variation in N content was more limited (~7–11%). Thus, body N:P varied from ~3.5 to 23, primarily due to differences in P content. Measured rates of P recycling declined significantly with increasing body P content, while N release was not correlated with body N content. Consistent with stoichiometrically explicit CNR theory, recycling N:P was a strong negative function of body N:P. These findings demonstrate once again a general feature of stoichiometric impacts on nutrient cycling and trophic dynamics: effects are strong when nutrient-rich consumers ingest nutrient-poor food items (such as algae and detritus, as in this study), in contrast to situations in which nutrient-rich consumers ingest nutrient-rich food (such as other animals, as in the study of Schindler and Eby). Finally, the study of Vanni and colleagues (2002) documented a strong taxonomic dependence of body stoichiometry, with major differences in P content and N:P at the level of taxonomic family (but not higher in the taxonomic scheme). Phylogenetic stoichiometric patterns have just begun to appear. Besides the fish study just mentioned, examples are zooplankton (Dobberfuhl 1999) and insects (Fagan et al. in press). These patterns are really intriguing and suggest many new questions. For example, "Is stoichiometry a slowly evolving trait associated mainly with higher taxonomic categories (as perhaps we just saw in fish)?" Better understanding of the phylogeny of stoichiometry will eventually tie directly back into the growth rate hypothesis of Chapter 4 and other evolutionary patterns such as the adaptive benefit of homeostasis.

Now we come to the third aspect relevant to fish CNR: how do fish maintain homeostasis in body C:N:P? Much less is known about the physiological mechanisms by which fish regulate intake, assimilation, and excretion of multiple substances to maintain stoichiometric homeostasis. We know of no detailed studies of physiological processing of C, N, and P by fish challenged with truly nutrient-deficient diets, as might be the case for fish consuming nutrient-limited algal biomass. However, some details of the physiological processing of multiple constituents in fish come from the field study of cyprinid fish previously described (Sterner and George 2000). The elemental composition of the contents of the foregut and hindgut were determined and compared to whole-body elemental composition. In general, gut contents were lower in nutrient content (% N and % P) than the fish themselves. When foregut % N was plotted against hindgut % N, a significant positive relationship was observed with a slope significantly lower than 1. The authors noted that this implies that animals with a high-N diet have higher N assimilation efficiency than those with a relatively low-N diet, the opposite of the negative feedback one would expect if regulation of digestive uptake was an important mechanism in C:N homeostasis. Thus, in this case homeostatic regulation must be taking place via physiological processing of nutrients that have already passed the gut wall. In any case, elemental imbalances between food and fish were weak in this study and thus fish were predicted to excrete N and P at relatively low ratio (~15), similar to the N:P of major diet items (zooplankton).

We have seen that fish CNR is potentially significant in the overall nutrient cycling regime in lakes. Its contribution will hinge on a variety of factors, including overall P availability, the relative abundance of zooplankton and other sources of internal nutrient supply, the relative abundance of fish and their overall consumption rates, and the degree of feeding on littoral zone prey items. In any case, the findings of those working on the influences of fishes on ecosystem nutrient cycling support one common theme of this book: it might be best to abandon the false dichotomy of "top down" versus "bottom up" in considering how nutrient supply and food-web dynamics interact. Instead, explicitly treating trophic interactions as biogeochemical transactions may yield the highest dividends.

CATALYSTS FOR ECOLOGICAL STOICHIOMETRY

Despite the considerable advances in understanding the stoichiometry of CNR in pelagic ecosystems that we have just described, substantive unresolved issues remain. These include the following.

- Since zooplankton release unassimilated nutrients in both soluble (excreted) and particulate (egested) forms, we need to know where homeostasis is maintained and if the stoichiometry of nutrient release differs for excretion versus egestion. If it does, does this affect nutrient cycling due to the fact that egested materials are likely to be lost from the water column by rapid sinking?
- Pelagic consumers include not only macrozooplankton and fish but also a wide array of rotifers and Protozoa. What is the nature of the stoichiometry of CNR due to the full suite of pelagic consumers in pelagic ecosystems? We have seen stoichiometric aspects of nutrient cycling of these groups individually, but to date no one has performed stoichiometrically explicit modeling of them all put together. Also lacking are comprehensive studies of stoichiometric aspects of CNR for a wide suite of consumers in a single food web.

In this chapter we presented evidence that stoichiometric mechanisms are involved in the regulation of nutrient recycling within food webs, with potentially important influences on biogeochemical cycles, trophic interactions, and community structure. We also summarized some theoretical approaches to analyzing these effects. We emphasized pelagic ecosystems but the law of conservation of matter holds for all trophic interactions and thus application of stoichiometric analysis to CNR should also be fruitful outside the pelagic zone. In particular, we wonder how such processes operate in other aquatic habitats, such as streams, and in the terrestrial realm. For application in lotic ecosystems, we are wondering about the following questions.

- Does consumer biomass represent a significant pool of nutrient elements (N, P) in streams?
- Do stream consumers differentially recycle N and P? If so, does differential nutrient recycling via the stoichiometry of CNR affect periphyton community structure?
- How does the very high C:nutrient of terrestrial detritus affect nutrient cycling in streams?
- Does the stoichiometry of CNR differentially affect the spiraling lengths (*sensu* Newbold et al. 1983) of N and P in streams?
- Does the food web, via alteration in the stoichiometry of CNR, affect nitrogen fixation or other N transformation processes in streams?

For application in terrestrial systems, including soil food webs, a similar set of questions comes to mind.

- Does differential recycling of N and P by above- or belowground herbivores affect plant growth status and perhaps the outcome of competition for various nutrients? During the growing season in soils, proliferation of low N:P detritivores or root predators might result in differential recycling of N relative to P and thus accentuate P limitation of plant growth. This could also increase nitrification or reduce soil N fixation due to enhanced NH_4 release.
- Does the structure of food webs alter the stoichiometry of CNR in terrestrial systems by altering the dominance of ecologically contrasting consumers that, as in lakes, differentially sequester or release key limiting nutrients during the growing season?
- In pelagic ecosystems, the dominant autotrophs (unicellular algae) have generation times considerably shorter than their metazoan grazers. In terrestrial systems, the dominant autotrophs (trees, shrubs, various annuals) have generation times similar to or considerably longer than their herbivores. How do these differences in time scales and time delays influence the nature of the feedbacks between CNR and autotroph dynamics?

SUMMARY AND SYNTHESIS

Awareness of the importance of consumers in regulating ecosystem processes such as nutrient cycling is on the rise. Because ecological stoichiometry expresses trophic interactions in biogeochemical terms, it seems that stoichiometric analysis will be a useful means of understanding this aspect of ecosystem function. In this chapter we synthesized recent conceptual advances applying stoichiometric thinking to nutrient cycling in pelagic food webs and suggested that it might be applied in other ecosystems. Here, stoichiometric theory focuses on both consumer and food elemental composition as critical parameters influencing rates and ratios of nutrient release by consumers, a process that generates feedbacks between the consumer and the autotroph food base.

Testing of stoichiometric theories of CNR is still somewhat preliminary, especially under field conditions. However, it seems clear that the ratios and absolute rates of nutrient release by animals have a strong stoichiometric dependence (Fig. 6.4) where variation in both food elemental composition and consumer elemental composition can have major effects. Studies have also supported the idea that relative rates of nutrient release (N vs. P) are influenced by stoichiometric balance, both via effects of food composition (Fig. 6.4A) and via consumer elemental composition (Fig. 6.4D, Table 6.1). These principles extend directly to microbial food webs and microbial processing of detritus (Fig. 6.5) as well. We will explore some of the consequences of this stoichiometry in Chapters 7 and 8 when CNR is considered in the context of larger-scale community and ecosystem studies and models.

The predictions of stoichiometric models of nutrient release are generally supported by experimental data. In the pelagic studies considered, N:P release ratios were primarily a function of algal N:P and secondarily a function of consumer N:P. Rates of P release by consumers were also strongly related to food P:C. We also considered some studies examining the role of stoichiometry in influencing how fish recycle nutrients. While the data indicate that fish are a potentially important player in nutrient cycling in aquatic ecosystems, because their growth efficiencies are low and the elemental imbalance between themselves and their prey items is generally small, differential nutrient recycling by fish is relatively weak. In addition, because fish often consume prey items of low N:P, their recycling ratios also tend to be low. However, new data for extremely P-rich algivorous and detritivorous tropical fish indicate that fish can generate very high N:P recycling ratios when they consume algae and detritus. The general features of the stoichiometry of consumer-resource interactions reflect fundamental biological processes linked to nutritional physiology of the consumer. Thus, the stoichiometric view of consumer-driven nutrient recycling should be easily transferred for use in understanding other ecosystems, including terrestrial and lotic food webs. There appear to be many avenues by which the basic components of stoichiometric analysis might begin to be assessed in such systems.

7

Stoichiometry in Communities:
Dynamics and Interactions

Prediction is very hard, especially when it's about the future.
—*Yogi Berra (1925–)*

In this chapter, we will see what stoichiometry has to say about collections of interacting species, in other words, communities. Some of the critical aspects of stoichiometry at lower levels of organization—namely, homeostatic regulation of chemical content, linkages between elemental content and the physical and chemical environment, and interspecific stoichiometric variability—will reappear. Some new patterns and feedbacks also will arise.

Our operational definition of a community will be a dynamic, biological system with a small set of interacting species or other players (some of which, like soil nutrient pools, may be abiotic). By "small set" we distinguish communities from ecosystems: in the latter, all possible interacting components, whether biotic or abiotic, are included. One of the major concerns of community ecology is to understand the dynamic behavior of systems with multiple interacting species. This can be accomplished with both theoretical and empirical approaches. (As long as we're quoting Yogi Berra, here is another one just for fun: "In theory there is no difference between theory and practice. In practice there is.") Ideas of coexistence, stability, and variability have been central to the study of community ecology. In this chapter we will see that stoichiometry influences both community dynamics and structure. We will see that homeostasis, food quality, and consumer-driven nutrient recycling generate structural changes and instability both locally (small changes near equilibrium) and globally (arbitrarily large changes). A major theme of this chapter is that stoichiometry has many consequences for the complexity of community dynamics.

Another emphasis of community ecology has been to describe species interactions in quantitative terms. Such a description has both a sign (whether one species is beneficial or detrimental to another) and a magnitude. We will see that stoichiometric mechanisms can change either of these. That is, even the qualitative nature of species interaction (whether species pairs are competitors, commensalists, or mutualists, for example) is at times under stoichiometric control. We will see this happening in some of the most abundant and biogeochemically significant organisms on Earth. Effects of stoichiometry on ecological interactions will also be manifested in food-web dynamics such as trophic cascades.

Why open this chapter with a logical conundrum from a former Yankee catcher? The role of prediction in ecology has been a subject of some epistemological discussion. A question we will consider in this chapter is, "To what extent does stoichiometry help us make ecological predictions?" One reason we want to emphasize this point here is that stoichiometric predictions are of a very particular and, we think, powerful type. They are not just extrapolations or interpolations, such as those emphasized by Peters (1991) and others. Instead, stoichiometry makes "first principle" predictions about qualitative and quantitative patterns in dynamic systems under conditions very different from those that have already been observed. An example from this chapter will be the interactions between herbivores and algae, which operate in one way at low light and in a qualitatively different way at high light. Stoichiometry correctly predicted the qualitative aspects of this novel set of observations before any experiments or observations under those conditions were undertaken. Regardless of whether you think prediction is the ultimate goal of ecology, or of science in general, successful *a priori* predictions such as these indicate a good understanding of how a system works.

This chapter begins by considering the role of nutrient content in modifying some well-known types of ecological interactions, including those between plants and mycorrhizal fungi, between algae and bacteria, between *Daphnia* and algae, and between insects and plants. In all these cases, we will see that stoichiometry influences the nature of the interaction in very powerful ways. We will then move into a detailed description of a stoichiometrically explicit model of nutrients, algae, and consumers. This model demonstrates some highly complex and seemingly esoteric dynamics. It shows, for example, chaos and multiple stable states. Nonstoichiometric models of the same system have much simpler dynamics; hence, one of the conclusions to emerge from this analysis is that stoichiometry may generate dynamical complexity. We then consider stoichiometric influences on trophic cascades (where upper trophic levels affect multiple lower trophic levels). We will see that stoichiometry can either constrain trophic cascades by diminishing the chances of success of key species, or be a critical aspect of spectacular trophic cascades with large shifts in primary producer species and major shifts in ecosystem nutrient cycling. At this point, we will consider a series of studies of light:nutrient balance that integrate a great deal of material from throughout this book. Effects of light on autotroph stoichiometry, along with effects of food quality on gross growth efficiency and population dynamics, are integrated in these experiments. Finally, we close this chapter with an example of a hypothesized major stoichiometric effect controlling community dynamics, in this case in tall-grass prairies.

SPECIES INTERACTIONS

Studies of strengths of species interactions have long been central in ecology (Connell 1961; Paine 1966). In particular, they have identified ecological interactions qualitatively by whether species have positive, negative, or neutral effects on one another and quantitatively by how strongly they influence each other. Competition, for instance, is defined as a "$-,-$" interaction of mutual inhibition. In this approach, interspecific interactions are defined by the resulting dynamics, not by the mechanisms of nutrient exchange, behavioral interference, consumption, etc. Community ecologists have also long appreciated the importance of context in determining the nature of dynamical interaction. Such phenomena include higher-order interactions in which the observed net outcome of a species interaction depends on the abundance of a third species in the community, such as if species B affects how A and C interact (Billick and Case 1994), or on other indirect effects (species A affects species C via A's effect on an intermediate species B: $A \rightarrow B \rightarrow C$) (Lawlor 1979). These ideas are engrained in today's community ecology.

Stoichiometry is a framework that can help us move these descriptive concepts into a predictive, mechanistic mode. Rather than defining species interactions by dynamics that have already occurred (a phenomenological, *a posteriori* approach), we can make predictions about dynamics from first principles, even in the absence of dynamical observations. The reason we can do this is that key trade-offs and feedbacks often involve strong stoichiometric mechanisms that affect the qualitative and quantitative nature of species interactions.

Stoichiometry and Competitive Ability

Community ecology seeks to understand the impacts of interspecific competition. One way to do this is with resource competition theory, which is a mechanistic, predictive approach (Tilman 1982). Resource competition theory has been subject to numerous successful tests with microorganisms, algae, higher plants, and even metazoans (Grover 1997). These studies have shown that the R^* concept is important in understanding the mechanisms and dynamics of resource competition. The R^* of a species is the concentration of resources necessary for growth to exactly balance a given mortality rate. At equilibrium, the species with the lowest R^* will win in competition for a single resource (Hsu et al. 1977). R^* is a measure of competitive ability; the lower the R^*, the better the competitor. As we will now see, there is a strong stoichiometric component to R^*.

Algal growth models allow us to explore the physiological basis of competitive ability. We will use again some of the growth models we described in Chapter 3. One can build upon the Droop relationship between cell quota (Q) and growth rate (μ) described there to find an expression for R^* as a function of several physiological parameters thought to be fundamental in algal ecology (Morel 1987). To do so, we first need to describe resource uptake. The Michaelis-Menten formula for uptake has the identical form to the Monod model of growth [Eq. (3.3)], but it describes uptake (v) instead of growth:

$$v = v_{\max} \overbrace{\left(\frac{R}{K_v + R}\right)}^{A},\qquad(7.1)$$

where $v_{\max}$ is the maximal uptake rate at high resource concentration, R is the resource concentration, and K_v is a half-saturation constant for uptake. The term marked A makes resource uptake zero at zero resource concentration; it makes it half of the maximal rate when resource concentration equals K_v; and it allows uptake to approach its maximal rate as R gets very large. The Michaelis-Menten and Monod models have identical mathematical form, but their parameters differ. Morel (1987) showed that the half-saturation constant for uptake is greater than the half-saturation constant for growth ($K_v > K_\mu$, where K_μ is as we defined in Chapter 3); part of the reason for this inequality is changed cell stoichiometry with growth rate.

Recall that the Droop relationship describes tight coupling between cellular nutrient quota and cellular growth rate. In its stoichiometric version [Eq. (3.4)], the Droop relationship relates biomass nutrient ratios to growth rate. The dynamics of the cell quota (Q) can be represented as

$$\frac{dQ}{dt} = \overbrace{v_{\max}\frac{R}{K_v + R}}^{A} - \overbrace{\mu'_m (Q - Q_{\min})}^{B},\qquad(7.2)$$

where the symbols Q (quota), μ'_m (growth at infinite quota), and $Q_{\min}$ (minimal cell quota) are the same as those in Chapter 3. The term marked A represents gains in quota from uptake [according to Eq. (7.1)] and the term marked B represents decreases in quota due to growth dilution of nutrients by biomass gain. By setting (7.2) equal to zero (i.e., assuming equilibrium) and substituting for Q using Equation (3.1), we can solve for R^*:

$$R^* = \frac{m\mu'_m \, Q_{\min} \, K_v}{(\mu'_m - m)v_{\max} - m\mu'_m \, Q_{\min}}, \tag{7.3}$$

where m is the mortality rate imposed on the population (Grover 1989). Although a bit of a mess, Equation (7.3) bears some scrutiny: it tells how certain physiological traits influence competitive ability. Factors that lower R^* and allow a species to be a better competitor include enhanced uptake kinetics, i.e., increased $v_{\max}$ or decreased K_v, either of which will allow a species to obtain a larger share of resources per unit time at a given resource concentration. The influence of stoichiometry on competitive ability is also apparent. Intuitively, we might expect that species with low cell quotas and thus high yield will efficiently turn resources into biomass and hence gain a competitive advantage. Equation (7.3) makes explicit how cell quota influences R^*. Decreased minimal cell quota ($Q_{\min}$) reduces R^* both by decreasing the numerator and by increasing the denominator. All else being equal, species with high yield (low nutrient content) would be expected to outcompete species with lower yield for that resource at equilibrium (Grover 1991c). In this single equation we see a major ecological influence of species stoichiometry.

For an example, recall the patterns in cell stoichiometry in a cyanobacterium and cryptomonad that we presented in Figure 3.5A. Imagine that we took a P-limited chemostat of fixed dilution rate and inflow medium containing only *Anabaena* and inoculated it with *Cryptomonas*. Theory predicts that *Cryptomonas* would outcompete *Anabaena* for P. It also says that the C:P of total phytoplankton biomass would increase, not because of a change in the dilution rate, nutrient supply, or light intensity but because of the displacement of a species with low C:P by another with high C:P. Thus, this example also illustrates the contribution of species-specific patterns in stoichiometry to community-level measures.

In spite of the importance of quota in determining R^*, this role seems not to have received as much attention as it deserves. An excellent example from the terrestrial literature is the comparative study of Wardle et al. (1998), who found that competitive ability varied inversely with plant N content. Since the elemental composition of an organism depends on its biochemical composition (Chapter 2) we can see another opportunity to connect events at the molecular or cellular scale to the level of community dynamics. For example, Lewis (1985) has speculated that haploidy in diatoms is an adaptation to increase competitive ability, because it reduces the amount of nutrients necessary to allocate to genetic material and thus could lower minimal cell quota, raise yield, and thus lower R^*.

Another place where stoichiometry influences competitive outcome is in nonequilibrium competition. As we discussed in Chapter 3, with temporal patchiness in resource concentration, the ability to acquire and store nutrients for later growth can be an important component of long-term success. We will now define a new parameter, the maximal achieved cell quota or Q_{max}, which is the cell quota at the true maximal growth rate μ_m. Nutrient storage creates a variable stoichiometry, and an index of storage ability is the ratio $Q_{max}:Q_{min}$ (see Chapter 1 for more on variable stoichiometry). Species with high $Q_{max}:Q_{min}$ ratios have been called "storage specialists" (Chapter 3; Sommer 1984). For competitors differing in Q_{max} only, Grover (1991a,b,c) has shown that competitive outcome depends on the variability of resource supply. As a competitive arena is changed from constant resources to patchy resources, the competitively dominant species change from ones with low Q_{max} to ones with high Q_{max}. Some storage competitors in fact can dominate temporally variable environments even if they have low equilibrium growth rates over all resource concentrations. In other words, even if they would lose in competition at any constant resource condition they may win in a variable world. These findings point to potential selective pressures on stoichiometric variability. It is a trait that should be under direct natural selection and might represent one of several types of potentially successful life history strategies.

We will not dwell further on considerations of resource competition, even though, as we have just seen, competition for inorganic resources can be considered a process with major stoichiometric overtones. The interested reader can consult recent reviews including Grover (1997) for a general treatment, Smith (1993a,b) for microbial examples, and Smith (1993c, 1996) for applications to host-pathogen interactions. Our attention now turns to a series of case studies on the stoichiometry of species interactions.

Mycorrhizal Fungi and Plants: Stoichiometric Boundaries for Mutualism?

Our first specific case study here is of a belowground mutualism. Mycorrhizal fungi dwell below ground in close association with plant roots. They obtain organic carbon from their plant hosts and transfer soil-derived nutrients to the plant host in a very widespread (Newman and Reddell 1987) and ancient (Pirozynski and Malloch 1975) symbiotic exchange. From the additional nutrients thus acquired, the host plant may grow faster (Jones et al. 1990). Plants generally interact with entire communities of mycorrhizal fungi across the whole root system; it is rare to have only one fungal species present (van der Heijden et al. 1998). However, we will see that many studies have adopted the simplification of a single fungal species, or one biomass pool of fungi. A mycorrhizal interaction can be viewed as the donation of C by a plant to a fungus, ultimately for the plant to obtain for itself a larger amount of C photosynthesized due to an increased acquisition of nutrient from the fungus. To aid in discussion, we will consider P to be the important soil nutrient. Other nutrients such as various forms of N, micronutrients, and in some cases even fixed carbon have also been shown to be delivered to plants via mycorrhizae (Simard et al. 1997).

This symbiotic interaction fascinates us because it has a very strong stoichiometric component involving the exchange of carbon for nutrients between two species. The benefits to host and fungus are highly dependent upon the C:P stoichiometry of the transaction. Schwartz and Hoeksema (1998) considered this symbiosis in the context of resource trading, a theoretical approach borrowed from economics. In their model, both species benefited from a trade when something they called the "C:P trade cost ratio" fell between other theoretical constructs called the "C:P isolation cost ratios" of the two species. The C:P trade cost ratio is equal to the units of C donated by the plant to obtain one unit of P from the fungus. Isolation cost ratios are hard to define concisely but they relate to the ability of a plant to harvest C relative to its ability to harvest P, as determined by both resource availability and adaptations for harvesting. For a photosynthetic green plant with leaves in the air, the C:P acquisition cost ratio is higher (C relatively cheap compared to P) than the C:P acquisition ratio of an entirely belowground mycorrhizal fungus (P relatively cheap compared to C). This model shows explicitly the stoichiometric bounds to the mutualism. If the C:P trade cost ratio is too high, the plant should not cooperate, having to give up too much C per unit P obtained in the trade. In contrast, if this ratio is too low, the fungus would have to give up too much P to obtain carbon in the trade. A wide difference in the two isolation cost ratios (e.g., by enhanced specialization of each species on its more easily obtained resources) generates a wider range of possible trade cost ratios for a mutually beneficial exchange. This interaction is beneficial when one species (the plant) is far better at obtaining C and the other (the fungus) is far better at obtaining P. Within this mutually beneficial range there is then a predicted C:P trade ratio for which it is advantageous for both species to specialize on a single resource, and trade for the other, rather than to generalize and attempt to obtain both resources on their own.

Other studies have arrived at similar conclusions from somewhat different approaches. Koide and Elliot (1989) couched their analysis of this symbiosis in terms of efficiency of acquisition and utilization of C and P, determining the circumstances where it is better for a plant to invest C in its own roots rather than in trading with fungi. They defined several efficiencies of C and P use by host plants and rewrote a fundamental C trade-off (whether plants invest in their own roots or in trading) as the product of two stoichiometric quantities, the efficiency of P utilization (increase in C

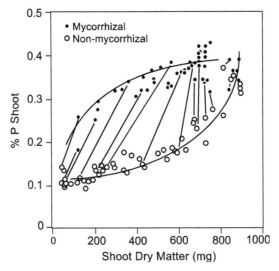

Fig. 7.1. Plant size (shoot dry matter) and P content in plants grown with and without mycorrhizal fungi, and grown on soils differing in P content. When mycorrhizal fungi (*Glomus mosseae*) are present (closed circles), leek plants (*Allium porrum*) grow larger and have higher phosphorus content, except at high fertility when plants with and without fungi are all large and have high shoot %P. Line segments connect plants grown on the same soil with and without mycorrizae. At low fertility, mycorrhizal fungi increase both shoot mass and shoot P content. Where growth is already high (high shoot dry matter), the fungus has little effect. Based on Stribley et al. (1980).

in whole plant divided by increase in P in whole plant, or the C:P of the plant) and the efficiency of P acquisition (increase in P in whole plant divided by total below-ground carbon expenditure). Jones et al. (1991) subsequently showed that willow plants with mycorrhizal symbionts had higher P acquisition efficiency than nonmycorrhizal plants (in their study, efficiency was measured as P uptake divided by C translocated below ground). It was "cheaper" to buy P from the mycorrhizae than to build the necessary roots.

There are some subtleties about using nutrients obtained via trading, however. It is interesting to note that the nutrient content of mycorrhizal plants is not identical to that of nonmycorrhizal plants of similar size. Stribley et al. (1980) showed that mycorrhizae-infected plants have three-fold higher shoot P content than plants lacking mycorrhizae (Fig. 7.1). This difference in P content indicates that the plants in this study were not as efficient at using the P obtained from mycorrhizae for growth as they were at using P obtained directly from the soil. In turn, this implies that plants participating in mycorrhizal exchanges may not obtain the full biomass growth potential out of the P they obtain in trade. The observed difference in P content between infected and noninfected plants may reflect the extra C drain of the fungi. It means that there may be a penalty involved in cooperating with mycorrhizal fungi: insect herbivores may grow better on cooperating plants than on noncooperators because of the higher nutrient content (Ayres et al. 2000). These data also show that the effect of the fungi is large at low soil fertility but essentially zero at high fertility. Earlier in the book we made the point that understanding ecological interactions from a stoichiometric standpoint is based on knowing the nutrient contents and variabilities of all the species involved. We have seen very little information on nutrient content of mycorrhizal fungi, and, given that much of the fungus often is contained within a plant's root system, it is easy to understand why. Nevertheless, if this nut could be cracked, it would be intriguing to develop a full stoichiometric picture of plant (above and below ground) and fungus (internal and external).

Factors that change the relative acquisition costs of C and P for either symbiont should change the ecological interaction between them. For example, increased pCO_2 (making C "cheaper" for the plant) has been shown to increase mycorrhizal association, although not necessarily plant P acquisition (Norby et al. 1986; O'Neill et al. 1987; Norby et al. 1996; Rillig et al. 2000). Other factors that increase photosynthetic rates, such as increased light, might be expected to do the same (Fitter 1991). Numerous studies have shown that, as expected, increasing soil fertility either decreases the extent of mycorrhizal associations (Mosse 1973; Menge et al. 1978; Thompson et al. 1986; Schwab et al. 1991) or reduces growth benefits of plants with mycorrhizae (Buwalda and Goh 1982).

In sum, this extremely important ecological interaction between plant and fungus clearly has a very strong stoichiometric aspect, the ecological and evolutionary dimensions of which are still being uncovered. It also offers us one of the few well-studied examples of stoichiometry in a mutualism.

Algae and Heterotrophic Bacteria: Competition Becomes Commensalism or Indirect Predation?

Our second example of stoichiometry in species interactions concerns algae and heterotrophic bacterial dynamics. For brevity, we will refer to the latter simply as "bacteria" in this section so that cyanobacteria do not complicate the discussion. We will see that the qualitative nature of their interaction (whether they help or harm one another) has a strong stoichiometric component. As we discussed in Chapter 6 when we discussed nutrient recycling in microbial food webs, pelagic bacteria were thought years ago to be generic remineralizers of nutrients bound in detrital organic matter. Bacteria in this role would benefit autotrophs by making nutrients more available. However, it is now understood that algae and bacteria in pelagic environments have a much more complex ecological interaction. Two recent reviews (Vadstein 2000; Cotner and Biddanda 2002) describe numerous advances made in clarifying the ecological role of microbes in the plankton. Bacteria and algae share part of their resource base and may frequently be simultaneously limited by the same resource (e.g., see Hessen et al. 1994). However, resource needs for the two differ in that bacteria but not algae have a high requirement for dissolved organic carbon (DOC). Further, algae themselves can be an important immediate source of DOC to bacteria. Hence, it is logical to deduce that that bacteria and algae may compete with each other (if both are limited by P), or algae may stimulate bacteria but may themselves not be much affected by the presence of bacteria (if algae are regulated mainly by P and bacteria regulated mainly by carbon). Because two fundamental constituents (organic C and P) are involved, there is a stoichiometric dimension to this interaction (Rothhaupt 1996).

Chemostat theory and experiments (see Chapter 3) have been applied profitably to these problems (Thingstad and Pengerud 1985). A simple model by Bratbak and Thingstad (1985) made predictions that compared favorably to results from growth experiments using combinations of algae and bacteria. The model is based on C and P dynamics. Bacteria rely entirely on organic C released by the algae. Both bacteria and algae require P from the same inorganic pool. In this model, coexistence of algae and bacteria is possible because both C and P are potential limiting factors. The authors solved the model for the relative proportion of algal and bacterial biomass as a function of dilution rate. They found that at low system dilution (i.e., slow growth, high algal C:P, and high algal exudation of organic C as a fraction of the algal C budget) bacteria should do well compared to algae. To test this, they used experimental systems consisting of three different continuous cultures: (1) algae alone; (2) algae and bacteria; and (3) bacteria alone. The alga was the diatom *Skeletonema costatum* and the bacterium was an unidentified vibrio-shaped marine strain. Cultures were continuously illuminated. The medium was nutrient-amended filtered seawater.

The results (Fig. 7.2) were consistent with the model's predictions. The presence of algae benefited bacteria, but only at low dilution rate where algal exudation of organic carbon should be great. The presence of bacteria, on the other hand, inhibited algae, particularly at low dilution where competition for P is most intense. The effects of these species on each other varied. They were largest at low dilution, where C:P stoichiometry departs from balanced, Redfield-like, proportions. Guerrini et al. (1998)

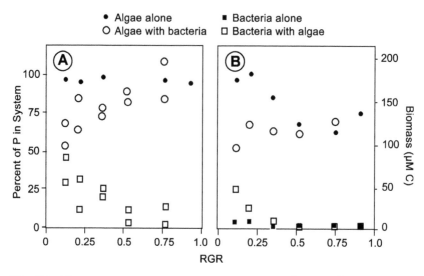

Fig. 7.2. Chemostat experiments examining interactions between bacteria and algae. A. When grown alone, algae contain all the chemostat P over a wide range of dilution rates (dark circles). In mixed culture, however, the relative dominance by algae and bacteria depends upon dilution rate. Bacteria (open squares) are relatively less important and algae (open circles) relatively more important as dilution is increased. B. In carbon terms, the biomass of algae is negatively affected by the presence of bacteria (compare open circles to closed circles) whereas bacteria are positively affected by algae (compare open to closed squares). However, at high dilution, algae and bacteria achieve similar biomass in mixed culture as when grown alone. The equilibrium effect of algae and bacteria on each other therefore has a stoichiometric component, driven by dilution rate. The interaction is strongest at low dilution rate where one typically finds large differences from Redfield ratios. Based on Bratbak and Thingstad (1985); to calculate biomass in carbon terms in B, values from the original Fig. 4 were used along with the information that reservoir P = 1 mM.

also observed a reduction in algal biomass in the presence of bacteria under conditions of low P concentration. Because chemostat experiments can be carried out over many population turnovers, these can be considered representative of equilibrium results and therefore indicate long-term dynamics analogous to "press" experiments (Bender et al. 1984). The equilibrium effects of algae and bacteria on each other therefore have a stoichiometric component, determined by dilution rate.

Another way to change the stoichiometric balance between C and P is to alter the balance of light and nutrients. At high light:nutrient ratios, algae fix (Chapter 3) and exude (Obernosterer and Herndl 1995) relatively more

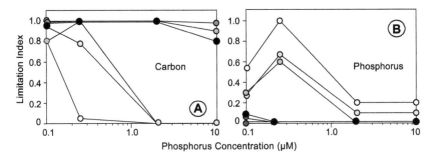

Fig. 7.3. Factors limiting bacterial growth in mixed culture with algae, and at varying light and P levels. Symbols from dark to light indicate increasing light levels. Bacteria exhibit greatest C limitation at low P concentration (A) and greatest P limitation at low P and high light (B). Based on Gurung et al. (1999).

C, thus increasing organic C flux to the bacteria. In contrast, at low light:nutrient ratios, algae have more balanced C:P metabolism. Gurung et al. (1999) cultured algae and bacteria under different combinations of light levels and total P concentrations. Following 14 days of incubation (enough for populations to stabilize), the authors measured biomass and assayed C versus P limitation in the bacteria with experiments using added glucose or phosphate. They found that algae were relatively more dominant (measured as cell numbers) at higher light levels. Increased P also favored greater algal dominance, particularly at high light. The identity of the resource limiting bacterial growth (whether organic C or P) also was very strongly dependent upon experimental conditions (Fig. 7.3). Somewhat paradoxically from a conventional standpoint, bacterial C limitation was greatest at low P. To resolve that paradox, we have to consider that low P also meant low algal biomass and thus overall low C exudation. Bacteria also were more limited by C at low light, where again total algal exudation was low. Bacteria were limited most by P at intermediate P and at high light where the stoichiometric imbalance of C:P was greatest. Gurung et al.'s results indicate clearly that light:nutrient balance can cause qualitative shifts in the ecological interaction between autotrophs and bacteria. We will consider other light:nutrient balance issues later in this chapter.

It is also interesting to note how fundamentally similar the above described dynamics are to the ones occurring on land. The interactions among plants, decomposers, and nutrient pools were summarized into a model by Harte and Kinzig (1993). The model is based on the suppositions that decomposers have superior abilities to obtain nutrients compared to plants and that they utilize nutrients in such a way as to maximize their own growth. Similar to the approach taken by Bratbak and Thingstad, Harte and Kinzig compared their model output to nitrogen fractions in plants, decomposers, and nonliving matter, finding good agreement. Again, competitive and mutualistic dynamics were possible.

These qualitative shifts in the interactions between autotrophs and microbes are interesting and approachable with stoichiometric modeling. They are also vitally important on land and in water, and they account for large fractions of the mass flow in ecosystems.

Daphnia and Algae: Intraspecific Competition Becomes Facilitation?

Ecological systems in general tend to be dominated by negative, rather than positive, intraspecific feedbacks. Positive feedbacks (higher population densities cause higher per capita rates of change) are destabilizing (DeAngelis et al. 1986). Occurrence of positive feedbacks has sometimes been a perplexing topic in population biology, as in early models of mutualisms (May 1976). At the community level, though, "indirect mutualisms"— where the enemy of my enemy becomes my friend—have been described (Vandermeer 1980) and may be quite common. We will now see a community-level change in dynamic outcome where normal (negative) density dependence becomes a positive feedback. Stoichiometry thus impinges on the fundamental mechanisms of population regulation.

In Chapter 5, we considered nutrient-limited herbivores and in Chapter 6 we discussed some nutrient recycling feedbacks. When assembling these ecological mechanisms into communities, some surprising things happen. For example, consider the potential dynamics occurring with a consumer consuming autotroph biomass of high C:P. With increased consumption, such as with an increase in herbivore biomass, autotroph biomass will decline. But this might also bring about an increase in the quality of autotroph biomass because a lower biomass of autotroph should result in a greater growth rate of the smaller remaining biomass due to simple density dependence. Urabe (1995), for example, observed that algal P content increased with herbivore density in a small, eutrophic pond. Recall that, in algae as in most autotrophs, higher growth is coupled to higher nutrient content [Eq. (3.1)]. A positive feedback loop may therefore result: increased herbivores lower autotroph biomass, increasing autotroph food quality, and thereby stimulating herbivore growth. This putative positive feedback loop should only be observed under conditions of poor food quality where herbivores are able to have a substantial impact on autotroph biomass and where autotroph biomass in the form of C is inhibitory rather than stimulatory to herbivore populations. Note that this feedback establishes the possibility of inverse density dependence, where more consumers are beneficial to the growth and reproduction of individual consumers.

Experimental evidence supporting such unconventional dynamics is seen in the study of Sommer (1992), where he cultured algae and *Daphnia* together in large (10 L) chemostats, altering the system dilution rate. Experiments lasted 60–80 d. At high dilution rate, algal biomass was low and their quality as food was high. *Daphnia* became abundant under those conditions. These were high-grazing, low-algae systems. In contrast, at low dilution rate algal biomass was high, their quality as food was low, and *Daphnia* remained rare. These were low-grazing, high-algae systems. When Sommer examined how herbivore vital rates related to herbivore density, he found a very strong positive relationship of herbivore density with herbivore birth rates and a strong negative relationship between density and death rates. In other words, the *Daphnia* exhibited the reverse of normal negative feedback density dependence. These trends were largest at low herbivore density. This positive feedback was a community-level "Allee effect," driven by stoichiometry. Allee effects (positive response of a population's per capita growth to increases in population density) are known to be destabilizing to population growth. Sommer also pointed out how extremely sensitive this simple community was to flow rate: a difference in flow of only $0.05\ d^{-1}$ divided communities that achieved a high-algae but low-zooplankton configuration from ones that achieved a low-algae, high-zooplankton configuration. There was a sharp threshold in dilution rates that separated two very distinct long-term outcomes.

Observations consistent with Sommer's have more recently come from Nelson et al. (2001), who maintained closed communities with *Daphnia* and edible algae for approximately 120 d. Their microcosms were 80 L in volume, and they contained either single clones of *D. pulex* (three replicates of three clones) or mixtures of all three clones (three replicate microcosms). Little stoichiometric data were taken; however, detailed demographic analysis gave clear evidence for two different system states. One state consisted of high herbivore and low algal biomass, and the other consisted of low herbivore and high algal biomass. Based primarily on demographic evidence, the low-herbivore state was suggested to be due to P limitation of the herbivores. For somewhat mysterious reasons, the state with low herbivores and high algae was observed primarily in the presence of multiple *Daphnia* clones. New experiments manipulating algal C:P by altering light intensity have also found strong signs of positive density dependence in *Daphnia*, mediated by nutrient recycling feedbacks on stoichiometric food quality (Urabe et al. 2002a).

In these studies, stoichiometric food-web feedbacks clearly shifted the ecological relationship between negative and positive density dependence in the herbivores, a switch between intraspecific inhibition (normal density dependence) and facilitation (positive density dependence). Very little attention in general food-web theory has been paid to the possibility that under poor food quality herbivore individuals do not compete with each other but instead may facilitate each other's success. Such facilitation may occur in intraspecific interactions (as seen in Sommer's study), but there is no reason that interspecific facilitation also should not occur under poor-food-quality conditions. Indeed, the study of Urabe et al. (2002a) demonstrates just such an effect for two species of *Daphnia* in their study. We note that other, nonstoichiometric, forms of facilitation by herbivores have been described, as where grazing by hares maintains dominance in the plant community by nonshrubby species that are preferred food for geese (Van der Wal et al. 2000). In addition, dynamical consequences of Allee effects, even in nonsocial animals, are poorly studied in community models. These effects bear further consideration.

Insects and Plants: Varying Predatory Effects?

Another stoichiometric influence at the community level is in tritrophic interactions, namely, plants being consumed by herbivores, which themselves are consumed by predators. Price et al. (1980) listed a number of ways predators influence the interaction between plants and herbivores. One, reduced plant nutritional quality, can result in prolonged development, increased feeding rates, or both. Either of these may influence the susceptibility of herbivores to attack by predators by changing the overlap in time and space between predator and prey. The dynamic effect of predators on prey depends in part on whether predation is met by "compensatory" responses of the prey (i.e., increases in reproduction as a result of increases in mortality; Sinclair and Pech 1996). In a recent example (Oedekoven and Joern 2000), the impact of predatory spiders on grasshoppers was studied in control and fertilized vegetation. In unfertilized vegetation, spiders did not affect grasshopper population size. However, in fertilized vegetation, spiders reduced grasshopper populations. Explaining this difference required invoking stoichiometric food quality effects. In the unfertilized vegetation (poor food quality), spider predation was met by compensatory growth by the grasshoppers. Grasshopper loss to spiders enhanced the survivorship of the grasshoppers not preyed upon. In contrast, grasshopper reproduction was high with and without spiders in the fertilized vegetation. Because of the different dynamics in fertilized and unfertilized vegetation, the authors conjectured a stoichiometric dimension to whether spiders influenced grasshopper population densities.

Oedekoven and Joern suggested that under poor food quality, grasshopper survivorship is determined mainly by the total N amounts present in the vegetation. Under unfertilized conditions, limited N meant that removal of some grasshoppers by spiders gave greater access to limiting N by the remaining grasshoppers. In this scenario, the N in vegetation controlled the number of grasshoppers; the effect of spider predation was merely to change which individuals in the population received adequate N. Note that this would be another, but weaker, form of herbivore facilitation as described above for *Daphnia* and algae. In the grasshopper case, rather than a net benefit, there was compensation eliminating the expected intraspecific detrimental effects (turning negative feedbacks to neutral). In contrast, under fertilized conditions, spider effects and overall survivorship were additive and conformed to conventional predator-prey dynamics. Grasshoppers had surplus N and so having spiders alter the number of grasshoppers removing N did not change the success of the remaining grasshoppers. Under one stoichiometric condition, spiders had a notable effect on grasshoppers. Under another, the population-level effect was immeasurable even though the predatory spiders were actively consuming grasshoppers. Ritchie and Olff (1999) also cited a number of very similar results to these in other terrestrial herbivory studies, and they emphasize that the strength of herbivory as an ecological force depends on the identity of the resource limiting plant growth, i.e., it has a stoichiometric component.

In sum, these theoretical and empirical studies on stoichiometry and species interactions demonstrate that qualitative and quantitative outcomes are controlled by stoichiometric influences.

POSITIVE FEEDBACKS AND MULTIPLE STABLE STATES

The dynamic behavior of continuous time models of more than two species (here "species" means a functional component, and may refer to an inorganic substance such as nutrients) can show many, sometimes complex, behaviors including multiple equilibria, stable limit cycles, or chaos (Yodzis 1989). We have already described positive feedbacks from a mechanistic standpoint; their existence suggests that some of these complex dynamics might occur in stoichiometrically explicit models. Therefore, seeing these specific empirical examples of community-level changes in interaction sign and strength brings us to a more general question of community dynamics, "Does stoichiometry lead to complex dynamics including multiple stable states?" The term "multiple stable states" refers to dynamic systems possessing more than one locally stable equilibrium point (we will not consider nonpoint attractors in any formal way). With multiple stable states, whether a two-species system ends up at attracting equilibrium point x_1, y_1 or x_2, y_2 depends on initial conditions. The "basin of attraction" for either point is that portion of the x, y plane where initial conditions within that basin lead the system to that point.

The existence of multiple stable states in ecological systems has been controversial (e.g., Grover and Lawton 1994). On the one hand, rigorous empirical documentation of systems with multiple stable states may be difficult (Sutherland 1974; Connell 1983), even though some diverse putative examples have been put forward (Gunderson 2000). On the other hand, nonlinear dynamic systems (which seems a reasonable characterization of ecological communities) are often multistable (i.e., possessing multiple equilibria). Therefore there seems to be a paradox: theory suggests that multistability should be the rule, but empirical observation has not yet borne that out in a rigorous way. Reconciling theory and observation in nonlinear systems, particularly with contributions of stochastic noise, is not easy (Kraut et al. 1999).

At several points in this book we have encountered a situation with different amounts of stoichiometric variability in resource and consumer. In virtually all such cases, the consumer has a greater degree of homeostatic regulation of stoichiometry and thus lowers stoichiometric variability in the higher trophic level. We saw this, for example, for autotrophs consuming inorganic nutrients: nutrient ratios in inorganic pools can vary essentially limitlessly (Chapter 1) but variation is more attenuated in the autotrophs (Chapter 3). We also saw this for herbivores consuming autotrophs, as in Chapter 5 and in the above discussion about *Daphnia* and algae. In this section we will explore the general idea that when there exist tight recycling loops and conservation of nutrients (recall Chapter 6), continual production of biomass of a constrained element ratio may generate positive feedbacks and multiple stable states.

Wedin and Tilman (1990) performed an experiment where five species of annual grasses were planted in monoculture on several soil types. After three years of growth, large differences in soil N mineralization had developed in response to the species of plant on the plot. Grass species with low C:N (in both above- and belowground biomass; the two were closely correlated) produced relatively small amounts of litter with rates of soil N mineralization much higher than in plots dominated by grass species with high C:N (Fig. 7.4). Differences in N mineralization across plant species in the fertile soils were particularly pronounced: a range of about 12-fold was observed. These results illustrate the importance of species identity in ecosystem processes, and they also indicate that stoichiometric relationships may be key in explaining those differences. In related experiments (Tilman and Wedin 1991), it was further observed that grass species with high C:N competitively displaced species with low C:N ratio in soils of low fertility (consistent with patterns predicted by resource competition theory described in the section "Stoichiometry and competitive ability," above). Hence, the good N competitors promoted their own competitive superiority, a form of positive feedback mediated by stoichiometric mechanisms.

Hobbie (1992) described a likely set of related positive feedbacks involving plant growth rate; we touched on these ideas also in Chapters 3 and 6.

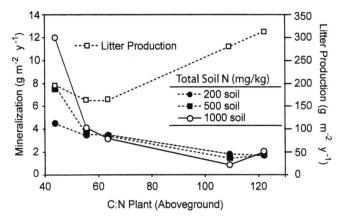

Fig. 7.4. Effect of differing plant C:N on soil N mineralization and litter production. Five grass species (plotted based on their low to high C:N ratios: *Agrostris scabra, Agropyron repens, Poa pratensis, Schizachyrium scoparium,* and *Andropogon girardi*) were planted in monoculture in several soil types (different in fertility, see legend). After three years of growth, the rate of litter production and the rate of N mineralization are strongly dependent on plant species, with plant C:N being a good predictor variable. Plants with low C:N produce less litter but are associated with higher N mineralization rates than plants with high C:N. Belowground C:N is ranked in similar order as aboveground C:N. Based on Wedin and Tilman (1990).

Plant species characteristic of nutrient-poor ecosystems exhibit slow growth, use nutrients efficiently (i.e., have high C:nutrient ratios), and, as we saw above, produce poor-quality litter that decomposes slowly. High C:nutrient ratios of plants and litter also mean that herbivores and detritivores should grow slowly and recycle nutrients sparingly. In contrast, plant species from nutrient-rich habitats grow rapidly, produce readily degradable litter, and sustain high rates of herbivory by consumers with high rates of nutrient excretion. These differences are evident in very broad-scale comparisons, such as between low-nutrient conditions typical of tundra or boreal forest and high-nutrient environments such as old fields or some grasslands. Hobbie stated this positive feedback as follows: "in general, plant species reinforce patterns of nutrient availability in natural ecosystems through their uptake and use of nutrients." Similarly, Shaver and Aber (1996) considered the literature on plant allocation of C and N, and wrote, "Since 1972, Mooney and colleagues have published a set of papers linking several previously distinct physiological, morphological and life-history characteristics into a unified theory based on the fitness of plants in sites of different resource 'richness.' . . . This theory can be represented as a circular set of interactions resulting in positive feedbacks between site quality, plant response, and future site quality." The nutrient cycling model of Sterner (1990b) discussed in Chapter 6 also exhibits positive feedback. By drawing off nutrients at a constant N:P from the resource pool through their growth, homeostatic consumers make a resource pool of moderately high N:P even higher (by retaining P and regenerating N) and, vice versa, make a resource pool of moderately low N:P even lower (by retaining N and regenerating P).

But before we jump to any conclusions, we need to note that the studies described in this section so far have not explored long-term stability. Short-term dynamics, as emphasized in these studies, often will not accurately predict long-term or equilibrium dynamics. (Whether natural systems ever reach these long-term equilibria, or are continuously buffeted by changing short-term forces, is another matter entirely.) Theoretical approaches contrast local stability, which is the tendency for a dynamic system to return to an equilibrium point after infinitely small disturbances, with global stability, which is the tendency of a dynamic system to return to equilibrium after arbitrarily large disturbances (e.g., Case 2000). This distinction sometimes makes a big difference to conclusions. For example, de Mazancourt and Loreau (2000) explored grazer optimization of primary production in models with a single limiting nutrient. Their results indicate that grazers at equilibrium may increase primary production in systems with nutrient recycling under specific conditions of (1) mass channeling through the herbivore pathway compared to other pathways in the ecosystem, and (2) nutrient loading. It was also shown (de Mazancourt et al. 1999) that effects of grazers on autotrophs can be very different at short and long time scales. In transient, short-term dynamics, simply increasing nutrient turnover rates can increase primary production. However, long-term grazer optimization requires these more specialized conditions about channeling and loading. Over the short term, physiological responses of autotrophs to changed nutrient regimes may be very predictable and cause a certain response of herbivores. In the long run, however, aspects such as retention of nutrients within the system, changed species abundance patterns, and other slow-cycling phenomena may become important. Short-term positive feedbacks do not establish the existence of the same over the long term or at equilibrium.

Tateno and Chapin (1997) constructed a model also showing that one set of anticipated positive feedbacks as discussed above did not occur in an equilibrium analysis, even though those dynamics between plants, litter, and N mineralization did occur in the initial transient dynamics. As empirical support for their conclusions, Tateno and Chapin pointed to an almost linear relationship between plant biomass and soil N. A positive feedback system, they argued, should demonstrate a curve with increasing slope as soil N increases (we will see in the next chapter that the usual trend is a decreasing slope with increasing fertility). Although the Tateno and Chapin model is lacking in certain features that may influence the presence of positive feedbacks (for example, making herbivory a function of plant nutrient content, or making plant growth rate depend on plant nutrient content), their work reminds us that both local and global dynamics need to be explored in order to gain a good appreciation of behavior. There has been much more discussion about short-term than about long-term feedbacks in models such as these.

Multiple Stable States in Grazer-Algae-Nutrient Systems

Positive feedbacks leading to multiple stable states in long-term equilibrium solutions have in fact been thoroughly examined in stoichiometric models of grazers, algae, and nutrients. Andersen (1997) modeled a community composed of one species of grazer, one algal species, and one nutrient within biomass pools indexed by C. Others have examined similar models. Loladze et al. (2000) presented a bifurcation analysis and numerical simulations of a similar model and Hessen and Bjerking (1997) also performed a simulation analysis of another. This stoichiometric system (algae, homeostatic zooplankton, nutrients) is perhaps unique in the degree to which it has been examined theoretically, with a complete analysis of equilibrium points, domains of attraction, and stability. The more mathematically inclined reader will want to read these sources in depth; here we will describe Andersen's model in enough detail to gain some insight into its solutions. We will assume the reader already has some acquaintance with isocline analysis of grazers and resources. Isoclines are sets of points of equal z value plotted in an x-y plane. In population models, nullclines on the x-y plane represent all combinations of x and y (e.g., *Daphnia* and algae) where growth of one or the other equals zero. For an introduction to this topic, see Case (2000). We already presented some of this model's components in the last chapter.

Before looking at the whole model, there are three main functional forms we must first describe. First, algae (A) grow according to Droop kinetics (see Chapter 3):

$$\mu_A = \mu_m' \left(1 - \frac{(\text{P:C})'_{A,\min}}{(\text{P:C})'_A} \right). \tag{7.4}$$

Here, μ_A is algal growth rate, μ_m' is the theoretical growth rate at infinite quota, $(\text{P:C})'_{A,\min}$ is algal P:C mass ratio at zero growth rate, and $(\text{P:C})'_A$ is algal P:C mass ratio at $\mu \geq 0$. Equation (7.4) directly compares to the stoichiometric version of the Droop equation [see Eq. (3.4)]. Second, the functional response of the grazer is "rectilinear" (McMahon and Rigler 1965), otherwise known as a Holling Type 1 function:

$$f' = f'_{\max} \overbrace{\mathrm{Min}\left(1, \frac{A_C}{F}\right)}^{A} \tag{7.5}$$

where f' is the ingestion rate, $f'_{\max}$ is a maximal ingestion rate, A_C is algal carbon biomass, and F is the incipient limiting concentration (the food concentration where feeding satiates). The minimum function labeled A makes ingestion maximal when $A_C > F$ and makes it a decreasing function of A_C when food levels are low ($A_C < F$). Third, as discussed extensively in Chapter 5, grazer growth (μ_G) may be limited by algal C (food abundance) or P (food quality):

$$\mu_G = \overbrace{(f' S_{C,\max} - r)}^{A} \overbrace{\mathrm{Min}\left(1, \frac{A_P}{(\mathrm{P{:}C})'_G A_C}\right)}^{B}, \tag{7.6}$$

where $S_{C,\max}$ is the maximum assimilation efficiency for carbon, r is the respiration rate, A_P is the concentration of algal-bound phosphorus, A_C is the concentration of algal carbon, and $(\mathrm{P{:}C})'_G$ is the grazer P:C. The term marked A gives the carbon balance of assimilation minus respiration. The minimum function labeled B makes grazer growth either strictly a function of carbon ingested if algal P:C is greater than grazer P:C, or gives it a linear dependence on algal P:C if the algal P content is lower than that in the grazer's body.

The three main governing equations then give the rates of change of algal biomass (A_C), zooplankton biomass (G_C) and phosphorus (P). Both algal and zooplankton populations are given in terms of their respective C mass. Algal biomass (A_C) dynamics are governed by the following equation:

$$\frac{dA_C}{dt} = \overbrace{A_C \mu_A}^{A} - \overbrace{A_C \sigma}^{B} - \overbrace{A_C m}^{C} - \overbrace{f' G_C}^{D}, \tag{7.7}$$

where μ_A is algal growth given by the Droop equation [Eq. (7.4)], σ is algal sinking loss rate, m is mortality due to hydraulic flushing, and f' is ingestion [Eq. (7.5)]. The term marked A gives the gains to algal carbon in the absence of losses, and the remaining terms give rates of carbon loss due to sinking (B), washout (C), and grazing (D). Zooplankton (G_C) dynamics are given by

$$\frac{dG_C}{dt} = \overbrace{G_C \mu_G}^{A} - \overbrace{G_C m_G}^{B} - \overbrace{G_C m}^{C}, \tag{7.8}$$

where μ_G is zooplankton growth [Eq. (7.6)], m_G is mortality loss specific to grazers, and m is the hydrologic loss (the same as for algae). The term A represents zooplankton growth (production minus respiration), which is either C or P limited, and the other terms represent zooplankton losses due to mortality (B) and washout (C). Finally, the model uses the following equation for the dynamics of phosphorus. Here, the variable A_P (algal P) is best interpreted as the sum of algal and inorganic P:

$$\frac{dA_P}{dt} = \overbrace{mP_L}^{A} - \overbrace{A_P\sigma}^{B} - \overbrace{A_Pm}^{C} - \overbrace{\mu_G(P:C)'_G\,G_C}^{D}, \qquad (7.9)$$

where m is the dilution (and hence mortality) rate, P_L is the concentration of phosphorus in loading, σ is the algal sinking loss rate, μ_G is grazer growth [Eq. (7.6)], $(P:C)'_G$ is the grazer P:C mass ratio, and G_C is grazer carbon biomass. The term marked A gives the rate of P input ("loading"). The term marked B gives the rate of loss of P due to sinking, and the term marked C gives the rate of loss of P due to washout. The last term, marked D, gives the net flux of phosphorus into zooplankton biomass. Phosphorus ingested but not incorporated into grazer biomass is not lost from this combined pool of algal plus inorganic P; hence, recycling is implicit. The fact that the model follows a single P pool (algae plus inorganic) makes it more analytically tractable, reducing the number of equations by one. The effect of this assumption is that P enters the algae directly without first occupying an inorganic pool. Equations (7.4) and (7.9) are consistent if phosphorus in algae is much larger than dissolved phosphorus (not an unrealistic assumption relative to the situations in most pelagic ecosystems). Where dissolved P affects dynamics, this simplifying assumption could create problems.

Figure 7.5 shows how equilibrium biomass levels of algae and zooplankton respond to input P concentration for one set of parameters in this model. Beginning at zero input concentration where neither algae nor zooplankton have positive equilibrium, there is a small range of P input concentrations ($P_I < P'_L$) where increased loading goes entirely into supporting a linear increase in algal biomass from a zero level. Zooplankton cannot be supported at these low algal concentrations, and in this range this is a mixed-reactor, algae-nutrient system. At a loading concentration of P'_L, algal density becomes sufficiently high that grazers can enter the system. Over the range of loading from P'_L to P''_L, grazers are linearly increasing with loading while algae, surprisingly, are linearly decreasing despite the added nutrients. Within this range, zooplankton biomass is determined by both C and P. Also, increased loading results in increased algal growth rate and (linked via the Droop equation) increased algal P content; zooplankton biomass responds positively. Finally, when the loading concentration is

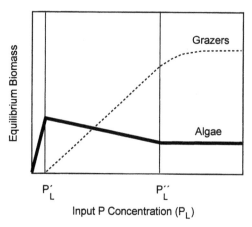

Fig. 7.5. Trophic-level response to P loading in Andersen's stoichiometric model of nutrients, algae, and grazers. Parameters correspond to Table 7.1.

greater than P''_L, zooplankton biomass smoothly approaches its asymptote (G'_C) and algal biomass is constant. In this highest range of loading, zooplankton are limited by food carbon concentration (A_C). There are some interesting subtleties at work, with increased loading generating higher zooplankton with constant algal biomass in this highest range of loading. The way this works, as in more conventional trophic models, is that algal production increases in this range even though algal biomass is fixed. The increased primary production at constant autotroph biomass can support greater heterotroph biomass. Thus, in a system with this behavior, increased loading results in increased algal P but not increased algal C. Hence, zooplankton biomass and algal P would be found to be positively correlated, and they are related in a certain sense (coupled through C), even if instantaneous zooplankton growth does not respond to increased algal P (zooplankton are instantaneously limited by C).

Figure 7.5 superficially resembles other, nonstoichiometric, food-chain models (e.g., Persson et al. 1996; Oksanen and Oksanen 2000). The similarity is that both types of model predict different functional relationships between grazers and resources in different ranges of loading (or productivity). The stoichiometric model is perhaps unique, though, in the prediction of decreased algal biomass with increased loading over the intermediate range of loading. Another difference between the stoichiometric model and others lies in the reasons why these functional relationships occur. In nonstoichiometric models such as those of Oksanen and Persson, changes in functional form occur when entire trophic levels are added to or subtracted from the community. In the stoichiometric model, shifts in functional relationships also occur as qualitative shifts in limiting resources occur.

Insights into the model's solutions are gained by studying its nullclines. Luckily for graphical analysis, the three-dimensional system of phosphorus, algae, and zooplankton can be projected into a two-dimensional system in the A_C-Z_C plane because any relevant equilibria will be located in that plane (Andersen 1997). This projection allows us to analyze this three-species model by consideration of only two nullclines, one for the algae (Fig. 7.6A) and one for the zooplankton (Fig. 7.6B). The algal nullcline has a "humped" shape similar to results of several other consumer-resource models (Case 2000) (Fig. 7.6A). Algae increase in concentration at all points between the nullcline and the A_C axis, and decrease outside of this curve. The piecewise functional response [Eq. (7.2)] causes a discontinuity at moderate and high loading at the point $A_C = C'$. At this particular point, increased phytoplankton biomass requires a higher zooplankton biomass to hold it at equilibrium due to the fact that the specific grazing rate drops with increased phytoplankton biomass once feeding saturation has been reached. Dynamical properties of nonlinear prey nullclines with maxima at intermediate prey densities have been thoroughly studied, beginning with Rosenzweig's (1969) seminal paper on this subject. At high algal biomass, as algal biomass increases further, fewer grazers are required to crop algal production because of the "normal" density dependence in the algae, lowering their growth rates due to P limitation as they use up P in the system.

The grazer nullcline in Andersen's model contains a new feature, stoichiometric in origin, which to our knowledge had never before been seen in predator-prey models. Like the algal nullcline, the grazer nullcline (Fig. 7.6B) also is concave and intercepts the x axis twice. Grazers increase in concentration at all points between the nullcline and the A_C axis and decrease "outside" this curve. In most classical predator-prey models, the predator nullcline is either vertical (no explicit density dependence or limiting factors other than prey density on predator growth), or a sloping relationship pointing upward and to the right (additional grazer density dependence, which might come from an intraspecific interference or competition term or from a carrying capacity representing other limiting factors), or it slopes upward and to the left (inverse density dependence). At low algal abundance in this model, there is a section of the nullcline that is vertical, and changes in algal density across that portion of the nullcline correspond to the classic situation of relief of food limitation in the grazer with no additional density considerations. The behavior of the model at high algal density requires some additional explanation.

The fact that the grazer nullcline in the Andersen model exhibits a hump and crosses the x axis twice is unusual and extremely interesting. This intersection of the nullcline with the C axis at high C implies that

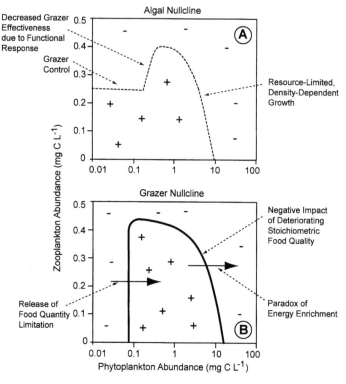

Fig. 7.6. Example nullcline results of the Andersen model. Nullclines (heavy lines) divide regions in space where a species increases (pluses) or decreases (minuses) in abundance. In the classical Lotka-Volterra predator-prey models, nullclines are horizontal and vertical lines (e.g., Case 2000). Regions similar to the classic Lotka-Volterra model can be observed in this more complex stoichiometric model. However, departures from the classic model are due to differing biological assumptions of the model. A. The algal nullcline for these parameters (Table 7.1) has a sharp bend and a hump. The bend occurs at the incipient limiting concentration C'. Humps in "prey curves" such as is shown in A are present in other models for similar reasons as they occur here (Rosenzweig 1969), and they tend to destabilize the dynamics. B. The zooplankton nullcline is also humped, but humped consumer nullclines have seldom been studied before. Zooplankton "starve" in all parts of the region marked with minus signs. On the left, this is because of energy limitation. On the right, they starve for nutrients because of high levels of bulk food with poor food quality. The hump in this nullcline is the direct result of a stoichiometric growth penalty: at high algal abundance (measured in carbon units), further increases in algal carbon negatively affect zooplankton growth. Loladze et al. (2000) termed this the "paradox of energy enrichment." Empirical observations consistent with this apparent paradox are discussed elsewhere in this chapter.

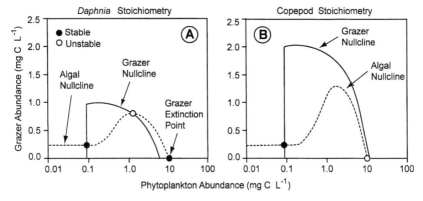

Fig. 7.7. Some of the community equilibria in the Andersen stoichiometry model, contrasting a low-C:P grazer (e.g., *Daphnia*) (A) with a high-C:P grazer (e.g., copepods) (B). For calculation of nullclines, only the grazer C:P was changed. Solid circles represent stable equilibria and open circles represent unstable equilibria. Multiple stable equilibria occur in the model with the low-C:P grazer.

there exists a point where positive grazer growth can be changed to negative grazer growth via increases in algal density! Previously, such dynamics have been surmised for prey exhibiting "herding" behavior, where, over some range of population density, prey enjoy increased predator protection through increased prey density. Dynamically, stoichiometric food quality effects have similar effects as the gathering together of musk ox to defend themselves against wolf attack (though in the case of food quality, the group defense occurs at high prey density, not low as would be the case for herding behavior). Increased algal C at constant algal P brings about reduced food quality and hence reduced zooplankton growth. At some point, reduced zooplankton growth can no longer offset any external mortality, and the grazer exhibits zero and then negative rate of change. To make a long story short, stoichiometric food quality effects bend the grazer nullcline into a hump. Loladze et al. (2000) characterized this transition from positive to negative grazer growth due to increased algal abundance as the "paradox of energy enrichment." In such a region, increased energy inputs (carbon or biomass) to the ecosystem in terms of solar radiation will increase yield of algae per unit nutrient (raise algal C:P, as discussed in Chapter 3). Thus, increased energy input has the paradoxical effect of crippling the trophic interactions at the herbivore-autotroph interface. This is a qualitative difference in a stoichiometric model compared to a non-stoichiometric one. In an upcoming section we will see empirical evidence for such nonintuitive dynamics.

Putting the algal and zooplankton nullclines together allows us to ascertain the qualitative nature of system dynamics. The community equilibria occur where rates of change of A_P, A_C, and G_C are all zero. Andersen (1997) showed that in general there are five possible community equilibrium points in this model. The actual number of equilibria and their stability properties for any parameter set depend upon phosphorus loading (P_L), dilution (m), and, as we will see, stoichiometric values. Example nullclines are plotted in Figure 7.6, and some equilibria are plotted in Figure 7.7. One of the five possible equilibrium points corresponds to an absence of both algae and zooplankton (the origin of the graph): algae can invade such a system as long as the dilution rate does not exceed their capacity to increase in the absence of zooplankton. This first equilibrium point is uninteresting to us and is not plotted in the figure. A second equilibrium point corresponds to nonzero phytoplankton but zero zooplankton; Andersen called this the "grazer extinction point." At this point, algae come into equilibrium with their sinking and dilution mortality rate, so that at equilibrium $\mu_A = \sigma + m$. This equilibrium point is a very interesting part of the model. Andersen showed that this equilibrium is unstable if zooplankton can have net positive growth when feeding on algae with this growth rate, and stable if not. An example of a stable grazer extinction point is plotted in Figure 7.7A and an unstable grazer extinction point is in Figure 7.7B. Whether grazers achieve positive growth around this point strongly depends on whether they can meet their biomass P demands given the imposed mortality rates. We discussed "hard TER's" in Chapter 5; these refer to food that is so poor in quality that positive growth is impossible even with zero mortality. These grazer extinction points are not exactly the same phenomena, but they have the same effect at the community level (there may be positive grazer growth, but it is not large enough to offset mortality). The presence of the grazer extinction point indicates that algal growth rate and P content are important in zooplankton persistence in this model, as they should be given the many studies we examined back in Chapter 5.

The three remaining equilibrium points are "internal," that is, they represent nonzero values of A_P, A_C, and G_C. They can be distinguished by the identity of the factor limiting zooplankton growth, whether it is food concentration (A_C), food quality (A_P), or both. The first of these, where grazers are limited by food quantity, corresponds to the situation where algal P content is sufficient for the requirements of the grazer [$(P:C)_A \geq (P:C)_G$] while algal concentration is subsaturating to zooplankton feeding ($A_C < F$). This equilibrium point corresponds to the classical situation of food-limited grazer growth and it is locally stable in this model. The second internal equilibrium point, where grazers are limited by food quality [$(P:C)_A < (P:C)_G$] but not by food concentration ($A_C \geq C'$), has the property that both algal and grazer biomass are directly proportional to the input P concentration (P_L), and hence are mutually correlated. This equilibrium point, shown in Figure 7.7A, is locally unstable however. The third internal equilibrium point, where grazers are limited both by food quality $[(P:C)_A < (P:C)_G]$ and by food concentration ($A_C < C'$), has the property that algal biomass is linearly decreasing with zooplankton biomass, and zooplankton biomass is a linearly increasing function of P loading. This last of the five total equilibrium points is locally stable.

Andersen (1997) contrasted model results at different dilution rates. At low dilution, he found a single internal equilibrium point as well as a single equilibrium with zero grazers but positive algal abundance. He was able to identify the basin of attraction of the internal, stable, equilibrium point. Initial conditions within this basin of attraction demonstrate counterclockwise spiraling into the equilibrium (damped oscillations, with the reductions in algae followed by reductions in zooplankton followed by increases in algae and then increases in zooplankton). Initial conditions outside this basin of attraction resulted in stable limit cycles. Both kinds of dynamics have been observed in real *Daphnia* populations (McCauley and Murdoch 1987), although there are many proposed mechanisms to account for them (McCauley et al. 1988, 1999). The other equilibrium point (with zero grazers) is locally unstable. At higher dilution rates the model has both a locally stable and a locally unstable internal equilibrium point. In addition, the equilibrium point with zero grazers becomes locally stable (similar to Fig. 7.7A). This model therefore unequivocally has multiple stable states. The dynamics of such systems can be highly complex and dependent on initial conditions. In this particular case, the model either demonstrates damped oscillations toward the internal equilibrium, or eventually ends up at the grazer extinction point.

We have calculated some Andersen nullclines in order to examine the effects of grazer P:C ratio on predicted system behavior (Fig. 7.7). The first of these two cases, with high grazer P:C $[(P:C)_G = 30]$, corresponds to Andersen's own example (parameters in Table 7.1), but with a dilution rate set at 0.07. The contrasting case has a grazer with a low P:C $[(P:C)_G = 15]$. Note how the isoclines and equilibria change with grazer C:P. With a high-P grazer there is a grazer extinction point, whereas with a low-P grazer there is not. In other words, the presence of multiple stable states depends on the body stoichiometry of the grazer, which itself is determined by major aspects of biochemical allocation and life history traits (Chapter 4). At an even more abstract level, this contrast suggests that systems with high-nutrient grazers overall might have more complex population dynamics (cycling, chaos, etc.) than ones with low-nutrient grazers.

The grazer extinction point is perhaps the most interesting feature of

TABLE 7.1

Summary of parameter values in model of Andersen (1997). In addition, the symbols A_P, A_C, G_C, and μ are variables

Variable or Parameter	Definition	Value	Dimensions
F	Incipient limiting food concentration	0.17	mg C l^{-1}
f'_{max}	Maximum ingestion rate	0.81	d^{-1}
m_G	Grazer mortality loss rate	0.02	d^{-1}
m	Dilution or mortality rate	0.01	d^{-1}
μ_m	Maximum grazer growth rate on high-quality food	0.4	d^{-1}
μ'_m	Algal growth rate at infinite cell quota	1.2	d^{-1}
P_L	Phosphorus loading concentration	40	μg P l^{-1}
$(P:C)'_{A,min}$	Minimum algal P:C ratio, by mass	3.8	μg P mg C^{-1}
$(P:C)'_G$	Grazer P:C ratio, by mass	30.0	μg P mg C^{-1}
r	Respiration rate	0.25	d^{-1}
σ	Algal sinking loss rate	0.0008	d^{-1}
$S_{C,max}$	Food carbon maximum assimilation efficiency	0.8	—

Andersen's stoichiometric model. Its existence arises from food quality constraints linked to stoichiometry. Zooplankton may be drawn to this point when phytoplankton biomass is large relative to total nutrient pools. Near this equilibrium point, food quality penalties suffered by zooplankton may be severe enough for zooplankton to decline even under high food quantity, further allowing algae to escape from grazing pressure, and further allowing algal carbon to increase. Under certain circumstances, this grazer extinction point can be a locally stable point with a large basin of attraction. But what meaning can this have in reality, in which lakes devoid of zooplankton do not exist? Although not yet examined theoretically, in the long run the importance of this grazer extinction point may lie more in explaining zooplankton community structure than in total extinction of herbivorous zooplankton as a whole. In other words, extinction of a high-nutrient grazer may create opportunities for low-nutrient grazers to be successful. This model thus may be a theoretical, dynamic underpinning for such community-level patterns as we saw in Figure 5.13.

TROPHIC CASCADES

The Andersen model describes a community with multiple equilibria. One is a "high-grazing" community with low algal biomass and the other is a "low-grazing" community (in fact, grazer extinction) with high algal biomass. Ecosystem alternations between high and low grazing, if not explicitly due to the same mechanisms as in this model, have been a major focus of limnologists for years. We know through this work that the structure of food webs (e.g., the number of trophic levels) can have large impacts on multiple ecosystem processes, such as primary productivity, autotroph biomass and species composition, gas fluxes, and microbial dynamics (Carpenter and Kitchell 1993; Pace et al. 1999). Effects of upper trophic levels on autotrophs and ecosystem processes are known as "cascading trophic interactions" (CTI) (for more on definitions, see Polis et al. 2000). In Chapter 6 we saw how food-web structure can alter the nutrient limiting phytoplankton production by altering the dominance of zooplankton taxa with contrasting body N:P (Fig. 6.1). Those fish manipulation experiments, performed in meso-oligotrophic lakes in northern Michigan, resulted in rapid (less than 1 yr) and large shifts in zooplankton and algal communities as well as changes in algal N versus P limitation. These were the first glimpses of the stoichiometric ramifications of CTI.

In turn, stoichiometry may influence the existence of CTI. Here we consider two recent whole-lake experiments that, on the one hand, continued to reveal the large stoichiometric rearrangements induced by CTI, and, on the other hand, indicated that stoichiometric mechanisms can interfere with the ability of CTI to produce major ecosystem changes. The studies involve a simultaneous pair of whole-lake food-web manipulations at the Experimental Lakes Area (ELA), Canada (Elser et al. 1998, 2000c). Fish were used to attempt to shift the food webs from high-N:P, copepod-dominated zooplankton to ones dominated by low-N:P *Daphnia*. A top carnivore (northern pike, *Esox lucius*) was added to two lakes contrasting in nutrient loading, productivity, and community structure but similar in having cyprinid minnows as the top trophic level. No other experimental manipulations were imposed besides the introduction of piscivores. One of these experiments was performed in Lake 227 (L227) where experimental fertilization at low N:P had produced annual dense blooms of N-fixing cyanobacteria for approximately 20 years (Hendzel et al. 1994). Stoichiometric CNR theory predictes that under N limitation, differential nutrient recycling (retention of P and release of N by low-N:P *Daphnia*) would raise the N:P of the lake's internal nutrient regime and thereby shift the lake toward P limitation, suppressing N-fixing cyanobacteria. An identical manipulation in P-limited oligotrophic Lake 110 (L110) was predicted to

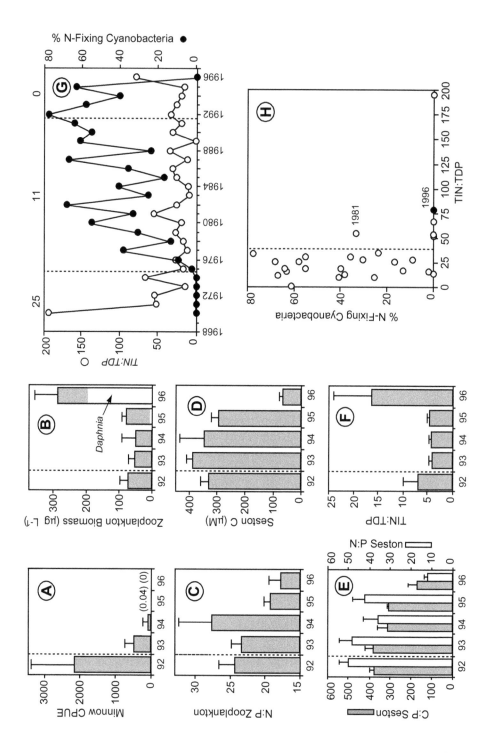

accentuate phytoplankton P limitation. The experiments involved a base-line year of monitoring (1992) followed by addition of piscivorous northern pike in year 2 ($\sim$60 fish per lake, 1993) and year 3 ($\sim$120 fish, 1994) and additional years of study (1995 and 1996) to evaluate ecological response. A reference lake (L240) with no experimental manipulations performed was also monitored for comparison with dynamics in L110.

Unlike the rapid response we described in Chapter 6 where the Tuesday Lake plankton reacted to a simultaneous manipulation of both piscivores (bass added) and planktivores (minnows removed) within a time scale of weeks (Carpenter et al. 1987; Elser et al. 1988), the L227 ecosystem re-sponded to pike on a time span of months to years. Minnows were elimi-nated in 2–3 years (Fig. 7.8A). *Daphnia* did not respond until the third year following initial fish manipulation, possibly due to the slowness of minnow response coupled either to delay in arrival of *Daphnia* propagules or to the extended time required to build a population from extremely low initial levels (Elser et al. 2000c). However, when *Daphnia* arrived, they did so with a vengeance. Large-bodied *Daphnia* were first observed in early spring, 1996, and by midsummer of that year, *Daphnia* very strongly domi-nated zooplankton biomass ($\sim$290 μg L^{-1}; Fig. 7.8B). Zooplankton N:P

Fig. 7.8. Stoichiometric parameters relating to a trophic cascade in a eutrophic lake. Top predators (northern pike) were added in 1993 and 1994. A. Cyprinid minnows, which before the manipulation were the top trophic level, decrease to immeasurable levels in terms of catch per unit effort (CPUE) by 1996. B. Zoo-plankton biomass shows a great increase in the same year that minnows disap-peared. The open portion of the bar represents *Daphnia* biomass in comparison to the whole zooplankton community. Previous to 1996, *Daphnia* are very rare. C. Zooplankton community N:P declines during the study, and is approximately equal to *Daphnia* N:P (see Fig. 4.2) by the end of the study. D. Seston C concentration shows a great reduction in 1996 compared to previous years. E. Seston C:P (dark bars) and N:P (light bars) show a large decline in 1996. F. Dissolved N:P [total inorganic N (TIN) divided by total dissolved phosphorus (TDP)] jumps in 1996. G. Long-term records (1970–1996) of dissolved N:P (open circles) and percent N-fix-ing cyanobacteria (closed circles). The numbers at the top of the graph give the experimental N:P loading ratios in force for the indicated periods (separated by broken lines). H. The contribution of N-fixing cyanobacteria to total phytoplankton abundance is related to the dissolved N:P; when this ratio is high, cyanobacterial prevalence is much reduced. In total, these observations are consistent with a stoi-chiometric shift of reduced zooplankton N:P and elimination of cyanobacterial competitive dominance by shifts in internal recycling weakening N limitation. Based on Elser et al. (2000d).

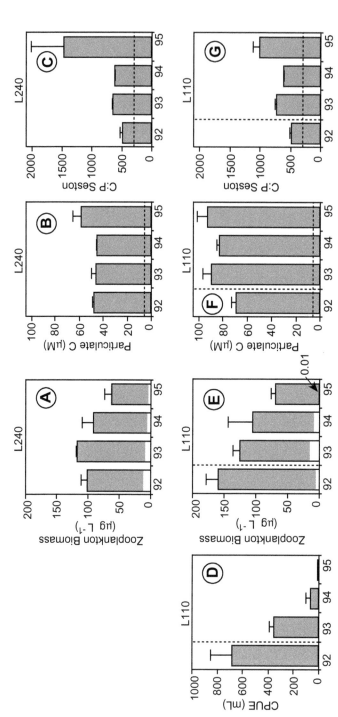

Fig. 7.9. Stoichiometric parameters relating to an absent or delayed trophic cascade in a lake with high seston C:P. Top row: unmanipulated reference lake (L240). Bottom row: L110 following northern pike addition. Dashed vertical lines separate pre- from postmanipulation conditions. In contrast to observations in L227 (see previous figure), zooplankton biomass and *Daphnia* biomass fail to increase during the study period (E) in spite of a greatly reduced minnow presence (D). Seston abundance increases (F) as does seston C:P (G). Annual trends in the manipulated lake parallel those of the reference system. These observations are consistent with the hypothesis that a stoichiometric constraint (high seston C:P; note values are above the TER, dashed horizontal line in G) prevented or delayed an increase in *Daphnia* following the minnow reduction. Seston quantity was abundant compared to feeding saturation levels (dashed horizontal line in F). Based on Elser et al. (1998).

(Fig. 7.8C) decreased to ~17 (similar to the N:P of *Daphnia* itself, Fig. 4.2). At peak abundance, the zooplankton contained ~30% of total water column P, a high percentage relative to other lakes (Andersen 1997). Phytoplankton responses also were large, with a sixfold reduction in total biomass on an annual basis (Fig. 7.8D) and a major reduction in particulate C:P and N:P (Fig. 7.8E). These proportional changes in zooplankton and algae are very large compared to those in other studies on CTI (Brett and Goldman 1996). Consistent with the predicted stoichiometry of CNR (Chapter 6), dissolved nutrient pools were also impacted, with disproportionate increases in inorganic N (especially NH_4-N) relative to dissolved P (Fig. 7.8F).

These changes resulted in a massive alteration in phytoplankton community structure. For the first time in 20 years in this lake, the L227 phytoplankton in 1996 lacked a significant contribution of cyanobacteria (Figs. 7.8G,H). This pattern is remarkably similar to the classic pattern of reduced cyanobacteria dominance at high TN:TP in the multilake data set analyzed by Smith (1983b). These changes also had ramifications at the scale of whole-lake N budgets, since heterocystous cyanobacteria had been a major source of N to the lake via N fixation (Findlay et al. 1994). Further supporting the potential role of differential N:P recycling by zooplankton CNR in regulating N fixation in this lake, MacKay and Elser (1998b) documented strong reductions in cyanobacterial N fixation in treatments grazed by *Daphnia* versus ungrazed or copepod-grazed treatments in mesocosm studies in L227. These results (Fig. 7.8) continue to show that large trophic cascades may induce strong stoichiometric shifts.

The rearrangement of community structure and ecosystem processes in L227 observed in 1996, although very significant in that year, did not continue thereafter. Subsequent sampling in 1997–2000 found that, although minnow populations remained apparently extinct, *Daphnia* populations did not continue to bloom. Dense cyanobacterial biomass returned, further supporting the idea that *Daphnia* presence eliminated the cyanobacterial bloom in 1996 through differential nutrient recycling of N and P. Food-web ecologists have recognized the difficulty of stabilizing cascade effects, especially in highly eutrophic lakes (Shapiro 1990; Scheffer et al. 1997; Gragnani et al. 1999). The L227 study adds to the list of examples of unstable trophic interactions under eutrophy. Perhaps stoichiometric food-web models (above) will be able to capture important aspects of these unstable dynamics.

Unlike the spectacular outcome of the L227 manipulation, in the case of L110 the experimenters were left with the challenging task of trying to explain (and publish) a *lack* of community response to fish manipulation (Elser et al. 1998). In L110 during the three years following pike introduction not only did *Daphnia* fail to increase in response to much-reduced minnow abundance (Fig. 7.9E), they went virtually extinct in the third year following pike introduction (Fig. 7.9E). A similar decline was also observed in an unmanipulated reference lake (Lake 240, Fig. 7.9A), suggesting a temporal coherence (Rusak et al. 1999), such as a regional climatic response. With CTI not explaining the zooplankton dynamics, Elser et al. (1998) considered several possible explanations for the lack of *Daphnia* response. Invertebrate predation was rejected because there was no major change in the abundance of such predators following minnow reduction. Insufficient food abundance was rejected as seston C concentrations were above levels normally accepted as sufficient to support *Daphnia* population growth (Fig. 7.9F). However, unsuitable food quality was not rejected; seston C:P was considerably above previously published TER values (Fig. 7.9G).

This hypothesized stoichiometric constraint suggested that *Daphnia* success (and thus the strength of CTI in lakes, as *Daphnia* is thought to be the key player) is jointly regulated by predation pressure and stoichiometric food quality (Fig. 7.10). *Daphnia* populations do best in lakes with low seston C:P (Fig. 5.14), and thus CTI should be dampened in lakes with high seston C:P. Further, factors that might affect phytoplankton nutrient limitation (Chapter 3), such as regional climatic variation, could impact on *Daphnia* independent of changes in predation pressure. This might account for the simultaneous reduction in *Daphnia* in 1995 in both L110 and L240 at the ELA described earlier, coincident with increased seston C:P in both systems. Others have observed region-wide shifts in zooplankton communities that might be climate induced (Rusak et al. 1999). Follow-up experiments in this same set of high C:P lakes at the ELA, including direct manipulation of phytoplankton P content, have supported the operation of stoichiometric constraints on *Daphnia* growth and reproduction (MacKay and Elser 1998a; Elser et al. 2001; Urabe et al. 2002b). Subsequent events have indicated that in L110 the stoichiometric constraint is not a hard barrier to this species' establishment but instead serves to delay the eventual success of *Daphnia* in the absence of predation pressure. Population recruitment rates, for example, were not unusually low (Sterner 1998). Three years after the funding for this study ran out, sporadic sampling revealed that *Daphnia* had finally appeared in appreciable numbers in L110 in 1997. Unfortunately, its eventual effects on the rest of the planktonic community remain unanalyzed.

It has been noted that the strength of CTI varies both among lakes (DeMelo et al. 1992; Brett and Goldman 1996) and among different types of ecosystems (Strong 1992; Polis 1999). One factor that might contribute to this variation is the suitability of autotroph biomass to support dense populations of efficient grazers. Schmitz et al. (2000) indicated that trophic cascades in terrestrial systems were attenuated whenever plants contained

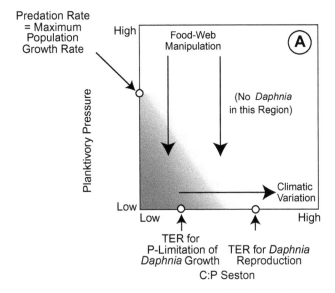

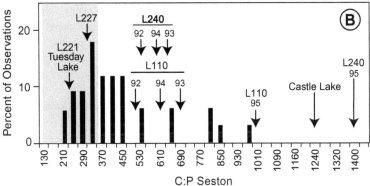

Fig. 7.10. Stoichiometric effects on trophic cascades. A. Schematic showing hypothesized boundaries for development of high *Daphnia* densities (dark shading). *Daphnia*, a keystone herbivore in CTI, will develop high biomass under conditions of low planktivory and low seston C:P. Variation in fish communities shifts the situation vertically. Factors that influence seston C:P, such as climate, create horizontal shifts. B. Four fish manipulation experiments, spanning a wide range of seston C:P ratios, produced very different patterns in CTI. At the low-seston-C:P side (L221, Tuesday Lake, L227), lake dynamics in response to fish manipulation generally follow CTI predictions. In contrast, where seston C:P ratios are high (L110, Castle Lake), *Daphnia* do not establish or are much delayed, and CTI predictions fail. A more complete histogram of lake seston C:P ratios is given in Fig. 3.13. Based on Elser et al. (1998).

antiherbivore defenses. We (Elser et al. 1998) and others (Strong 1992; Hessen 1997; Polis 1999; Müller-Navarra et al. 2000) have hypothesized that much of the variation in the strength of CTI among ecosystems is associated with food quality.

Data to test for the role of C:P stoichiometry in trophic cascades at the whole-lake level are rare, but several studies report enough information to begin such an analysis, and together they produce some intriguing patterns. Figure 7.10B places several CTI experiments in the context of observed variation in seston C:P in lakes. Examples of relatively strong and rapid trophic cascades in Tuesday Lake (Carpenter et al. 1987), L227 (Elser et al. 2000c), and L221 (Vanni and Findlay 1990) are on the low-C:P side of the histogram. Two examples of delayed or unusual trophic cascades in L110 (Elser et al. 1998) and Castle Lake (Elser et al. 1995d) are on the high-C:P side. In the case of Castle Lake, cessation of stocking of the lake's dominant planktivore (rainbow trout) was followed by near extinction of *Daphnia*, increased phytoplankton productivity, and decreased water transparency, all trends the opposite of CTI predictions. Seston C:P in Castle Lake is among the highest yet reported (Elser and George 1993). While these analyses are necessarily anecdotal, they do suggest that as one moves from marine to lentic to terrestrial ecosystems (i.e., left to right on a histogram of autotroph C:nutrient ratio, Fig. 3.17), the relative strength of "top-down" trophic coupling weakens, supporting arguments that food quality influences the strength of trophic cascades. In particular, it might explain a difference in strength of CTI between aquatic and terrestrial systems, the latter of which are notable for high autotroph C:nutrient ratios (Chapter 3). The generality of strong CTI has recently been called into question (Chase 2000; Polis et al. 2000; Schmitz et al. 2000), and hopefully this set of questions will be sorted out soon. One possible factor in explaining variation in CTI strength across ecosystems is that the keystone herbivores that are necessary to impose trophic cascades onto their food base may need to be fast-growing generalists. As such, they likely have nutrient-intensive bodies (Chapter 4) and thus should be more strongly constrained by insufficient nutrient content in their food (Chapter 5). Here we see suggestions that the stoichiometry of the ribosome (Chapter 2) impinges all the way up to food-web and ecosystem dynamics.

LIGHT:NUTRIENT EFFECTS AT THE COMMUNITY LEVEL

In Chapter 3, we described the response of autotrophs to gradients in light and nutrients. We saw how biomass C:nutrient ratios respond positively to changed balance between light and nutrients due to adaptive adjustments among rates of photosynthesis, nutrient uptake, and growth. We argued that light:nutrient balance was a key factor determining autotroph elemental composition in the field. We have also discussed how C:nutrient balance influences secondary production (Chapter 5) and nutrient recycling (Chapter 6). In communities, these ecological interactions are combined and operate simultaneously. Here, we will describe some of the food-web implications of light:nutrient balance by examining a series of experimental studies at the community level.

Let us begin by first considering energy input and energy transformations in simple systems. Take your TV or stereo as an example, and consider what happens when you turn up the volume. If the device is switched on and everything is functioning properly, when you turn up the volume, the useful work done (speakers pushing air, causing disturbances that your ear detects as sound) increases and the sound gets louder. The volume control is an electronic gatekeeper, adjusting how much energy goes into the amplifier circuit. Now, replace the amplifier with a food web. Solar energy is the relevant energy input. What kinds of useful work do food webs do that might be like the sound produced by an amplifier? One thing we might consider to be useful work is the rate of production of higher trophic levels. This is a particularly apt comparison when those populations are being harvested. We can easily call that situation "useful work." Some would refer to this as an "ecosystem service" (Costanza et al. 1997). Even when top trophic levels are not being harvested, the conditions promoting top predators in ecosystems have important ecological consequences (see previous section).

So, do food webs behave like your stereo? With more energy in, do we get more useful work out? At some level, the answer is "yes." Several studies have estimated both primary and secondary production across many ecosystems, and have found the two to be correlated across very broad ranges. McNaughton et al. (1991), for example, observed such an increasing trend when comparing primary and secondary production in many diverse terrestrial ecosystems (Fig. 7.11). Downing et al. (1990) also described positive relationships between fish production and primary production. Cyr and Pace (1992) estimated grazing losses across a range of productivity. Although there are clear positive trends, the range of secondary production supported at a given level of primary production is wide. For example, in the work of McNaughton and colleagues, at intermediate primary production, the spread in the data ranges over approximately two orders of magnitude, or about 100-fold! One cannot parse out "real" variation from methodological and sampling error from such a data set, but there seems to be much variation in secondary production left to be explained once primary production is accounted for. Among the possible explanations are factors associated with particular systems. In McNaughton's work (Fig. 7.11), tropical grasslands tended to have low secondary produc-

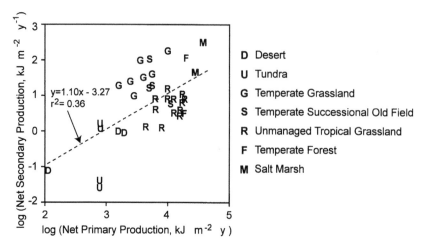

Fig. 7.11. Relationship between secondary (herbivore) production and primary production in a diverse set of terrestrial ecosystems, including tundra, forest, grasslands, and others. The slope is statistically different from zero but there is much variation left to be explained. Based on McNaughton et al. (1991).

tion and temperate grasslands tended to have high secondary production relative to the overall trends. Invertebrate herbivores tended to fall above the line and vertebrate herbivores tended to fall below the line.

But, at a deeper level, the comparison between stereos and trophic interactions is not so clear. What if, instead of looking at relationships between primary and secondary productivity, we look at the relationship between secondary productivity and solar energy input? In other words, how efficiently do food webs convert solar energy to useful work as we give them higher and higher solar energy? Light impinging upon terrestrial systems is affected primarily by latitude and climate. Over at least regional scales, there is not much geographic difference in solar energy input to the surface of the earth and, by implication, to plant communities dense enough to have near-100% basal area coverage. In aquatic communities latitude and climate are still factors, but additional factors come into play. Streams receive varying degrees of solar energy input due to different degrees of shading from riparian vegetation. For example, Hill et al. (1995) examined periphyton production and snail growth in a stream in Tennessee. With increasing light, periphyton growth showed an increasing, saturating function (a *P-I* curve), as did snail production. In water columns in lakes and oceans, mixing patterns and light absorption by colored compounds may profoundly affect light climate (see Chapter 3). Hence, two lakes that are side by side may have highly contrasting solar energy inputs to autotrophs. If one lake had colored water and deep mixing and the other had clear water and shallow mixing, light available for epilimnetic photosynthesis would be very different. Several studies have now explored the stoichiometric aspects of how secondary production relates to light levels in planktonic systems. Our discussion will go from the small scale and short term in artificial environments to larger and longer experiments under more natural conditions. The bottom line from these experiments is that the story we tell about energy flow and trophic dynamics in our basic biology textbooks may be quite misleading.

Urabe and Sterner (1996) performed experiments in 1 L of growth medium in which algae (*Scenedesmus*) were allowed to grow at a range of light intensities for 6–8 d, after which young *Daphnia* were added. After an additional 6 d of incubation, the biomass and chemical composition of algae as well as the biomass of zooplankton were determined. This experiment assessed the short-term response of herbivore production to a light intensity gradient in a single trophic interaction. The results indicate that responses to light were not all linear, but they could easily be explained from stoichiometric considerations. First, mean algal biomass over the duration of the incubation was an increasing function of light (Fig. 7.12). This suggests that at some level it is correct to consider algal biomass to be light limited in these systems. Second, as expected from what is known about changing allocation patterns versus light (Chapter 3), because the quantity of phosphorus in the communities was fixed, increasing algal carbon with light generally meant that algal P:C decreased with light (Fig. 7.12). Third, and finally, zooplankton biomass peaked at intermediate light levels (Fig. 7.12). A change from positive response to light to negative response to light occurred at roughly the same P:C in the different treatments—one which matched *a priori* expectations based on simple TER calculations (see Chapter 5) of when this herbivore would switch between C and P limitation.

The authors interpreted this unimodal pattern of herbivore production versus light intensity as resulting from energy limitation of both algae and zooplankton at low light, but P limitation at high light where even the animals were P limited. In fact, at high light, carbon was inhibiting to growth (more algal C lowered animal growth). This strange thermodynamic pattern is very different from what your stereo does and from what we tell our introductory ecology students about food chains. In these flasks, turning up the volume (increasing photon input) at first generated more sound (secondary production) at the low end of the light gradient, but further increases caused the useful work to decrease. At high light and high production, primary production inhibited secondary production because of accentuated C:P imbalance. Stoichiometry has imposed some strange nonlinearity on the thermodynamics of these ecosystems. For the

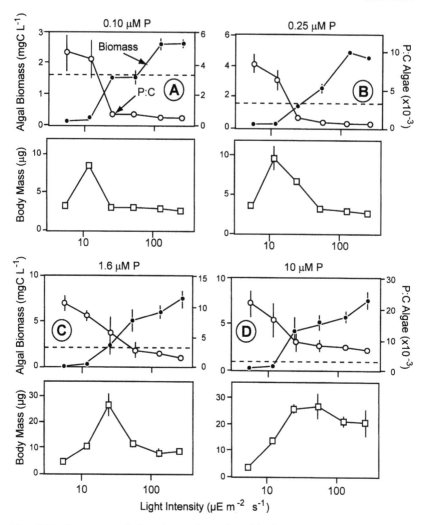

Fig. 7.12. Responses of algae (*Scenedesmus*) and herbivores (*Daphnia*) to varying light intensity at different concentrations of phosphorus. The four pairs of panels represent four different levels of P. Error bars indicate ±1 s.d. Algal biomass (solid circles) increase with light intensity while P:C ratios (open circles) decrease. Herbivore growth is given as body mass at age 6 d (open squares). Herbivore growth is highest at intermediate light, and the light intensity resulting in maximum herbivore growth increases with increasing phosphorus (compare location of peak in different panels). The dashed horizontal line is the P:C TER below which the growth rate of the herbivore is expected to be limited by P rather than by C. Note that in each case *Daphnia* growth declines precipitously when algal P:C first crosses the TER. Based on Urabe and Sterner (1996).

Daphnia in the high-light beakers, one could just as easily say that they suffered (indirectly) from too much light as one could say they had too little P. This echoes our discussion in Chapter 3 of how light:nutrient balance affects autotroph nutrient limitation, in which we implied that severe nutrient limitation of autotroph growth is not simply the result of low nutrient inputs (supply-side limitation) but also a result of elevated light intensity which raises demand for nutrients. Decreased herbivore performance in the presence of increased algal abundance is empirical evidence for the paradox of energy enrichment (Loladze et al. 2000) referred to earlier in this chapter.

Small containers allow for a large number of treatments to be examined efficiently and with replication. However, small scale also imposes artificialities and necessarily limits the temporal scale of the study (Hairston 1989). In another study, Sterner et al. (1998) further explored the response of simple communities to light, but these experiments were conducted in large, indoor "plankton towers" (Lampert and Loose 1992) of ca. 1 m diameter and 11 m height, which allowed for establishment of a thermal gradient (warm upper layer and cool bottom layer). The experiments were similar to the beaker studies of Urabe and Sterner in that simple communities consisting of *Scenedesmus* and *Daphnia* were used. However, the total duration of the experiment was lengthened to 26 d, and some new realities such as settling of algal particles and potential vertical migration of the animals were included. The expanded experimental scale reduced the possible number of treatments, and only two light intensities at a single P level could be run. Two experimental runs were performed in the pair of towers, with one tower having high light and the other low light. The experimental setup (incident light, mixing depth, initial P levels, etc.) was selected so that the high-light treatment would produce unfavorable food quality as seen in the right-hand side of Figure 7.12. The experimental conditions were chosen to put the high-light systems in high-light:nutrient conditions. The results (Fig. 7.13) confirmed the observations of Urabe and Sterner (1996). High light resulted in greater algal growth (Fig. 7.13 A), lower algal P:C (not shown here), and reduced zooplankton production. At high light, zooplankton biomass did not measurably change over the 26 d, meaning herbivore production approximately balanced natural mortality, which in the absence of predators we can assume to be low. Thus, inhibition of secondary production by elevated primary production also occurs over lengths of time approximately equal to the generation time of the herbivores, and in larger and more realistically structured habitats. This is further evidence for a seemingly surprising but actually predictable stoichiometric influence on ecosystem energetics, a paradox of energy enrichment. Urabe et al. (2002a) also used indoor artificial ecosytems to examine effects of light intensity over the longer term (90 d). As men-

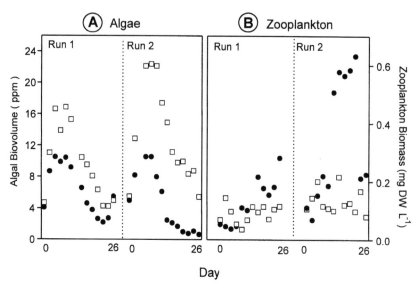

Fig. 7.13. Algal (A) and zooplankton (B) dynamics differ between high-light (open squares) and low-light (closed circles) communities in two different runs in large-scale experimental plankton towers (separated by dotted lines). In comparison to low light, algal biovolume is higher at high light but zooplankton biomass is lower. In fact, at high light, net zooplankton biomass change over time is effectively zero in spite of the presence of a large quantity of algal food. Based on Sterner et al. (1998).

tioned earlier in the chapter, this experiment also documented the detrimental effects of high light and elevated algal C:P on herbivore growth, but in this case the nutrient recycling feedbacks during slow population buildup were able to eventually ameliorate poor food quality conditions.

A third study was performed in enclosures at a similar spatial scale as the towers (1-m diameter, 4-m deep enclosures) but with a much more species-diverse community from a P-limited lake at the Experimental Lakes Area, Ontario (Urabe et al. 2002b). Recall that the ELA is the location of L110, where we saw evidence for stoichiometric constraints on the trophic cascade and possible climatic variation impacting on *Daphnia* success (Figs. 7.9 and 7.10). Twelve enclosures were filled with ambient lake water and various zooplankton including large *Daphnia* were added at low levels to all to assure the presence of a large herbivore. Half of the enclosures were enveloped in thick shade cloth that reduced light intensity to ~7% of its level in the remaining unshaded bags. Each set of enclosures, shaded and control, received a gradient of P fertilization from 0 to 15 μg L^{-1}. The investigators examined the joint effects of shading and P enrichment on seston quantity, C:N:P stoichiometry, and zooplankton production after 5 weeks. Thus, this experiment represents a further step in testing the light:nutrient hypothesis at even longer temporal scales and under conditions closer to those found in nature, including a much more diverse set of species of algae, zooplankton, and microbes.

Despite the fact that lakes of the ELA in general are P limited (Schindler 1977, 1978, 1991; Hecky et al. 1993), P enrichment had no effect on seston abundance in either control or shaded enclosures (Fig. 7.14A). However, at each level of P enrichment, shaded enclosures had lower seston levels than controls. P enrichment had large effects on seston C:P, which declined from ~500 in unenriched enclosures to ~150 in enclosures receiving 15 μg P L^{-1} (Fig. 7.14B). Shading also affected seston C:P, which was lower in the shaded enclosure than the control enclosure at each level of P enrichment. These changes in the quantity and quality of seston due to light and P manipulation affected zooplankton production in what should be, by now, an expected direction. Zooplankton biomass in both control and shaded enclosures increased with P enrichment (Fig. 7.14C). Consider especially the data at low levels of P enrichment, where zooplankton biomass in shaded enclosures considerably exceeded levels achieved in the fully illuminated controls (note the logarithmic scale). Thus, at low levels of nutrient input (similar to natural conditions at the ELA), a decrease in light intensity (energy input) resulted in an *increase* in secondary production (useful work). However, at high P input, elemental imbalance was modest and increased zooplankton production was achieved in fully illuminated enclosures that could sustain high food concentrations. All these trends are consistent with the stoichiometric effects of food C:P on zooplankton community dynamics and production that we have highlighted in this chapter and throughout this book.

This series of studies from flasks, artificial ecosystems, and field enclosures all support the idea that light:nutrient balance regulates trophic transfer efficiency at the base of pelagic food webs due to effects on the stoichiometry of phytoplankton-grazer interactions. They all produced a set of internally consistent, strange, but simple patterns. Others have observed improved herbivore growth rates when feeding on autotrophs from lower light levels, for example in *Manduca sexta* consuming tomatoes (*Lycopersicon esculentum*) (Jansen and Stamp 1997). In another example, a community model of plant-herbivore interactions along productivity gradients, which includes an assumption of reduced forage quality in plants favored by high light, emphasizes that the identity of the limiting resource (whether it be nutrients or light) plays an important role and provides a slightly different mechanism for reduced herbivore productivity at high light (Huisman et al. 1999). The latter model is motivated in part by successional trends in salt marshes, where certain herbivores show maximal produc-

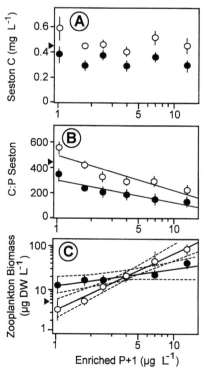

Fig. 7.14. Light-nutrient interactions in a field enclosure experiment at the Experimental Lakes Area. Effects of light and nutrient balance on (A) seston quantity and (B) C:P stoichiometry during the 4.5-week experimental run, and (C) mean zooplankton biomass for the final three sampling dates. Closed circles indicate the shaded enclosures, open circles denote unshaded (control) enclosures. Dark triangles on the y axis indicate values for the lake. Temporal variation is represented by standard error (vertical bars) on the mean among sampling dates. Regression lines are inserted when relationship is significant at $p < 0.05$. For zooplankton biomass (C), 95% confidence intervals are denoted by dotted lines to evaluate significant differences between control and shaded enclosures. Note that reducing light to less than 1/10 of its ambient value in unenriched low-P enclosures leads to an approximately fivefold increase in zooplankton biomass (C) due to improved stoichiometric food quality. Based on Urabe et al. (2002b).

tivity at intermediate successional stages although primary production is greatest late in succession (van de Koppel et al. 1996). In sum, under certain stoichiometric conditions, food webs may produce more animal biomass with a reduction in energy input. These effects are repeatable across scales and predictable from a few stoichiometric relationships. As in

Chapter 5 where a multiplicity of internally consistent results lent strength to stoichiometric theories, the convergence of results on light:nutrient balance at the community level is quite striking and we think deeply revealing. We suggest that it is time to tell our beginning students the truth about the nonlinearities and stoichiometric constraints impinging on how food webs really work. Thermodynamics impinge heavily on trophic dynamics but much of this impact follows from the law of mass conservation as captured by stoichiometric theory, and not from entropy and the second law, as textbook ecology generally insists.

FEEDBACKS OWING TO THE "CONSTRAINTS OF STUFF": C:N RATIOS IN TALL-GRASS PRAIRIE

Wedin (1995) wrote of the importance of plant C:N in determining the structure and characteristics of "humid" grasslands, such as the much endangered tall-grass prairies of North America. His arguments integrate many of the themes we have covered in this chapter and so we highlight them here. The key community players are the plants [C_4 grasses with low foliar N content (see Fig. 3.9) and which produce low N litter (Fig. 7.4)], mammalian and invertebrate herbivores, and fire. Wedin proposed that there are three feedback loops acting to maintain the characteristic vegetation of the native prairie. The first of these is the previously discussed feedback between low-N plants, low-N litter, low N mineralization, and enhanced or maintained competitive dominance of low-N plants. The second is related more to the topics of Chapter 5, namely, the poor quality of low-nutrient autotrophs for herbivores. The native tall-grass prairie, Wedin argued, is inadequate forage for vertebrate and invertebrate herbivores, thus lowering the animal density. Unfortunately, he did not specify exactly how lowered herbivore density helped maintain tall-grass ecosystems; one presumes that a high density would lead to an "overgrazed" condition with altered vegetation dominated by grazing-resistant plants. The third feedback involves the buildup of low-N litter, which promotes fire, a necessary factor in the maintenance of these grasslands. Fires remove both C and N from the ecosystem (C as CO_2 and N in different volatile compounds). By removing vegetation and nutrients, fires create high light:nutrient ratios; these conditions favor competitive dominance by the native tall-grass C_4 grasses.

Comparison of the three "consumers" in this system—decomposers, herbivores, and fire—is aided by reference to TER theory (Chapter 5). Ranked in order from low to high N requirement (high to low C:N TER), these consumers are (1) fire (zero N requirement), (2) decomposers (low

N requirement), and (3) herbivores (high N requirement). Wedin argued that tall-grass prairie is dominated by the consumer best suited to the stoichiometry of the vegetation. That "best-suited" consumer for vegetation with high C:N is fire, which, evidence suggests, was a more important consumer in tall-grass ecosystems than in other grasslands. These "constraints of stuff" (Wedin's term) interacted to mutually stabilize the tall-grass ecosystem in its native state. In the past century, however, "overgrazing, anthropogenic N inputs, and the exclusion of fire have all led to the displacement of native tall C_4 grasses by non-native C_3 grasses" (Wedin 1995, p. 255). Knowledge of these hypothesized stoichiometric thresholds may be critically important in managing these threatened ecosystems.

In our final major chapter now to come, we shall broaden the perspective once more, building on themes from this chapter to evaluate stoichiometric patterns at ecosystem and global scales.

CATALYSTS FOR ECOLOGICAL STOICHIOMETRY

The hallmarks of ecological stoichiometry at the community level are positive feedback, thresholds, and multiple attractors. These phenomena are well defined at a theoretical level. Empirically, one surprising aspect—the paradox of energy enrichment—is now well supported. Most of our catalysts for ecological stoichiometry in this chapter concern complex dynamics. But first, here are a few other suggestions.

- Resource competition has a strong stoichiometric component. All along, we have stressed the many implications of organism nutrient content (from genomics to life history evolution and beyond). Surprisingly, studies testing the stoichiometric determination of competitive ability are still needed. Questions needing answers include, "To what extent does competitive ability in constant or patchy environments depend upon stoichiometric patterns of nutrient content?" Expressed another way, "What is the role of stoichiometric yield in resource competition?"
- Stoichiometric intersections with community ecology are still poorly understood. We are suggesting that high-nutrient habitats promote rapidly growing consumers (liberated from mainly P limitation, and allowing their high-ribosome lifestyle to flourish). Data coupling life history strategies, even as simply defined as r- versus K-selected species, to ecosystem nutrient content and function are needed.
- We have argued that the strength of ecological interactions is affected by stoichiometric considerations. Additional studies of these ideas, particularly in poorly studied systems, would be very welcome. Do the

ideas we presented about autotroph-microbe interactions carry over to other systems, such as soils? Trophic cascades are a special case in that a well-developed body of literature suggests that food quality (C:P in part) influences the response of ecosystems to top predators. Explicit tests at the ecosystem level are needed but will be difficult and expensive to perform.

- There are several general approaches to understanding plant interactions with mycorrhizae; these all have a strong stoichiometric dimension, with particular C:P ratios necessary for a mutually beneficial interaction. To make a complete stoichiometric model of this interaction, however, requires knowing the nutrient requirements of mycorrhizae. This information could then be related to trading values within a multiple-currency framework.
- The concept of resource trading might well be profitably applied to the evolution of tissue specialization and perhaps of multicellularity. That is, within a single plant for instance, the leaves are swapping C with the roots for P. Other good examples come readily to mind.

We encountered one good example of intraspecific facilitation under poor-food-quality conditions (Sommer's *Daphnia*) as well as several where effects of predators varied with food quality. We also noted that within the context of tritrophic interactions, terrestrial herbivores are thought to exhibit interactions between food quality and mortality patterns. Although food quality bears most directly on growth rates (Chapter 5), these studies point to a surprising degree of complexity in the relationship between food quality and population dynamics.

- Cross testing of ideas in different systems would be a productive way to proceed.
- Certainly we need more theoretical and empirical work on the role of stoichiometric food quality in mediating interactions among multiple herbivore species.

In general, we need a great deal more information on the question of how stoichiometry contributes to the temporal variability of populations.

- Does stoichiometry promote cycling? Chaos?

These questions should continue to be pursued at both the theoretical and empirical levels. A major gap has already been stressed: we need good empirical studies testing multiple stable states driven by stoichiometry. Although positive feedbacks, thresholds, and multiple stable states seem to be the hallmark of homeostasis in stoichiometric models, to date we have no unambiguous experimental test supporting the existence of these exotic dynamics in a natural system. So far, there are no empirical studies that can be said to validate the existence of multiple stable states in nature. The closest demonstration so far is a laboratory study of zooplankton in which alternation between a high-grazing and low-grazing state was observed (Sommer 1992). However, in this study, the dilution rate of the system was varied, so this shows sensitivity to dilution rate, not multiple stable states of a single system per se. This gap in fact is the same as in other nonlinear models. There is a plethora of these threshold effects, and multiple basins of attraction have been discovered in many kinds of simple nonlinear systems, but there is a lack of empirical demonstrations of the same.

- If nature is governed by nonlinear dynamics, why are the hallmarks of nonlinear systems so difficult to observe? We leave it to others to answer that question, and we prefer to close on a more positive message.

Bridging the gap between short-term and long-term dynamics is an important priority (Hastings 2001). Some have argued that long-term positive feedbacks driven by stoichiometry do not occur (Tateno and Chapin 1997). Yet we have seen them rigorously demonstrated in mathematical models (Andersen 1997; Loladze et al. 2000), and certain key features were maintained in a series of studies on light at expanding temporal scale.

- What we don't yet know is what characteristics are necessary in order to find positive feedbacks and multiple stable states. These "interactions" between short-term and long-term outcomes are not just of interest in the context of theory. Ecological experimentation is typically conducted at short to medium time scales, while comparative observational studies involve patterns that have been produced over the long term.

Stoichiometry has great potential to predict broad ecosystem patterns and could give large-scale ecosystem study a theoretical basis it currently lacks.

Finally, additional studies of light:nutrient effects in real ecosystems over long time scales with adjustments of species would extend the continuum from beaker to bag that we discussed earlier.

SUMMARY AND SYNTHESIS

As promised, we have now seen the implications of the lower-level processes encountered in previous chapters for community structure and dynamics. Ecologists recognize that species interact to varying degrees; sometimes interactions are strong, sometimes they are weak, and sometimes even the qualitative nature of the interactions changes (DeRuiter et al. 1995; Laska and Wootton 1998; Ruesink 1998). The context of ecological interactions clearly is important to determining interaction strength, but how do we advance from there? How could we build a predictive theory of species interactions? How could we know, before we observe them together, when species will interact strongly or weakly, or even change their interactions from beneficial to inhibitory or neutral? In some cases, stoichiometry can help with this goal. In this chapter we saw several examples where even the nature (sign and magnitude) of ecological interactions changed according to stoichiometric balance.

We also saw a number of examples where factors such as system dilution rate or the light:nutrient balance had major effects in determining how species interacted with each other, and, as a consequence, how matter and energy moved through trophic links. These variables set an overall stoichiometric context within which species interact. We saw that communities with components generally close to Redfield C:N:P proportions (e.g., at high dilution rates in the Andersen model) behaved differently from those with marked imbalances in nutrient ratios. Imbalances seem to accentuate certain interactions, such as those between autotrophs and microbes; they also create surprising outcomes such as intraspecific facilitation within herbivore populations and inhibition of herbivores by increased autotroph biomass.

Another community property we saw in these examples is the operation of stoichiometric thresholds. In Sommer's *Daphnia*, small differences in culture dilution rate tipped the ecological dynamics strongly between high and low grazing with concomitant low and high autotroph biomass. In Wedin's analysis of grassland stoichiometry, small differences in plant C:N ratio were surmised to be associated with large differences in community structure, such as between long- and short-grass prairie. Stoichiometric thresholds are a fundamental discovery about communities and may have great importance for management.

We also saw how stoichiometry generates multiple stable states in communities. Strong feedbacks often induce highly nonlinear responses once thresholds in the levels of driving variables are crossed (DeAngelis 1992). Positive feedbacks have their own unique impact on system dynamics (DeAngelis et al. 1986). Theoretical analysis of stoichiometry in communities indicates that homeostasis should generate complex dynamics in food webs, including sensitivity to initial conditions and multiple stable states. On the other hand, low-quality food may generally help stabilize dynamics, dampening predator-prey cycles (Hessen 1997). The general dynamical and stability properties of models with balanced and imbalanced foods are still not well characterized.

To many, the overriding goal of science should be to be able to make successful predictions, either about the future of systems as we know them today, or about systems responding to perturbations. To some, extrapolation or interpolation of preexisting trends is the best way to predict (Peters 1991). However, some other epistemological framework is needed to establish what variables scientists think should be related to each other in a mechanistic or statistical sense. Perhaps the most striking aspect of the set of studies on light:nutrient balance we encountered in this chapter was that a few simple statements about C:P balance produced consistent patterns, ones that were accurately predicted from basic principles such as TER theory (Chapter 5). It is wonderfully exciting that one could use information about physiological and cellular processes (Chapter 2), autotroph growth patterns (Chapter 3), and herbivore growth efficiency (Chapter 5) to first understand and then correctly predict how secondary production will behave in a light gradient! It is one thing to predict one variable from another in a statistical sense. It is quite another to make *a priori* predictions about system behavior without a preexisting trend line. In a complex world of nonlinear interconnections among species and ecosystem components, such a prediction is a success indeed. Perhaps the idea of a predictive ecology based on first principles is not completely hopeless.

8

Big-Scale Stoichiometry: Ecosystems
in Space and Time

What can be said with assurance is that there is a unique and nearly ubiquitous compound, with the empirical formula $H_{2960}O_{1480}C_{1480}N_{16}P_{1.8}S$, called living matter. Its synthesis, on an oxidized and uncarboxylated earth, is the most intricate feat of chemical engineering ever performed—and the most delicate operation that people have ever tampered with.—Deevey (1970)

Our journey from molecules to ecosystems is almost complete. What remains is to examine the highest levels of biological organization at the largest spatial and longest temporal scales. Here, we are returning to the intellectual roots of ecological stoichiometry. Ecosystems were the subject of many of the seminal studies we have already relied heavily upon, including those of Lotka and Redfield and others. Even these historical figures had predecessors pointing to the importance of consideration of multiple substances in biomass. For example, consider this quotation dating from the first decade of the twentieth century, years before Redfield's and even Lotka's writings:

> Chemical analysis shows that the animal and plant body is mainly built up from the four elements, nitrogen, carbon, hydrogen, and oxygen. Added to these are the metals, sodium, potassium, and iron, and the non-metals, chlorine, sulphur and phosphorus. Calcium or silicon are also invariably present as the bases of calcarious or siliceous skeletons. All these, with some others, are indispensable constituents of the organic body, and in an exhaustive study of the cycle of matter from the living to the non-living phases, and *vice versa*, we should have to trace the course of each (Johnstone 1908).

And before that, from the middle part of the nineteenth century, Liebig himself can be seen as perhaps the founding figure of ecological stoichiometry.

Quite a few stoichiometric properties of ecosystem processes have already been examined in this book (for example, consumer-driven nutrient recycling in Chapter 6). We have made numerous observations about the behavior of organic matter, tying it to the balance of carbon and other essential nutrients. Hopefully by now we have made a convincing case that stoichiometry plays a large role in explaining much variability in many of these individual fluxes of matter!

Now let us expand the scale. Entire ecosystems are comprised of almost countless individual fluxes. One fundamental question that arises therefore is, "When one adds them all up, integrating interactions, dynamics, and abundances across all species in a large habitat, is there still stoichiometric variability that guides whole ecosystems in one direction or another?" Alternatively, do other controls take over at this very large scale? In other words, are stoichiometric effects numerous, but often opposite in sign, so that when one considers entire trophic levels or entire ecosystems they tend to cancel and all or most ecosystems are stoichiometrically alike? Our approach in this chapter is to take the broadest perspective in terms of spatial and temporal scale, considering such issues as stoichiometric patterns across habitat gradients and habitat types, in successional time, and up to the global scale.

Ecosystem ecology is concerned with the fluxes of matter and energy in specified habitat units. Up to now, in describing the scope of ecological stoichiometry, we have asked which of the myriad of all ecological phenomena should be included. In the biogeochemical analysis of ecosystems, however, the situation is reversed: so much of what we know about the fluxes of matter and energy in ecosystems is based on mass balance of multiple substances, one must almost wonder what in this field *isn't* ecological stoichiometry. Biogeochemists often refer to the "coupling" of element cycles. Element cycles are coupled to each other in that there must be definite or constrained proportions at certain points in ecosystems. This, of course, is stoichiometry. So our challenge in this chapter is somewhat different from before; we do not intend to review all of ecosystem ecology in one chapter! Rather, we will touch on some of the most relevant high points and attempt to connect some of these familiar patterns with their underlying biological mechanisms identified in previous chapters.

With all the complexity that they subsume from lower levels of organization, ecosystems may have in a single location at a single time most of the various phenomena we have already discussed. This functional richness may be either a curse or a boon, depending on your perspective. The nutrient contents of the living and nonliving components within a single ecosystem will potentially show many different patterns. In the opening chapter, we considered C:N:P cycling in the oceans as described by A. C. Redfield. Oceanic ecosystems have relatively balanced pools and fluxes of C, N, and P in dissolved, autotroph, and consumer components over vast areas. We also saw in later chapters that many species, and many habitats, have imbalanced C:N:P ratios, and we discussed some of the implications of such imbalances. By adopting a broadly comparative approach here, by the end of this chapter we will develop some major contrasts between

stoichiometrically balanced ecosystems and ecosystems that are stoichiometrically imbalanced.

In this chapter, we begin with raw data and then move into more conceptual territory. The first part will discuss several patterns in C:N:P stoichiometry at the ecosystem level. We will consider patterns in nutrient
content across trophic levels within habitats and within trophic levels
across habitats. The balance of N and P in ecosystems has been thoroughly
studied since the time of Redfield, and we will consider some of the modern approaches to this topic. In doing so, we can see some of the broad-
scale patterns in cycling of multiple elements and gain some insight into
biological-chemical couplings at large scales. Other topics we will consider
include the C:N:P stoichiometry of the world's major rivers and stoichiometric patterns in successional time. We should note here that many of the
same themes we develop in this chapter also are discussed in a general
review by Harris (1999). Harris's theme that "the biogeochemistry of [the]
elements can be explained by the physiology and stoichiometry of the major functional groups of micro- and macro-biota in these systems" strikes a
familiar chord! We close this "observational" section with some ideas on
global oxygen cycles and the maintenance of a relatively homeostatic element content in this abiotic pool. The conceptual material to follow integrates many of the ideas we have considered up to now. We will develop
an extremely simple mathematical equation that relates stoichiometry and
growth (a theme of Chapters 3 and 4). Ecosystem ecologists have used the
concept of "nutrient use efficiency" as a guiding principle, and as this is a
stoichiometric concept, we will make heavy use of it here too.

In the last part of this last chapter, our focus is on the planet Earth as a
stoichiometric system. This focus is consistent with one of ecology's current frontiers: understanding the cycles and feedbacks associated with accelerating human impact on the Earth. Concepts we have used—homeostasis, feedback, coupling of element cycles—are a major part of this active
field. Ecologists and earth system scientists are still grasping for an understanding of the regulatory forces on major biogeochemical cycles. Elu-
cidating the "biotic feedbacks in the global climatic system" (Woodwell and
Mackenzie 1995) has been a priority. Negative feedbacks at the global
scale, in some respects the very reason for the persistence of life on earth,
inspire awe and have been likened to mystical gods ("the Gaia hypothesis":
Lovelock 1988; Lenton 1998).

EMPIRICAL PATTERNS IN ECOSYSTEM STOICHIOMETRY

To address ecosystem-level questions in ecological stoichiometry, we need
to know the distributions of nutrients within ecosystems, i.e., the amounts
and ratios of nutrients within such things as autotroph tissue, animals at

different trophic levels and in different life stages, decomposers, soils, etc. These amounts and ratios of nutrients in different components of ecosystems are determined by all of the processes shaping ecosystem structure as classically defined (which determine biomass patterns) as well as the differences in stoichiometry among species (which determine relative changes in nutrients). Although it may seem as though there should be many good descriptions of nutrient distributions in ecosystems in the scientific literature, sadly this is not the case. In fact, it has been pointed out that even biomass distributions within trophic levels have rarely been reported in the literature, even if the trophic pyramid is one of the cornerstones of ecology (del Giorgio and Gasol 1995). A comprehensive stoichiometric ecosystem analysis would require simultaneous measurement of biomass and several chemicals in several ecosystem components (perhaps measurements of C, N, and P in 5–10 major biomass and inorganic pools would be a reasonable start). This level of analysis either has rarely been attempted or has rarely been published, once done. Hence, it is beyond our scope (and far beyond the available data) to attempt an encyclopedic description of ecosystem stoichiometry. Instead, here we will review a set of diverse studies to see what a survey of the available information has to say.

First we must grapple with the fact that ecosystem-level stoichiometry is multivariate at several levels: at the level of the measurements themselves (e.g., C, N, and P) and at the level of the descriptor categories (autotrophs, decomposers, etc.). How can we best organize and think about such information? Tables of numerical entries convey the appropriate information, but they are not terribly useful as an analytical tool or to identify patterns. A means to express several chemical substances simultaneously within food webs of interacting species was suggested by Sterner et al. (1996) in a graphical device they referred to as "trophochemical diagrams." Trophochemical diagrams plot abundance of three substances using Cartesian coordinates for two of the substances and the area of a circle centered on those coordinates for the third. Then, indications of mass flow (or species interactions) are given by connecting these "bubbles" with arrows. A generalized trophochemical diagram is given in Figure 8.1A. Note how this diagram shows stoichiometric relationships among species in a food web. Species with balanced stoichiometry (similar $Y:X$ ratio) will be located on a single ray emerging from the origin (Fig. 8.1B). Species with large relative elemental imbalance will be dispersed across the plot. Three trophochemical diagrams for simplified lake food webs were also presented by Sterner et al. (1996) (Fig. 8.1C). Carbon, being closest to a measurement of biomass, was chosen for the circular area and N and P were plotted on the axes.

Ecological interactions with great stoichiometric imbalance, generating food quality constraints (Chapter 5) or indicating potentially large skewing

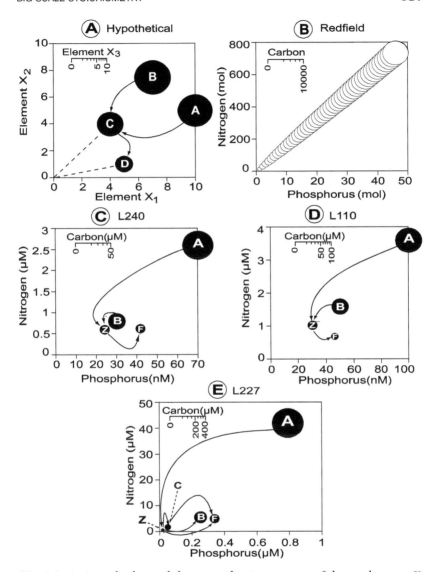

Fig. 8.1. A. A trophochemical diagram indicating amounts of three substances X_1 and X_2 (plotted on axes), and X_3 (circular area) within four hypothetical species or ecosystem components A–D. Two X_2:X_1 ratios are given by the slopes of the two dashed lines, and elemental imbalance between species C and D is given by the angle between the lines. B. The Redfield ratio on a trophochemical diagram. C–E. Example trophochemical diagrams for three lake ecosystems. For C–E, A is algae (large seston), B is bacteria (small seston), C is *Chaoborus*, F is fish, and Z is zooplankton. Based on Sterner et al. (1996).

of consumer-driven nutrient recycling (Chapter 6), will be easily identified in such a diagram. Although we cannot present them here, we have also produced animations showing successional patterns of C, N, and P with time. Consistent shifts in stoichiometry with increasing trophic level would generate similar patterns of nutrient fluxes on the diagrams (e.g., right-hand curves progressing from large biomass of autotrophs in the upper right to smaller biomass of higher trophic levels near the origin, if N:P ratios tend to decline at higher trophic levels).

Mental image is a powerful aid in organizing one's thoughts about complex problems. Examples in ecology include Eltonian pyramids, Odum's energy flow diagrams, and Lindeman's picture of the Cedar Bog Lake ecosystem. Trophochemical diagrams provide one tool to visualize ecosystem-level stoichiometric patterns. Now, let's see what some of those patterns are.

Trophic Levels

As Hairston et al. (1960) famously observed, the biosphere is, for the most part, green (for a more multihued perspective see Polis 1999). In other words, a very large fraction of the living biomass on Earth is to be found within autotrophs. Besides being biomass dominants, by taking up inorganic elements from their surrounding environment, autotrophs are major entry points for inorganic substances into biotic ecosystem processes. Hence, patterns at the whole-ecosystem scale (including all living things within any given geographic area) may often strongly reflect what the autotrophs are doing. We have already seen that photosynthetic organisms have an elemental composition that reflects the balance of materials and energy impinging upon them (Chapter 3). But variation in the stoichiometry of living biomass—even of the relatively variable autotrophs—is more restrained than in the nonliving world.

Although ecologists recognize a generally decreasing amount of biomass within trophic chains (highest in autotrophs, lowest in top consumers), the stoichiometric aspect concerns how much nutrient is to be found in these upper trophic levels. One aspect of this question is the ratio of autotrophic to heterotrophic biomass, which was considered by del Giorgio and Gasol (1995). These two argued that heterotrophic biomass is a decreasing fraction of total biomass as productivity increases (see also our discussion of the work by Cebrià, below), or, in other words, that more productive ecosystems are relatively "greener." Supportive evidence for this trend is also presented by Biddanda et al. (2001). According to del Giorgio and Gasol, at low autotroph biomass, lake ecosystems have a greater biomass of heterotrophs (including microbes and larger organisms) than of autotrophs.

So now what kind of stoichiometric signal might we expect to see across trophic levels? Because heterotrophic biomass is more stoichiometrically constrained than autotrophic biomass, we might expect larger and more variable C:P and C:N ratios as production increases. That would be logical given this information, but as we will see it is not what is observed. We will see that there is more to this pattern than just the relative biomass within trophic levels.

A few specific ecosystem-scale studies involving higher consumers consider nutrients in consumer trophic levels explicitly. Hessen et al. (1992) estimated the contribution of crustacean zooplankton to particulate phosphorus in lakes, using survey data from 45 Norwegian lakes of varying trophic status. They found that, on average, total zooplankton biomass contributed 20% of particulate P and the variation around this average was very high. Similar results were reported by Hassett et al. (1997) in a survey of 25 lakes in north central North America (zooplankton contained ~17% of particulate P). In the study of Hessen and colleagues, zooplankton were a particularly important component of ecosystem P at low productivity, reinforcing again the suggestion of increased importance of heterotrophs at low production. Along these same lines, a startling calculation relevant to the terrestrial realm comes from Moen et al. (1998, 1999) who estimated that the amount of P in just the antlers of a moose (shed annually) is equivalent to 10% of the P that cycles in the surrounding hectare of boreal forest!

It is clear by now in this book that various forms of heterotrophic consumers from small to large should generally have a more stoichiometrically fixed and more N- and P-rich composition than autotrophs. As a conceptual aid, consider the patterns sketched out in Figure 8.2A. The expected ranges of C:P in various pelagic trophic levels indicate a narrowing of ranges and a shift in mean toward more nutrient-rich biomass as we progress from the basal trophic level (i.e., algae) to different forms of consumers. We plot heterotrophic bacteria as a trophic level above phytoplankton because some of the carbon used by pelagic bacteria is from phytoplankton exudation and detritus. The actual trophic position of bacteria will vary with many factors. Note too that this stoichiometric trend illustrates a potential major loss of carbon relative to nutrients with increasing trophic position. As we have described elsewhere (Chapter 5), our view is that this consumption and loss of carbon relative to other elements in large part is the outcome of stoichiometric effects on trophic transfer efficiencies and represents a new explanation for trophic pyramids.

Ideal data to describe these ecosystem patterns would include direct measurements of C, N, and P in multiple ecosystem components across a great diversity of habitat types. As we have already bemoaned, such data are rare. However, we combined the compilations of data on algae and zooplankton from Elser et al. (2000b) with those of fish from Sterner and

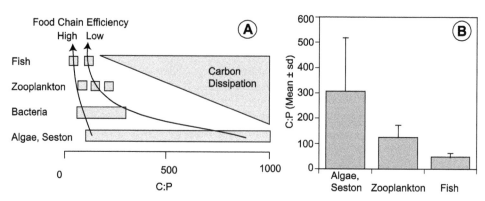

Fig. 8.2. A. Schematic of hypothesized range in C:P in several pelagic trophic levels. This figure represents an integration of many studies from the laboratory and the field, and is not based on any one particular data set. With increasing trophic level, organisms become richer in P, and the variance among species decreases. Note that elemental imbalance at the ecosystem level may vary widely. Predictions regarding food-chain efficiencies are discussed later in this chapter. B. Measured variability in C:P ratios across lake trophic levels. Figures for algae and zooplankton are from Elser et al. (2000c) and for fish are from the data set compiled by Sterner and George (2000) (see their Fig. 9).

George (2000) to arrive at one direct view of these trends (Fig. 8.2B). These compilations clearly show a decrease in the mean C:P and a diminishing of its variability (standard deviation) across trophic levels in going from algae to zooplankton to fish. Standard deviations are only one of many possible ways to measure variability; however, expressed also as coefficients of variation (the standard deviation as a percent of the mean) variability diminishes in the same direction (algae, 69%; zooplankton, 39%; fish, 29%). A similar pattern, although with different absolute values, would be expected for C:N. In fact, just such a pattern has recently been documented in terrestrial food webs: biomass C:N appears to decrease progressively as one moves from foliage to folivorous insects to predatory and parasitoid insects (Fagan et al. in press). Food chains consisting mainly of arthropods (plant–herbivorous insect–parasitoid insect–hyperparasitoid insect) likely will have different stoichiometric patterns than, say, short food chains containing large vertebrate herbivores. Here, we are hypothesis rich but data poor.

Comparisons across Habitats

Another way to gain some understanding of ecosystem-level stoichiometric patterns is to examine specific contrasts, such as those between marine and

TABLE 8.1
Total C, N, and P content (μg/g) and appropriate ratios in boreal forest soils and
lakes from the same region. Soils were sampled in the organic LF horizons
(partially decomposed forest litter). Compiled by Hecky et al. (1993) from other
sources

Ecosystem	C	N	P	C:N	C:P	N:P
Boreal soils	460	11	0.73	52	1700	35
Boreal lake sediments	200	12	2.0	12	270	23

freshwater ecosystems or between aquatic and terrestrial ecosystems. Several studies that we have already mentioned fall into this category. Patterns specific to autotrophs were covered in Chapter 3 and herbivores were considered in Chapter 4. Here, we will highlight some ecosystem-level distinctions that have been found.

Hecky et al. (1993) presented data on C, N, and P ratios of particulate mater for 51 lakes extending from arctic to tropical climatic regions, including small lakes as well as some of the largest lakes of the world. This paper was significant in part because of its conclusion that "Redfield ratios are the exception not the rule." They also reported that C:P and N:P ratios in freshwater systems were higher and more variable than in marine systems. Further, they found that within freshwater systems, C:N, C:P, and N:P ratios were all lower in subarctic lakes than in tropical or temperate lakes. These particulate composition ratios, they said, imply that there exist a wide variety of lake conditions, including N and P deficiency, as well as N and P sufficiency. They also compared terrestrial soils to lake sediments, showing that, in the boreal region they sampled, soils were much higher in C:N and C:P than were lake sediments (Table 8.1). Finally, they showed that particulate matter in meteorological precipitation and in streams was similar in composition to that of the nearby lakes and very different from surrounding soils. We will review some more patterns from running water below, but the major result from this study was the emphasis on the diversity of stoichiometric patterns within freshwater systems.

Elser and Hassett (1994) and Hassett et al. (1997) compared N:P in seston and bulk zooplankton communities in approximately 30 freshwater and 20 marine sites; seston data were also discussed in Chapter 3. They found that seston N:P was higher in lakes than in marine sites, but N:P of zooplankton was lower in lakes than at marine sites. Coupling these two trends, elemental imbalance between seston and zooplankton was shown to be very different in freshwater and marine sites. Freshwater particles were relatively P deficient compared to the composition of consumers, whereas in marine systems, particles were relatively N deficient. They also found a significant correlation between P content of particles and zoo-plankton community structure; namely, *Daphnia* were less abundant at high seston C:P and N:P (see also Chapter 5 for similar trends). Regional or site-specific differences in stoichiometric couplings may also explain differences among studies.

Elser et al. (2000b) compared autotroph and primary herbivore stoichiometry in terrestrial and freshwater aquatic systems, showing both similarities and differences in C:N:P stoichiometry across these habitats. As previously discussed (Chapter 3), terrestrial leaves were much higher in C:P and C:N but similar in their N:P to phytoplankton. Terrestrial herbivores (insects) and zooplankton were relatively nutrient rich in comparison, and were indistinguishable from each other (see also Chapter 4). Thus, stoichiometry at the base of freshwater and terrestrial food webs is primarily distinguished on C:nutrient axes rather than N:P. Such analyses are made possible because all biota rely on the same key suite of elements (C, N, and P), permitting us to examine disparate ecosystems in a common scheme and to better integrate knowledge of diverse habitats.

Nitrogen Debt and Organism Chemical Content

In Chapter 1, we described balanced cycling of N and P in the world's oceans and related it to the Redfield ratio, one of the cornerstones of ecological stoichiometry. This model of the world's oceans suggests that N comes into balance with the P cycle so that neither one alone should be expected to be strongly deficient for phytoplankton growth. We saw there that Redfield hypothesized that P ultimately limits marine production but that various mechanisms bring N into balance with P supply. In its most basic features, Redfield's view of the "evolution" of nutrient cycles to take on ratios similar to those of life and life's requirements contrasts with the models of variable abiotic and constant biotic stoichiometry emphasized in Chapter 6 where divergences in nutrient ratios were emphasized. Those models were driven by a stoichiometrically constant draw of mass from the abiotic pool. Let us examine some of these larger-scale "balancing" mechanisms a little more thoroughly.

A great deal has been written on biogeochemical balancing mechanisms for the N:P ratio in ecosystems. A key aspect is the ability of N fixation to make up an N debt. Organisms may tap the superabundant source of N_2 when they need it to make up for N deficiency. Well-known examples come from fertilized lakes (Schindler 1977; Hendzel et al. 1994) and from the general increase in cyanobacterial dominance at low N:P in lakes (Smith 1983b). Nitrogen fixation occurs at high rates when N is limiting, as can be ascertained from a plot of N fixation vs. N:P (Fig. 8.3). As discussed in Chapter 7, stoichiometric food-web feedbacks generating shifts in recy-

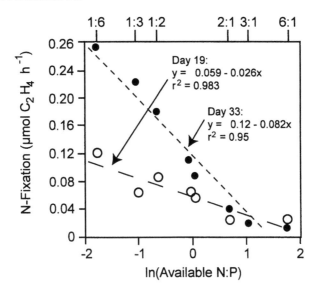

Fig. 8.3. Nitrogen fixation (measured by acetylene reduction) in soil as a function of soil N:P. Nitrogen fixation was linearly related to the logarithm of soil N:P, indicating that fixation generally makes up for a deficit of N relative to P. Nitrogen fixation rates increased with incubation time. Based on Eisele et al. (1989).

cling N:P can also impact N fixation (MacKay and Elser 1998b; Elser and Urabe 1999; Elser et al. 2000c). There are additional, more complex, controls on N fixation. Fixation of a single molecule of N_2 requires 16 molecules of ATP and the reducing power of eight electrons. Breaking the N_2 bond alone requires 226 kcal/mole (Schlesinger 1997). In other words, N fixation is energetically expensive and may not be a viable strategy when sufficient energy sources are lacking. Interactions of N fixation with trace-element availability have also been suggested. There is evidence that either molybdenum (Howarth and Cole 1985; Howarth et al. 1988a) or iron (Falkowski 1997; Karl et al. 1997) may exert some control over N fixation rates in systems where their concentrations are low. In the subtropical Atlantic, however, N fixation seems to be controlled more by availability of light and P than by metals (Sañudo-Wilhelmy et al. 2001). At the opposite end of the spectrum, biogeochemical mechanisms for "removing" N may also keep N and P in balance. Perhaps the best example is marine coastal denitrification. Most of the N reaching the ocean from land is converted to N_2 and ends up in the atmosphere rather than being transported offshore (Seitzinger 1988).

These biogeochemical mechanisms where the N and P (and possibly other) cycles intersect provide a different perspective on the way that living organisms alter their environments from the consumer-driven nutrient cycling models we saw in Chapter 6. Over the large temporal and spatial scales relevant to Redfield's hypothesis, biological production may well have a relatively constant stoichiometry, but life's ability to tap and drain different environmental pools of elements is still critical. Instead of driving the abiotic world further and further into N deprivation under N deficiency, an entirely new ecosystem feedback, i.e., N fixation, kicks in. Other examples with similar characteristics include microbial production of exoenzymes, which liberate necessary substrates from external abiotic pools. The inherent instability of systems with homeostatic consumers we saw in Chapter 6 is a powerful evolutionary and biogeochemical force. Only by overcoming this inherent instability can stable biogeochemical systems persist. Each of these separate regulatory mechanisms, from bacterial enzyme production, to consumer physiology and growth, to seasonal and longer periods of geochemical processing, occurs in some sort of nested hierarchy. Understanding the homeostatic regulatory points at each of these possible temporal and spatial scales will be a key focus for future study.

Nitrogen versus Phosphorus Balance in Aquatic Environments

The view of N coming into balance with a P-limited ocean contrasts with the classic view that marine production is primarily N limited (Ryther and Dunstan 1971), a view many modern studies continue to embrace (Falkowski 1997). In contrast, freshwaters have classically been considered to be mainly P limited (Schindler 1977). However, as we will now see, patterns in N:P stoichiometry in aquatic environments and resulting patterns in nutrient limitation are more complex than this. A clear-cut difference in nutrient limitation across marine and freshwaters has been called into question by several authors (Hecky and Kilham 1988; Howarth et al. 1988b) and a great deal of debate has ensued. The starkness of the contrast in viewpoints on the role of various nutrients in limiting oceanic primary production for examples could not be greater:

- "There is no evidence that phosphorus significantly limits primary production in coastal or open oceans on a global scale" (Falkowski et al. 2000).
- "Nitrogen is probably rarely the principal element limiting phytoplankton in the open oceans because its supply seems ample relative to phosphorus" (Downing 1997).
- "The relatively high N:P ratios in the near surface waters suggest that P, not N, is the production rate limiting nutrient in (the North Pacific subtropical gyre)" (Karl et al. 2001).

- "[D]ifferences between views of marine and freshwater scientists on nutrient limitation are perhaps based more on technique and inference than on any fundamental difference in ecology" (Hecky and Kilham 1988).

Let us now see what order can be brought to this seeming chaos. Let us consider freshwaters first. Elser et al. (1990) surveyed the literature on short-term nutrient addition experiments in lakes and found that 45% of experiments showed N stimulation of algal growth, similar to the ~50% that showed stimulation by P addition. Combined N + P enrichments yielded consistently larger stimulation than N or P alone in a great frequency (~90%) of experiments. Downing and McCauley (1992) examined N:P of potential nutrient sources for freshwaters (Table 8.2) and found that relatively undisturbed sources such as runoff from unfertilized fields, meteorological precipitation, and groundwater had high N:P while anthropogenic sources exhibited low N:P. As mentioned in Chapter 3, these shifts in lake TN:TP were correlated with shifts in algal response to N and P enrichment in bioassays. Their findings suggest a large stoichiometric difference in nutrient balance between background natural processes and human-impacted ones, with impact on the nature of autotroph nutrient limitation in lakes. Consistent with this difference in the N:P of nutrient sources, lake N:P decreased with increasing productivity (Fig. 8.4). This analysis suggests that oligotrophic lakes with primarily natural nutrient sources may be mainly P limited while those receiving considerable anthropogenic nutrients will tend toward N limitation. Figure 8.4 also illustrates a broad-scale stoichiometric constraint on N and P. Note that these two variables are mutually correlated across ecosystems. Some combinations of N and P are not observed. Why? This is perhaps just like Redfield's original observations of one element coming into balance with another (Fig. 1.11A). As we previously described, nitrogen fixation and denitrification may be key factors. Operating over a large scale, homeostasis of organism chemical content produces an ecosystem-level signal in total nutrient pools. It is somewhat tautological to say that organisms make the nutrient composition of ecosystems more like biota and less like the inorganic world. However, as we have seen, organism activities make ecosystems more biotic in ways beyond just adding their own biomass; their actions in tapping and draining biogeochemical pools may make total nutrient pools similar to biotic signals.

In marine waters, nutrient concentrations (Downing 1997) and nutrient addition experiments (Downing et al. 1999b) indicate that there are consistent differences in N versus P limitation patterns in different habitats. Specifically, nearshore waters close to anthropogenic sources as well as deep

TABLE 8.2

Average N:P ratios in potential nutrient sources in freshwater lakes. See Downing and McCauley (1992) for original sources

Source	N:P Mass	N:P Molar
Runoff from unfertilized fields	247.4	547.8
Export from soils, medium fertility	75.0	166.1
Export from forested areas	71.1	157.4
Export from rural and croplands	60.1	133.1
Export from soils, fertile	33.3	73.7
Groundwater	28.5	63.1
Precipitation	25.4	56.2
Runoff, tropical forest	23.5	52.0
Precipitation	23.2	51.4
Export from agricultural watersheds	20.0	44.3
River water	18.9	41.9
River water (Mississippi)	12.2	27.0
Sewage	10.0	22.1
Seepage from cattle manure	8.9	19.7
Zooplankton excreta	8.9	19.7
Fertilizer, average	7.9	17.5
Precipitation, tropical	7.7	17.1
Redfield ratio	7.2	16.0
Feedlot runoff	6.4	14.2
Sediments, mesotrophic lake	6.3	14.0
Urban stormwater drainage	5.8	12.8
Sewage	5.3	11.7
Zooplankton excreta	5.0	11.1
Pastureland runoff	4.8	10.6
Urban runoff	4.7	10.4
Sediments, oligotrophic lakes	3.3	7.3
Sewage	2.8	6.2
Septic tank effluent	2.7	6.0
Sediments, eutrophic lake	2.8	6.2
Gull feces	0.8	1.8
Rocks, sedimentary	0.8	1.8
Rocks, felsic	<0.1	<0.2
Earth's crust	<0.1	<0.2
Rocks, mafic	<0.1	<0.2

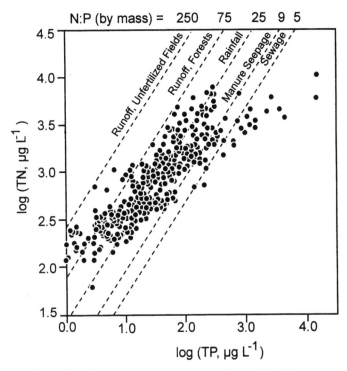

Fig. 8.4. Relationship between mean summer total N and total P concentrations in epilimnetic waters of world lakes. With increasing fertility, N:P ratios decline and more closely resemble animal waste products than natural background. From Downing and McCauley (1992).

TABLE 8.3
Frequency of total N:P ratios (TN:TP) greater or less than the Redfield ratio in different marine habitats. From Downing (1997)

	Percentage of Total N:P Ratios	
	<16	≥16
Open oceans, upper (< 50 m)	12	88
Open oceans, surface films	33	67
Estuaries, harbors, bays	39	61
Open oceans, lower (> 50 m)	64	36

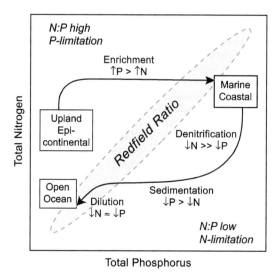

Fig. 8.5. Schematic of hypothesized shifts in total N:P balance from upland epi-continental systems (low-order streams, headwater lakes) subjected mainly to non-anthropogenic nutrient loading through enrichment producing lower N:P and emptying into marine coastal environments. There, high rates of denitrification coupled to other processes again raise N:P such that it is somewhat higher than the Redfield ratio (indicated by oval area). Moving offshore, sedimentation and dilution lower N and P, and increase N:P. For comparison to real data, see Figures 1.11 and 8.4. Figure drawn (with some artistic license) to illustrate the patterns described by Downing (1997).

waters in the offshore more often show ratios of total nitrogen to total phosphorus below the Redfield ratio compared to shallow samples from the deep open ocean (Table 8.3). Perhaps this is the source of the prevailing paradigm of dominant N limitation in the oceans: bioassays are frequently done nearshore and thus will consistently yield a response to N fertilization because of the low TN:TP in shallow marine systems. If you refer back to Figure 1.11C, which also comes from Downing's work, you will see that samples of the open photic zone (shallow depths in deep waters) cluster at the lower left of a plot of TN versus TP. This plot also resembles the one for lakes (Fig. 8.4) in that the slope indicates reduced N:P with increasing fertility. These systematic differences in N versus P balance in marine and freshwaters were synthesized into a conceptual scheme for global cycling (Downing 1997), which we compare graphically to the classical Redfield view in Figure 8.5. In this figure, the narrow oval-shaped region outlined with a dashed curve indicates the classical view of a Redfield ocean where a consistent balance between N and P is maintained.

This view may be appropriate for labile fractions of N and P. In contrast, the cycling of total N and total P seems to occupy a broader region of the graph, and one that has a shallower slope described by the circuit of solid arrows in Figure 8.5. From nutrient-poor upland freshwater systems of high N:P, enrichment lowers N:P until runoff is discharged into bays and estuaries. There, denitrification removes the great majority of N, further lowering N:P. Subsequent sedimentation and dilution reverse the trend and N:P increases as we move offshore. Both oligotrophic freshwater and marine systems occupy high-N:P regions where P limitation should prevail, but high-nutrient areas of both freshwater and coastal marine systems occupy low-N:P regions.

So perhaps the obvious distinction in nutrient limitation of marine versus freshwaters is not the appropriate one at all. One further reason the classical view may need to be reexamined is that micronutrients are now much better understood than in the past, driven in large part by advances in sampling and measurement techniques (Bielefeld et al. 1996). Although enough evidence has accumulated to indicate an important role of iron in regulating marine production, comparisons to freshwaters are still difficult due to the fact that relatively few inland systems have been studied with modern metal-clean methods.

Rivers of the World

The world's large rivers link inland and oceanic systems. The concentrations of nutrients observed at the outflow of major rivers are a result of meteorological precipitation chemistry, weathering, human nutrient transports and uses, uptake and retention within watershed, and *in situ* sedimentation and conversions such as denitrification. Major rivers drain large continental regions and thus integrate a vast biological, chemical, and geological territory. These are excellent systems to look at integrations of cycles across large land areas. Perhaps this integration results in enough averaging of different processes that major rivers are similar in their elemental composition. Large rivers also exhibit different degrees of pollution, from systems such as the Yukon and Mackenzie Rivers that drain regions with low human population density to ones such as the Mississippi and Rhine Rivers that drain highly industrialized and intensively cultivated regions (Turner and Rabalais 1991; Cole et al. 1993b; Justic et al. 1995). Therefore, they also can be used to see whether there is a common anthropogenic stoichiometry signal.

Meybeck (1982, 1993) considered the concentrations and biogeochemical controlling forces on C, N, P, and S in rivers (Table 8.4). He argued for a relatively constant value for riverine particulate C:N of about 10:1 (original data from Ittekot and Zhang 1989). It proves to be difficult to evaluate

TABLE 8.4
Global average of C, N, and P (mg/L) in unpolluted world streams and rivers
(Meybeck 1993). Values in the middle column are weighted by discharge volume.
Those in the right column are time and space averaged without considering
discharge

Chemical Species	Water-Discharge-Weighted Concentration	Most Common Natural Concentration
Dissolved inorganic carbon	10.2	6.0
Dissolved organic carbon	5.75	4.2
Particulate organic carbon	4.8	3.0
Total organic carbon	10.5	
$N\text{-}NH_4^+$		0.015
$N\text{-}NO_3^-$		0.10
Dissolved organic nitrogen		0.26
$P\text{-}PO_4^{3-}$	0.025	0.010

constancy of stoichiometric ratios across rivers in part because of an absence of data on some important chemical fractions. Most of Meybeck's analyses concern dissolved available nutrients such as nitrate and phosphate. Summaries of TN and TP are surprisingly hard to find. Redfield's hypothesis about P controls on oceanic productivity in geologic time would lead us to expect high N:P (total fractions) in the outflow of major rivers. Similar to the scenario sketched by Downing, Justic et al. (1995) suggested that increased eutrophication has had a large and consistent effect on river stoichiometry, often making proportions of dissolved N and P closer to the Redfield ratio than they are in unperturbed systems. An exception would be the lower Mississippi River, which has high TN:TP due to N leakage from the Corn Belt. It seems likely that the question of large-scale integration by rivers will have a complex solution, with some parameters demonstrating a "central limit theorem" of stoichiometry while others are much more spatially and temporally variable.

Stoichiometry in Time

Another place to look for stoichiometric patterns is along a time axis. Succession is one of the oldest concepts in ecological science. Ecosystem-level nutrient dynamics during primary or secondary succession relate to many processes, such as weathering, nitrogen fixation, dryfall, evapotranspiration, soil development, and biomass accrual (Gorham et al. 1979). Few, however, have taken an explicitly stoichiometric approach to this subject.

Over the long time spans comprising primary succession in terrestrial ecosystems, soil development is a major structuring process. Primary succession includes an overall shift from a strongly abiotic stoichiometric signal to one dominated by biology. Walker and Syers (1976) said that early in the development of a new soil from parent material, most soils already contain all of the P that they will ever have. This hypothesis considers atmospheric P input to ecosystems to be minor. In contrast, most parent materials contain little or no N. The early stages of succession are marked by rapid accumulation of N from the atmosphere via fixation. In this view, the biota act to bring the ecosystem's nutrient content closer to a biotic balance point. Later in succession, P is gradually lost or bound in insoluble or biologically unavailable forms. High C:P ratios in foliage in forests on highly weathered tropical soils are consistent with P scarcity in these systems. This model has held up to tests involving experimental nutrient additions (Vitousek et al. 1993; Vitousek and Farrington 1997; Hobbie and Vitousek 2000). Such patterns in plant C:N:P ratios along successional gradients likely have predicable consequences on secondary production, consumer-driven nutrient cycling, and other processes we have discussed.

Secondary succession of forests following clearing is typified by a rapid flush of nonwoody and shrubby vegetation followed by slower return of dominance by large trees. Examination of nutrient distributions during this secondary succession provides a context where we can see big shifts in biomass dominance from photosynthetic to structural tissues, which should have strongly divergent stoichiometry (Chapter 3). Reiners (1992) described the patterns in elemental composition in an experimental watershed at the Hubbard Brook, where all woody vegetation was cut and left in place and vegetative regrowth was suppressed with herbicides for an additional three years. During regrowth, samples of different vegetative biomasses were taken and total element budgets estimated. The results (Fig. 8.6) demonstrate the large buildup of nutrients in vegetative tissues over the almost 20-year study period. Both N (Fig. 8.6A) and P (Fig 8.6B) remained relatively low in herbs and shrubs, but trees became major nutrient pools over time. The N:P ratio was not greatly dependent on these successional dynamics (Fig. 8.6C). The similarity in N:P of these major autotroph types reminds us of the similarity in N:P ratios between microscopic pelagic autotrophs and leaves (Chapter 3). Reiners also calculated the amounts of different elements within different tissue types (new wood and bark, old wood and bark, and leaves; Figure 8.6D). These data are also important to us from the standpoint of how representative foliar nutrients, a variable we have often made reference to, are compared to total forest ecosystem nutrient budgets. Foliar nutrients were the greatest fraction of total vegetative nutrients approximately ten years after succession began rather than immediately. By the last sampling period, foliar nutrients were on the order of 25–35% of total vegetative nutrient masses. Individual elements also demonstrated some unique dynamics; for example, there

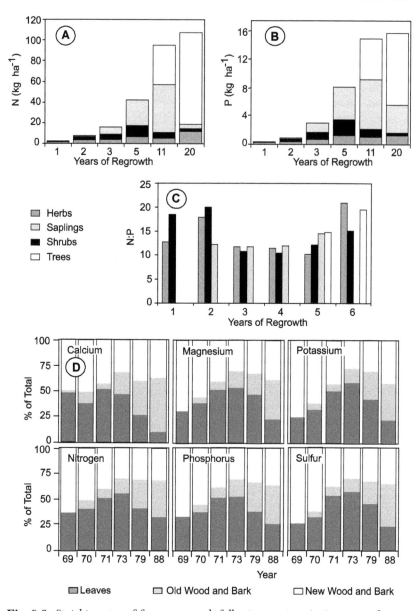

Fig. 8.6. Stoichiometry of forest regrowth following cutting. A. Amounts of nitrogen in different vegetation types as a function of years of regrowth. B. Same as A, but for phosphorus. C. Same as A and B, but for N:P. D. Concentrations of various elements in different tissue types. Based on Reiners (1992).

was a particularly large depletion of foliar Ca by the last sampling. Reiners suggested that elements most directly involved in metabolism and less in structure accumulated rapidly relative to biomass.

Stoichiometric shifts in geological time were the subject of a study by Murphy et al. (2000), who hypothesized a startling connection between stoichiometric uncoupling and mass extinctions. These workers examined C:N:P stoichiometry in organic matter buried in about 100 m of sedimentary deposits which temporally bracketed the Late Devonian mass extinction. During this approximately three-million-year period, there were wholesale losses of marine taxa, particularly from shallow tropical environments. Estimates are that as many as 82% of species were lost during this event (Labandeira and Sepkoski 1993). The study of Murphy and colleagues is a rare consideration of nutrient stoichiometry in time scales of this length. The authors observed variable C:N:P; excursions in marine sedimentary material from nearly perfect Redfield values of 100:15:1 to eye-popping values of 5000:170:1 were seen. Apparently the C:N:P stoichiometry of long-term burial of organic matter is highly variable. Factors that promote nutrient recycling rather than burial at the sediment-water interface include anoxia, which may be promoted by increased production (discussed further below). Furthermore, these departures from Redfield proportions were approximately coincident with increased $\delta^{13}C$ in the organic matter. One mechanism for increased ^{13}C is increased productivity (increasing CO_2 drawdown). Hence, both the stable C isotope data and the C:N:P data suggest that the Devonian mass extinction occurred during a time of increased productivity. While the authors interpreted the data largely from a diagenetic (sediment formation) perspective (i.e., they assumed that organic matter entered the sediments at the Redfield ratio and C, N, and P were selectively regenerated or retained during diagenesis), we wonder whether some deviations in the C:N:P of primary production in overlying waters may have somehow been involved in triggering these excursions in sedimentary C:N:P.

Homeostasis of the Abiotic Oxygen Pool

In this book we have argued that there is generally an overall regularity in the chemical content of living things that exists against a backdrop of wide abiotic variability in the same parameters. We have discussed from a variety of theoretical and empirical angles how the "drain" of a constant chemical formula from a coupled stoichiometric system (i.e., during the production of a new biomass by a homeostatic organism) creates instabilities in abiotic nutrient cycling, as in the models of CNR in Chapter 6. It is time now to consider a contrary set of patterns and see what can be learned about their causes.

The concentration of oxygen in our atmosphere has remained within the bounds of 15 and 35% over the past 500 million years (figures on the oxygen cycle come from Schlesinger 1997). Since the (modern) residence time of oxygen in the atmosphere is about 4000 years, its concentration has held relatively steady for more than 100,000 turnovers. What can account for this remarkable constancy of a dynamic abiotic pool? Globally, the oxygen cycle is driven in part by the buildup and breakdown of organic matter, and the creation of one mole of glucose generates as much oxygen as its breakdown uses [Eq. (1.1)]. Hence, a constant biomass of organic matter built up by photosynthesis and broken down by respiration would not perturb the standing concentration of atmospheric oxygen. But that is not the end of the story. Total global biomass over long time scales is not constant. Furthermore, the relatively rapid biological processing of oxygen is combined with a slower geological-scale processing, consisting of organic C burial (in oxidized form and hence taking with it some oxygen), and oxidation of uplifted iron and subsequent oceanic burial. Other forces acting on atmospheric oxygen include production of water by photolysis and oxidative destruction by such things as methane and ammonia. Taken separately and added up, there is no particular reason why all these inputs and outputs have to balance. Furthermore, too much of a deviation from today's level of oxygen would have catastrophic consequences. Oxygen levels higher than 30% would cause global wildfire; levels below 10% would not support large animals (Kump and Mackenzie 1996). Several explanations have been hypothesized to explain this crucial pattern of the constant redox state of the atmosphere.

One explanation for the homeostasis of atmospheric oxygen was suggested by Van Cappellen and Ingall (1996). Their model connects atmospheric oxygen levels to the efficiency of burial of phosphorus and level of productivity in the oceans. Over geologic time scales, they consider marine productivity ultimately to be P limited (see arguments in Chapter 1 and in the section above). To understand the basis of homeostasis of oxygen in their model, consider what to expect under higher than "normal" atmospheric oxygen. If respiration were unchanged, high atmospheric oxygen would eventually cause increased oxygen in deep marine waters. High oxygen promotes efficient P burial due to enhanced binding with Fe hydroxides at high oxygen (Mortimer 1941, 1942; Ingall et al. 1993). Efficient P burial would mean low C:P in marine sediments and reduced delivery of P back to upper waters, lowering oceanic productivity and thereby slowing marine oxygen production. In the section above, we saw how variable the C:N:P stoichiometry of organic matter sedimentary deposits can be. These processes constitute a negative feedback loop on atmospheric oxygen. The reverse situation (inefficient P burial and high sediment C:P) can be expected under low atmospheric oxygen. Van Cappellen and Ingall quantified these relationships and showed with numerical simulations that reasonable parameter values could stabilize atmospheric oxygen at appropriate time scales.

It is interesting to compare these processes to those we have encountered before, which generate unstable and variable abiotic pools. In the Van Cappellen and Ingall model, the driving process is the geochemistry of buried sediment. The stoichiometry of the "drain" of material in buried marine sediments is not tightly constrained; instead, the C:P of organic matter burial varies highly according to oxygen availability. In homeostatic nutrient recycling models, the constancy of nutrient content in one part of the ecosystem generates instability at another part. In contrast, in this model of atmospheric oxygen, it is the stoichiometric variability of deep ocean burial that produces constancy elsewhere in the system. It is a sobering thought that life as we know it may depend on the "lucky accident" that the C, P, Fe, and O cycles interact in particular ways at the bottom of the ocean. Here is another connection between the physical chemistry of elements (as we saw in Chapter 2) and the functioning of living systems. Is this what the face of "Gaia" looks like?

The differences in these examples seem to relate well to time scales. At the time scale of summer growing seasons, the life spans of individual zooplankters and hence the time scale of zooplankton population increases are meaningfully long and thus can drive variation in abiotic C:N:P by drawing C, N, and P into a homeostatic pool. However, over geological time scales, the life spans of individuals, or of population change, are insignificant and are averaged out and only the stoichiometry of long-"lived" pools associated with biology can have a role: e.g., the effects of detrital pools with very long turnover. Since detritus does not have the strong biochemical or cellular constraints of living systems (e.g., Chapters 2–4), it can draw C:N:P at widely divergent ratios, which under particular conditions can create constancy in the remainder.

At a more abstract level, this example suggests that in a dynamically coupled stoichiometric system, constancy of one pool (whether biotic or abiotic) must be accompanied by variability in another pool. Let us call this a new stoichiometric theorem. We are assuming that some potential pools of nutrients are stoichiometrically different from others, where, even with dynamic coupling, all elements could move around at the same rates and a trivial equilibrium could exist.

This finishes our survey of large-scale stoichiometric patterns. These large-scale trends are a backdrop against which many of the dynamical processes we have described throughout the book must operate. In the remainder of this chapter, we will consider several important ecosystem-level patterns in stoichiometry in greater depth. To begin, we will present a mathematical model linking several of the concepts from elsewhere in the book. Once we have these linkages established, we use them to analyze some specific patterns of C:nutrient linkages in ecosystems.

LINKAGES IN THE STOICHIOMETRY OF BIOMASS YIELD: USING ONE SUBSTANCE TO OBTAIN ANOTHER

In its most basic form, primary production is a physicochemical reaction of the form

$$\text{Light Energy + Inorganic Carbon + Nutrients} \rightarrow \text{Biomass + Heat.} \quad (8.1)$$

All beginning biology students learn the reactions of photosynthesis with fixed stoichiometric coefficients [Eq. (1.1)]. Although correct as far as it goes, it seems to us that this is not a good overall impression of biomass accretion by autotrophs. Neither photosynthetic rate per unit of carbon in the plant cell nor per unit nutrient in the cell is constant. As we have seen, plant biomass C:nutrient and N:P ratios vary widely within and across species (Chapter 3). What we miss by looking only at fixed stoichiometric equations is an impression of the flexibility of the chemical reactions of biomass gain by primary producers. Even for carbon alone, the specific activity of plant carbon (the light-saturated amount of C fixed per unit biomass, or what we will refer to here as photosynthetic efficiency) varies substantially (Fig. 8.7).

The mechanism underlying the positive relationship between the two variables plotted in Figure 8.7 is that leaves with high concentrations of photosynthetic enzymes (RUBISCO) and pigments simultaneously have high photosynthetic efficiency and high N content. The driving factor in the stoichiometric flexibility is changing plant allocation of matter to N-rich photosynthetic enzymes. This figure harkens back to the autotroph growth models covered in Chapter 3. Because we can assume constancy of C:biomass ratios, the vertical axis of Figure 8.7 is closely related to specific growth rate and hence this plot is an empirical demonstration of the theoretical model of Ågren (Fig. 3.11, where $\Gamma_n < \Gamma_{n,\text{opt}}$). Another interesting aspect of Figure 8.7 is that it visualizes the use of one substance (in this case, N) to gain another (in this case, C). A measure of the efficiency of plant N use for carbon fixation is obtained if we divide the measurement on the vertical axis by the measurement on the horizontal axis. A parameter with the following general dimensions results:

$$\frac{\text{C fixed}}{\text{leaf N} \cdot \text{time}}. \quad (8.2)$$

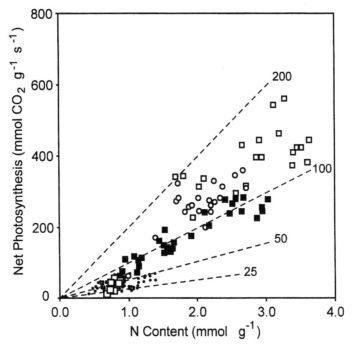

Fig. 8.7. Light-saturated photosynthetic efficiency (carbon fixed per unit biomass per time) relates strongly to the N content of C_3 leaves. This is because the N and C cycles are linked through N-rich RUBISCO and other enzymes that make up a large fraction of leaf biomass. Dashed lines indicate different values of nitrogen productivity ($\Phi = \dfrac{\text{mmol } CO_2}{\text{mmol N s}}$). Based on Field and Mooney (1986).

The ratio of the vertical and horizontal measurements (values indicated by the four rays emerging from the origin in the figure) has been called the "photosynthetic nutrient use efficiency (PNUE)" (e.g., Poorter and Evans 1998) or (as we described in Chapter 3) the "nutrient or nitrogen productivity" (Ingestad 1979b; Ågren and Bosatta 1996). Recall from Chapter 3 that we use the symbol Φ'_C to indicate the mass of new carbon production per unit plant nutrient. Nitrogen productivity in this data set varies approximately from 25 to 200, an eightfold range. Note too that this plot indicates a stoichiometric constraint: the points are not scattered all over the plot. Given information about N in biomass, something about carbon-specific rates of light-saturated photosynthesis is known. The carbon and nitrogen cycles therefore are linked at the leaf level through stoichiometric constraints captured by Φ'_C.

Another trend is visible in Figure 8.7. Plants with high N content not only have higher light-saturated photosynthesis, they also have relatively high Φ'_C compared to leaves with lower N content. This is apparent in that leaves with the highest N content have Φ'_C in the 100–200 range while leaves of lowest N content have Φ'_C in the 25–50 range. In other words, nutrient-rich leaves are more efficient at using N to gain C, most likely because at high N relatively more N is located in the photosynthetic machinery, compared to structural proteins, defensive compounds, etc.

There is much more we can do with this concept of using one substance to obtain another. First, consider the question, "How does Φ' relate to biomass nutrient ratios?" To answer this, we will consider an analysis conceptually similar to one given by Berendsee and Aerts (1987), although our means of presenting these concepts differs and we derive more generalized conclusions. We begin by making the definition for PNUE more mathematically precise:

$$\Phi = \frac{dC}{Ndt}, \tag{8.3}$$

where C stands for plant carbon and N stands for plant nitrogen (molar units). To simplify things, for the time being we will ignore loss terms such as respiration and exudation, although these may be important considerations in various situations (Siddiqi and Glass 1981; Chapin 1991a; Aerts 1995). Next we recognize that at stoichiometric equilibrium (balanced growth), the specific rate of growth (dM/Mdt where M is the mass of an element) of all elements is identical, or in other words,

$$\frac{dC}{Cdt} = \frac{dN}{Ndt}. \tag{8.4}$$

In fact, at stoichiometric equilibrium, biomass itself (M) has the same specific growth rate as any individual component of biomass:

$$\frac{dC}{Cdt} = \frac{dN}{Ndt} = \frac{dM}{Mdt} = \mu, \tag{8.5}$$

where μ is the specific growth rate as usual.

Rearranging (8.4), we get

$$\frac{dC}{Ndt} = \frac{C}{N} \frac{dN}{Ndt}. \tag{8.6}$$

Note that the first term (to the left of the equals sign) is Φ (rate of C gain per unit N, reflecting the use of N to gain C). On the right side of the equation we see the molar ratio of carbon to nitrogen and the specific mass flux of N, which under balanced growth is the same as the growth rate of biomass. Equation (8.6) decomposes Φ into a biomass ratio and a specific growth rate. Equation (8.6), simple as it is, contains a spectacular insight; if the term to the left of the equals sign is constrained (as it was in Fig. 8.7), a relationship between biomass composition and growth rate is defined. For generality, we will revert to symbols X_1 and X_2 to indicate molar quantities of two substances (which may be C and N, or C and P, or perhaps even proteins and ribosomes). Also, let us consider the efficiency of X_2 in gaining X_1 to be as fixed as it can be; in other words, it is a constant (Ψ). This gives us

$$\Psi \cdot \frac{X_1}{X_2} = \mu. \tag{8.7}$$

This is a grand result: one equation with two assumptions—balanced growth and constant efficiency of X_2 in gaining X_1—links biomass composition and growth rate by a direct proportionality. Under these assumptions, biomass composition and growth rate must be related (as we saw they were in Chapter 4 for animals using P to obtain mass or for cells using ribosomes to obtain proteins). A fundamental insight gained by Equation (8.7) is that stoichiometric biomass ratios should be directly linked to growth (a major theme of this book) whenever the efficiency of using one substance to get another is constant. A corollary is that they will be unlinked when that efficiency is unconstrained.

A second fundamental insight from this equation comes when we consider that C:nutrient ratios relate directly to competitive ability (Chapter 7). All else being equal, a species with high biomass C:nutrient ratio is a superior competitor for that nutrient than other species [Eq. (7.3)]. Rearranging (8.7), we get a clear picture of a trade-off in competitive ability (biomass ratio) and growth rate:

$$\frac{X_1}{X_2} = \frac{\Psi}{\mu}. \tag{8.8}$$

With constant Ψ, the higher the value of $X_1{:}X_2$ the greater the competitive ability for X_2, but the lower growth must be because of the proportionality with Ψ.

Trade-offs in growth and competitive ability have long been a part of thinking in evolutionary ecology (MacArthur and Wilson 1967) and have also been related to general power-efficiency trade-offs (Odum and Pinkerton 1955; Smith 1976; Arendt 1997). We will mention two recent examples that seem highly relevant. Tessier et al. (2000) recently reported that there was a trade-off among zooplankton between growth rates on rich and poor resources. They reported that their results indicate that "*Daphnia* trade off high maximum growth rates for low minimum resource requirements," a perfect fit to this general theory. For the second example, Wardle et al. (1998) suggested that certain plant traits were closely associated with leaf N content. These included palatability to herbivores (higher with high N content) but also competitive ability (lowest with high N content).

Earlier, we left aside the subject of nutrient losses and how they enter into our analysis of efficiencies. However, mass balance may be altered by patterns in both gains and losses. In terrestrial plants, loss of dead and scenescent tissues is a major feature of the life cycle (Thomas and Sadras 2001). One way to think about losses is simply to redefine $\frac{dX_1}{X_2 dt}$ as an expression of net change (gains vs. losses). This would change the physiological mechanism being described by the equations, but it would not change any of the more abstract insights we have derived. Another approach is to turn the whole analysis into one of losses. That is, by symmetry, we treat matter gains as irrelevant (constant) and think about $\frac{dX_1}{X_2 dt}$ as a description of losses instead of gains. At the level of mass balance of the entire organism, both losses and gains enter into our consideration of growth and efficiency trade-offs.

A third fundamental insight from Equation (8.6) comes when we consider more about biomass ratios. In terrestrial ecosystem studies, the use of a nutrient to gain biomass appears in the concept of "nutrient use efficiency." There are various definitions of NUE, but it is always an amount or flux of carbon divided by an amount or flux of nutrients. Chapin (1980), for example, defined nutrient use efficiency simply as biomass C:N. In its various guises, the numerator and denominator have been measured as mass quantities (e.g., C or N in leaf litter, C or N in plant tissue, N in soil), or as fluxes (net primary production, N mineralization). We will review some of the empirical studies of NUE below, but for now, we want to point out that the equations we have just covered can be used to relate three very fundamental ecological parameters, Φ, μ, and NUE. Under stoichiometric equilibrium, these three are related in a very simple way:

$$\mathrm{NUE} = \frac{\Phi}{\mu}. \tag{8.9}$$

A similar derivation has been around for a long time (Hirose 1975), but to our knowledge it has been used only in plant physiology literature dealing with C and N; we believe these trade-offs are far more general.

In this section we have brought together a number of major concepts from the book. Among these are growth-nutrient couplings, competition-nutrient couplings, and growth efficiency. We have provided a model that links a lower-level process (nutrient productivity Φ, or its generalized form Ψ) to higher-level ones (biomass ratios, growth, and NUE). This theory says that these couplings and linkages should be most prevalent when, for some physical, chemical, or biological reason, there are constraints within organisms on how one substance determines the rates of change of another substance. Another application of this thinking could be to N fixation, which one can think of as use of Fe to gain N (see Sañudo-Wilhelmy et al. 2001). Chapters 2–4 give us some inkling of the biological basis of these constraints for C, N, and P: the coupling of elements in key molecules (proteins, nucleic acids) and the dominant role that major biochemicals play in particular biological functions (e.g., RUBISCO in photosynthesis, rRNA in protein synthesis). These "transforming" substances need to be present in sufficient quantities or they themselves will regulate growth, which is an important aspect of these kinds of constraints. One key feature related to the growth rate hypothesis is the protein synthesis rate per ribosome: if this value is constrained, then the hypothesized connections linking C:N:P stoichiometry, ribosomal content, and growth rate should be valid and widespread. The closer values such as these come to a global constant, the tighter the growth-stoichiometry linkages.

NUTRIENT USE EFFICIENCY AT THE ECOSYSTEM LEVEL

In nature, nutrients are often the "limiting reagent" in primary production [Eq. (8.1)]. Primary productivity is generally positively related to measures of nutrient availability or fertility in lakes (Vollenweider 1968), grasslands and wetlands (Vermeer and Berendse 1983), temperate streams (Van Nieuwenhuyse and Jones 1996), and forests (Vogt et al. 1986). Primary production is usually increased when the system is fertilized, a fact the global human population now relies on for its subsistence (Loneragan 1997). In fact, the increased yield of primary production in ecosystems in response to added nutrients is one of the most repeatable and predictable features of nature, and thus these facts are some of the most basic ones in all of ecological science. However, although biomass and nutrients generally are positively related across systems, the mapping from one to the other is often not a perfectly straight line. There often is considerable scatter in plots of biomass versus nutrients (Shapiro 1980) and the relationships are often not linear (Smith 1982; McCauley et al. 1989). In other words, the stoichiometry of conversion of nutrients to biomass production in ecosystems is highly variable. Two major factors helping to explain this conversion are trophic structure (Carpenter et al. 1985; Sarnelle 1992; Mazumder and Lean 1994) and plant physiological response (Chapter 3; Droop 1970; Chapin 1991a).

In Fig 8.7, we saw that leaves with high N content were more efficient at using N to obtain C; this is because relatively more N is allocated to

C-obtaining machinery at high N content. By implication, then, plants with high growth rates, or plants on highly fertile sites, might be expected to be more efficient at using N to obtain C. However, this seems backward. Shouldn't efficiency be greatest when resources are scarce, not when they are highly abundant? Equation (8.9) related NUE to PNUE and growth. Consider what to expect under fertilization based on that equation. Berendse and Aerts (1987) pointed out that, with increased fertility, NUE tends to increase from increased PNUE but it also decreases from increased μ. Ingestad (1979b) showed that in birch seedlings grown at a range of fertilities, "nitrogen efficiency" (dry matter produced per unit nitrogen) decreased with increasing nitrogen status, but "nitrogen productivity" (dry matter produced per unit nitrogen and time) increased up to optimum. Hence, from mechanistic principles alone, it may not be easy to know how NUE and fertility truly are related.

Terrestrial Examples

Empirical patterns of NUE at the ecosystem level were first examined by Vitousek (1982) who studied carbon and nutrients in a wide diversity of terrestrial forests from temperate to tropical. Vitousek used steady-state assumptions of zero interannual change in carbon stocks as well as balanced growth (stoichiometric equilibrium) and suggested that annual net primary productivity could be measured as the quantity of carbon in litterfall per year. Also, N lost from litter at steady state equals N uptake by the plant community. A plot of litterfall biomass versus litterfall N mass (Fig. 8.8A) shows an increasing but saturating trend. All of this together implies that there is diminished biomass production per unit N uptake as productivity increases. This was interpreted as a changing NUE with fertility; ecosystems at low fertility showed higher NUE. Another point we need to make here is that, although the curvilinearity in this plot is not visibly striking, small degrees of bend can indicate large differences in the efficiencies involved. More than a fourfold range of efficiencies is represented by the envelope surrounding these data (see the indicated rays in Fig. 8.8A).

Vitousek's work spawned a great deal of further study on NUE at the ecosystem level, only some of which we have space to cover here. An interesting example comes from Pastor et al. (1984), who examined patterns of productivity in eight temperate forest stands as a function of soil processes, including N mineralization. Their results also indicated that, with increased fertility, ecosystem productivity increased in hyperbolic fashion, demonstrating reduced growth efficiency with greater growth rates (Fig. 8.8B). Bridgham et al. (1995) proposed a different decomposition of NUE, arguing that patterns of growth as a function of nutrient uptake should be different from patterns of growth as a function of nutrient availability in soil, particularly at very low fertility (see also Pastor and Bridgham 1998). They presented empirical patterns of production as a function of soil nutrients in peatlands that were consistent with the general pattern of diminished growth efficiency at greater productivity (Fig. 8.8C). Note the effect of a positive x intercept in Fig. 8.8C. Given this intercept, this measurement of NUE is maximal at intermediate soil P; however, a decrease of NUE with increased fertility still occurs at the high end of the soil P range, for the reasons we have already discussed. Our final example concerns N and P growth efficiencies in several dominant temperate tree species on two neighboring sites with contrasting fertilities. Boerner (1984) showed that in nearly all comparisons statistically significant differences in growth efficiency across sites were consistent with the general trend of lowered NUE at higher growth rate (Fig. 8.8D). Although in all these cases the precise definition of NUE varied, from these and other empirical studies it is apparent that terrestrial systems generally show reduced efficiency of converting nutrients into biomass at increased fertility. In Chapter 3, we highlighted fundamental similarities in the physiological coupling of stoichiometry and growth for terrestrial and aquatic autotrophs. Let us now see if cross-habitat similarities are apparent at the ecosystem level as well.

Aquatic Examples

R. A. Vollenweider published a series of papers over a time span of about 20 years exploring the relationships between nutrient supply and both primary productivity and biomass, and considering the role of factors such as lake hydraulic residence time and sedimentation (Rodhe et al. 1958; Vollenweider 1965, 1968, 1970; Vollenweider and Dillon 1974; Vollenweider 1975, 1976). These were highly influential papers; they combined an ecosystem approach with an engineering one and contributed greatly to the restoration of culturally eutrophic lakes. Early on in this work (Rodhe et al. 1958), the suggestion of reduced ecosystem efficiency at high fertility was already evident (Fig. 8.9A). This log-log plot of areal photosynthetic rate versus areal phytoplankton biomass shows our familiar increasing but saturating shape. This figure suggests that dense phytoplankton aggregations fix less carbon than would be expected from extrapolation from sparser populations. This figure does not represent nutrient use efficiency per se; rather, it is an expression of the efficiency of biomass at fixing carbon. Nevertheless this trend also carries the same message about changed efficiency of biomass creation with changes in fertility (or biomass).

A second example, more directly comparable to Figure 8.8A, comes from Smith (1979), who plotted areal photosynthesis against total phos-

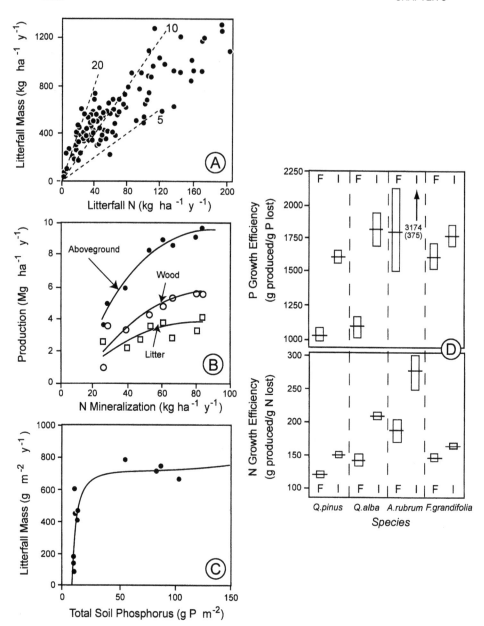

Fig. 8.8. The efficiency of terrestrial ecosystem biomass production per unit nutri-ent diminishes with increased productivity. This figure shows four somewhat differ-ent approaches to measuring that efficiency. A–C show hyperbolic functions indica-tive of lowering efficiency with increased values. A. The relationship between the

phorus. Smith's data come from a single lake and represent seasonal variability. This study and several to follow are based on total phosphorus, which is the most widely used measure of lake fertility. Areal photosynthesis reached a plateau at high fertility (Fig. 8.9B). Data published by Megard (1972) show the same trends when plotted on these axes (Fig. 8.9C). Thus, it appears that areal photosynthesis as a function of total phosphorus in lakes has a strongly saturating relationship. In fact, there are theoretical models of self-shading that say that there is a true maximal areal photosynthesis in water columns (Talling 1957; Bannister 1974b,a; Megard et al. 1979).

Finally, because production and biomass are linked (e.g., Krause-Jensen and Sand-Jensen 1998), it makes sense in this context also to examine plots of phytoplankton biomass versus nutrients. A number of authors (Smith 1982; McCauley et al. 1989; Prairie et al. 1989) have commented upon the shape of such plots, which—no surprise by now—exhibit positive, saturating shapes. An example of such a plot is given in Figure 8.9D, which also points out that consideration of the balance of N and P helps predict ecosystem behavior. Lakes on the saturating portion of the curve (right-hand side) are characterized by low TN:TP ratios. These are where we expect N rather than P limitation (Guildford and Hecky 2000).

Reasons for Changed Nutrient Use Efficiency with Fertility

Having now seen an overall pattern of diminished production efficiency with increased nutrients in many studies, let us consider the reasons why such a pattern would exist. Although NUE is often described as an increased efficiency at low nutrient levels, which appeals to our common sense, low efficiency per se is never favored either by competition or by natural selection. Even at high fertility, an autotroph with high growth efficiency should be more successful than one with low efficiency, all else

---

amount of nitrogen in fine litterfall and the dry mass of that litterfall in terrestrial forests. Each symbol represents 1–6 years of measurement in a forest stand. Ecosystem types range from temperate coniferous forests and Mediterranean-type forests at low productivity to evergreen tropical forests at high productivity. B. The relationship between production of several biomass forms and N mineralization in eight temperate forests. C. The relationship between litterfall and total soil phosphorus for 12 North Carolina peatlands. D. Contrasting growth efficiencies in four species of terrestrial trees from two neighboring sites with different fertility (F, fertile; I, infertile). Panel A is based on Vitousek (1982), panel B is based on Pastor et al. (1984), panel C is based on Bridgham et al. (1995), and panel D is based on Boerner (1984).

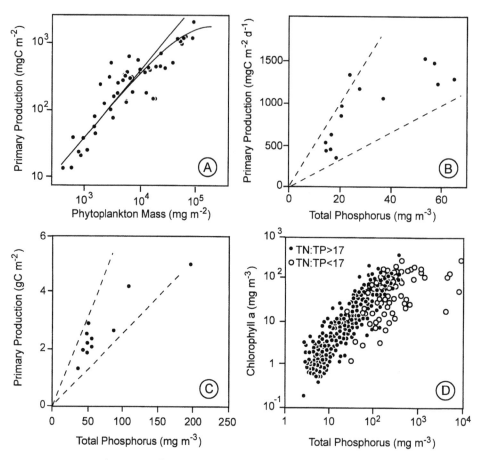

Fig. 8.9. Analogous results to Figure 8.8, but for lakes. A. Areal primary production vs. areal algal abundance, with a slight tendency for saturation at high phytoplankton biomass. Curve fits from original source. B. Areal production vs. lake fertility (TP) for Lake Washington, Seattle. C. Areal production vs. lake fertility for Lake Minnetonka, Minnesota. Each point represents a mean value for a single bay in this geometrically complex lake. D. Algal abundance vs. lake fertility. Open and closed circles denote different N:P ratios as indicated in the legend. Panel A is based on Rodhe et al. (1958), panel B is based on Smith (1979), panel C is based on Megard (1972), and panel D is based on Smith (1998).

being equal. We saw this in the previous chapter when we discussed the stoichiometric determination of competitive ability. Hence, perhaps the way to discuss these general patterns of NUE and fertility is to ask the question, "Why do autotrophs become less efficient at converting nutrients into biomass at high fertility?" We think the answer to this question comes from appreciating the multiplicity of potential limiting factors to plant growth and the trade-offs they impose (Chapin et al. 1987). In Chapter 3, we presented evidence that in lake ecosystems, biomass nutrient ratios change with changing patterns of light:nutrient balance. Those patterns can now be expressed simply as follows: NUE is highest at high light:nutrient ratios. In all ecosystems, whether terrestrial or aquatic, increased fertility generates increased biomass of primary producers. Increased biomass means increased self-shading. Thus, there is a fundamental and basically inescapable reduction in light per unit plant biomass with increased fertility. At low fertility, plants with high growth:nutrient relationships are favored. At high fertility, plants with high growth:light relationships are favored. This realization lifts us out of the need to explain why efficiency is reduced at high nutrients. It is caused not so much by a reduction in efficiency, but by a change in what currency is being optimized by the organisms. Plants do not really become less efficient, rather there is a changing optimization being performed (Sims et al. 1998), so that when one focuses on any particular component of the process, apparent shifts are observed.

The other major factor that may influence NUE in ecosystems is trophic structure and feedbacks from the food web. Earlier in this chapter, we discussed papers dealing with relative abundances of autotrophs and heterotrophs. Those papers indicated that infertile habitats were dominated more by heterotrophs, but heterotrophs generally have low C:N and C:P. Hence, heterotrophs should introduce a contrary signal (higher nutrient content at low fertility). However, trophic structure can contribute in a major way to stoichiometric biomass yield by modulating plant loss rates. Indeed, the term "trophic cascade" was originally coined to refer to the variance in primary productivity explainable by trophic structure and not explained by nutrients (Carpenter et al. 1985). By cropping autotroph populations and thus maintaining them in a state of more rapid growth (with consequent increased nutrient content, Chapter 3), herbivores may lower biomass yield at a given nutrient level. Differential nutrient use efficiency because of trophic structure is a part of some of the trophic-level models we described in the previous chapter (those by Oksanen, DeAngelis, and others). In these models, increased fertility has several impacts. It allows additional trophic levels to be added to the system. It also creates a stair-stepping pattern of autotroph biomass where, in some ranges of nutrient loading, autotroph biomass increases with nutrients, while in other ranges autotroph biomass is constant with increased nutrients. Where autotroph biomass is constant with increased nutrient loading, NUE as measured by standing biomass decreases (although NUE may not decrease if measured by productivity). The models of homeostatic consumers we also examined have similar, but more complex dynamics.

All in all, data on primary production as well as biomass standing stocks indicate that the stoichiometric yield (carbon or biomass per nutrients) of autotrophs consistently relates to ecosystem fertility: at high fertility, yield is reduced. This occurs both in forests and in lakes and therefore indicates a fundamental linking of carbon and nutrient cycles across diverse kinds of ecological systems. We will consider the possible mechanisms for such a repeatable pattern below, but here it is worth noting that these fundamental similarities between planktonic and terrestrial systems have not yet drawn much attention. Nowhere in the terrestrial NUE literature that we examined was there a single citation of an aquatic study. Similarly, aquatic scientists continue to explore production-nutrient issues apparently without recognizing the very similar patterns in NUE observed by terrestrial ecologists.

Having now considered how upper trophic levels may contribute to stoichiometric yield of primary producers, let us see if some of the concepts of "use efficiency" we have considered can be applied directly to heterotrophic growth.

THE STOICHIOMETRY OF FOOD-CHAIN PRODUCTION: A NEW TERM, CARBON USE EFFICIENCY

Nutrient use efficiency is a botanical concept (a rare application to animal growth can be found in Ayres et al. 2000). It refers to how autotrophs use nutrients to fix carbon. Because autotrophs often comprise the majority of biomass of ecosystems, NUE might also legitimately be considered to reflect an efficiency of the whole ecosystem. But NUE says nothing about upper trophic levels. Further, it may not be a terribly useful concept there. If a consumer is strictly homeostatic, its C:nutrient ratio is constant, and so necessarily would be its NUE. Instead, a meaningful measure of efficiency of upper trophic levels is the classical concept of trophic transfer efficiency (rate of production of one trophic level divided by rate of production of the next lower trophic level). However, to emphasize the correspondence to NUE, let us call this efficiency of transfer between trophic levels the carbon use efficiency, or CUE.

One form of CUE is secondary production (output) divided by primary production (input). For now we will leave it open whether we are talking about a single consumer population at a single trophic level (such as the

TABLE 8.5

Rates and ratios of primary and secondary production in five tropical lakes. See original paper (Hecky 1984) for primary references

	George	Chad	Lanao	Malawi	Tanganyika
Gross primary production ($g\ O_2\ m^{-2}\ d^{-1}$)	11.1	4.0	8.7	3.1	3.5
Net primary production ($g\ C\ m^{-2}\ d^{-1}$)	1.5	1.0	1.7	0.7	0.8
Fish yield ($kg\ ha^{-1}\ y^{-1}$)	136	14	60	40	125
Carbon transfer efficiency (% annual basis)	0.25	0.038	0.097	0.16	0.43

top consumer in a food chain), or if we are talking about an entire trophic level such as all the herbivores in an ecosystem, or in fact if we are talking about all the secondary production fueled ultimately by the primary production in that ecosystem. The CUE varies widely in nature. Hecky (1984) provided a set of examples in his discussion of several tropical lakes. He compiled data on fish production and primary production and calculated their ratio, referring to it as "carbon transfer efficiency." The end members of this comparison were Lake Tanganyika, which has extremely low phytoplankton abundance, and Lake Chad. Approximately 0.4% of primary production was transferred to fish production in the former but only 0.04% was transferred in the latter (Table 8.5). Similarly, Downing et al. (1990) observed that fish production as a fraction of primary production ranged from 0.002% to 1% in a sample of 20 lakes primarily from the north temperate zone.

A fundamental result of ecological stoichiometry is that as autotroph C:nutrient ratios increase, herbivorous animals utilize nutrients for growth very efficiently even if the absolute rate of animal production might be low (Chapter 5). Simultaneously, they use carbon very inefficiently, as they must release excess carbon to maintain their homeostasis. So, as long as consumer C:nutrient threshold element ratios (TER's, or the stoichiometric ratio in the food causing the consumer to be variously limited by different elements; Chapter 5) are sufficiently low, consumer CUE should be low where autotroph NUE is high. Of course there are many other factors besides nutrient ratios that determine food-chain production, such as size structure, plant defenses, and the number of trophic levels from plants to the consumer of interest. But, all else being equal, CUE should be inversely related to NUE. As far as we know, this hypothesis has not been advanced before. Given the generalization discussed above that NUE declines with fertility, there is a corollary hypothesis that fertile habitats should be inefficient at making plant biomass out of nutrients, but they should be efficient at making animals out of plant carbon. A second corollary comes into play when we consider the light:nutrient relationships discussed elsewhere in the book (Chapters 3 and 7). Recall that C:nutrient ratios increase as the light and nutrient balance shifts in favor of greater light. Hence, as light increases at constant fertility, NUE in autotrophs should increase and CUE in the food chain should decrease, reflecting the operation of fundamental stoichiometric mechanisms at the autotroph-herbivore interface. Therefore this is a good solution to the paradox of energy enrichment—a decrease in herbivore performance with increased energy input—we discussed in the previous chapter. These are a new set of *a priori* stoichiometric predictions awaiting empirical test. These insights can help us understand some key factors governing the ultimate fate of primary production in ecosystems, a topic to which we now turn.

THE FATE OF PRIMARY PRODUCTION

Understanding the fate and processing of carbon at large scales has become increasingly important in recent years due to carbon's role in climate change. Ecosystems are highly variable in their patterns of flux and storage of organic carbon. A compelling series of papers by Cebrián, Duarte, and colleagues (Cebrián and Duarte 1994, 1995; Cebrián et al. 1996; Duarte and Cebrián 1996; Cebrián and Duarte 1998; Cebrián et al. 1998; Cebrián 1999) explored patterns in the fate of carbon in ecosystems of widely different type. From over 200 published reports, Cebrián (1999) compiled a data set containing information on autotroph nutrient content, biomass, detrital mass, and magnitude and trophic fate of primary production (consumption by herbivores, detrital production, export, decomposition, and refractory accumulation) for several types of communities characterized by their dominant autotrophs. The community types were marine and freshwater phytoplankton, marine and freshwater benthic microalgae, marine macroalgae, freshwater macrophytes, seagrasses, brackish and marine marshes, grasslands, mangroves, and shrublands and forests. Few studies reported entire C budgets, so sample sizes differed for individual comparisons. Most of the terrestrial data referred only to aboveground biomass. Also, a number of assumptions had to be made about certain fluxes. In spite of these caveats, this effort represents the most complete comparative analysis of the gross features of C budgets in diverse ecosystems available today.

Autotroph nutrient content varied considerably across these community types (Figs. 8.10A,B). Plankton communities, composed of small autotrophs lacking in support tissues, were relatively high in both N and P.

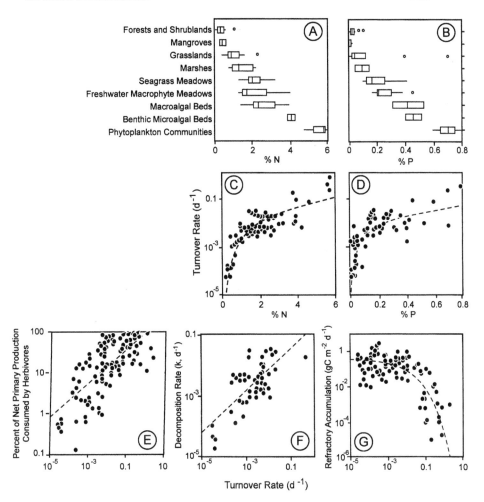

Fig. 8.10. Carbon flux in ecosystems as related to autotroph stoichiometry. A, B. Nutrient content of autotrophs in different habitats. Boxes represent 25 and 75% quartiles and the central line represents the mean. Following the original figure, bars encompass a range related to the 25 and 75% quartiles (see original). Dots indicate individual points outside these ranges. C, D. Biomass turnover rate as a function of autotroph nutrient content. E–G. Carbon dynamics as a function of autotroph turnover rate. Loss of primary production to herbivores (E) to decomposition (G) and to refractory accumulation (G). Curve fits from the original source. Based on Cebrián (1999).

Recall that nutrient concentrations generally decline with increasing size of autotroph from algae to trees (Fig. 3.6). Mangroves, along with forests and shrublands, had the lowest N and P content. Clearly, carbonaceous woody support tissues have a large influence on overall autotroph nutrient content in different ecosystems. A major life history trade-off between support and growth is also apparent in that biomass turnover rate (d^{-1}) was positively related to plant nutrient content (Figs. 8.10C,D). Biomass in support tissues turns over less rapidly than in high-nutrient photosynthetic ("protoplasmic") tissues. Biomass turnover rates ranged approximately 1000-fold, with biomass doubling times (the reciprocal of turnover rate) ranging from a low of several days (means for phytoplankton and benthic algal turnover) to a high of several years (means for mangroves as well as forests and shrublands). Values for N:P are not given; however, Elser et al. (2000b) showed that terrestrial and aquatic photosynthetic biomass has approximately the same N:P, again suggesting that the primary factor accounting for the differences across community types (Figs. 8.10A,B) is the amount of carbon supporting a core protoplasm (this fits with the logical framework of Reiners, Chapter 1). Cebrián (1999) also showed that net productivity (g C m^{-2} d^{-1}, using standard terminology where net autotroph production equals gross autotroph production minus autotroph respiration) was *unrelated* to autotroph nutrient content. That is, stoichiometry was strongly related to specific rates (i.e., normalized for biomass), but not to absolute production rates. Congruently, turnover rate was not related to net production rate.

Now, what is the fate of this autotroph production in the ecosystem? Net primary production can be consumed by herbivores, can enter detrital biomass pools and be broken down or be exported, can be stored as living biomass within the ecosystem, or can be buried long term in a nonliving state. At biomass steady state, storage is zero, leaving only three of these fates. There was a very strong stoichiometric signal in the gross features of C flux in these diverse ecosystems. The pathway of carbon through the different ecosystems was strongly dependent on autotroph nutrient content (Figs. 8.10E–G). The percent of primary production eaten by herbivores ranged from a low of about 5% (mean for forests and shrublands) to a high of 40–50% (mean for phytoplankton and benthic microalgae). Notably, these were the communities that were most different in their nutrient content, and a highly significant relationship between the percentage of primary production consumed by herbivores and autotroph turnover rate (which relates to both N and P content) was observed (Fig. 8.10E). Empirically, these data seem to suggest a curvilinear relationship, not a linear one, although Cebrián used a linear fit. The trend in Figure 8.10E indicates that, in ecosystems with high-quality autotrophs (see Chapter 5), herbivores were major conduits for organic matter but were only minor players where autotrophs were low in quality. Things were different for absolute consumption of plants by herbivores. Absolute consumption was best predicted by absolute productivity (McNaughton et al. 1989; Cyr and Pace 1993), not turnover rates (Cebrián 1999).

The importance of detritivory was opposite to that of herbivory, as one would expect. Ecosystems where grazing was a major carbon flux (those with microscopic autotrophs as the base of the food web) generally channeled only about half of their primary production through detritus. In contrast, ecosystems where grazing was a minor flux channeled more than 80% of net primary production through detritus. Moreover, the rate of detrital breakdown was strongly positively correlated with autotroph turnover rate (Fig. 8.10F), consistent with the work of Enriquez et al. (1993) on detrital mineralization discussed in Chapter 6. At this very broad level of analysis, it is useful to think of autotroph biomass in terms of an overall heterotrophic potential. Autotroph biomass as a substrate for heterotrophic growth, whether detritivores or herbivores, depends fundamentally on the balance of carbon with other essential elements.

Why do ecosystems behave this way? Why is herbivory favored more by rapidly turning over plant matter than detritivory, since as we have seen both may be positively stimulated by increased plant nutrient content? Both decomposers and herbivores have high requirements for nutrients and generally perform better when living on high-nutrient substrates. This overall trend is opposite to some general theories of herbivory, specifically the "plant apparency" theory (Feeny 1976), which stresses how large, slow-growing plants are more easily discovered by mobile herbivores. The relationship between plant nutrient content and the importance of herbivory implies a large-scale, "bottom-up" control for herbivores. Another possible factor for the difference in C fluxes is that there may be a major difference in TER between herbivores and detritivores, with a larger nutrient requirement exhibited by herbivores (see the discussion of tall-grass prairie in the last chapter) and thus herbivores are excluded from ecosystems with high C:nutrient ratio in autotroph biomass, leaving more material available for processing by detritivores with higher TER's. However, perhaps one big part of the answer to this question is that herbivores "get first crack at" plant biomass while it is still living. Living biomass of rapidly growing, nutritious autotrophs is first fed upon by a well-nourished herbivore biomass that can then proliferate and further consume more autotroph production, resulting in a large carbon flux from autotrophs to herbivores. Detritivory consumes the leftovers. The amount of leftovers is a function of how successful the herbivores are. If this hypothesis is correct, herbivore-nutrient interactions are key to understanding broad patterns of C flux in ecosystems, even though in many ecosystems herbivory seems to be a minor part of the story.

The final possible fate of primary production in an ecosystem at biomass steady state is a buildup of refractory carbon. This flux amounts to any difference between net production and the sum of detrital breakdown, detrital export, plus herbivory. Cebrián's analysis indicated a rapid falloff of refractory accumulation rates as plant turnover rate and nutrient content both increased. He also found a large influence of stoichiometry on the quantities of carbon in various ecosystem pools: with increasing turnover rate and autotroph nutrient content, both autotroph biomass and detrital biomass decreased.

Cebrián's (1999) study suggests that, at this very large scale of comparison, the overall pathways of carbon flow in ecosystems are strongly related to stoichiometry, in particular to autotroph nutrient content and its effects on herbivore success. He concluded,

> Communities composed of more nutritional plants (i.e. higher nutrient concentration) lose higher percentages of production to herbivores, channel lower percentages as detritus, experience faster decomposition rates, and, as a result, store smaller carbon pools. These results suggest plant palatability as a main limiting factor of consumer metabolical and feeding rates across communities. Hence, across communities, plant nutritional quality may be regarded as a descriptor of the importance of herbivore control on plant biomass ("top-down" control), the rapidity of nutrient and energy recycling, and the magnitude of carbon storage.

Having now examined stoichiometry in carbon budgets, let us close the chapter with a more expanded discussion of how stoichiometry fits into planetary-scale biogeochemistry.

GLOBAL CHANGE

Human domination of Earth's ecosystems has become a major societal concern and research focus (Vitousek et al. 1997b). Human impacts are a result of both the population density of our species and its high per capita control of resources (Tilman et al. 2001). Anthropogenic influence on climate has of course been one major focus, but many other global-scale human impacts are biotic, such as homogenization of species distributions and increased extinction rate. Still other human effects are best understood as major perturbations in global biogeochemical cycles, including those of carbon, nitrogen, and phosphorus. The couplings of biogeochemical cycles of C, N, and P to each other and to climate dynamics involve a highly complex web of interactions and feedbacks between the geosphere, atmosphere, and biosphere (Falkowski et al. 2000). These interactions operate across diverse time scales and also involve poorly understood interactions between terrestrial and marine ecosystems. Although much has recently been written about this topic, this chapter provides an opportunity to explore stoichiometric shifts at these scales, as our species realigns various global cycles. Stoichiometry comes into play in terms of how the biota couples the different cycles together. Further, changed balances and ratios of nutrients at huge scales may have a major impact on the distributions of plant, animals, and microbes at smaller scales. Let us first consider how much humans have altered global cycles of C, N, and P.

Earlier we considered the homeostasis of atmospheric oxygen. In contrast, other atmospheric gases have varied manyfold over the same period of geologic time. Atmospheric CO_2 today is at much higher levels than in preindustrial days. From the perspective of past changes coupled to glacial-interglacial cycles, the rapidity of this change has been extremely swift (Falkowski et al. 2000). Although we are a long way from an accurate, balanced global C budget (Schindler 1998), current estimates are that the combination of fossil fuel combustion and changed land use contributes about 8 billion metric tons per year against a background flux of 61. This is an increase above a natural background from global terrestrial respiration of 13% (Table 8.6).

Human impacts on the nitrogen cycle come from nitrogen fertilizer use, from fossil fuel combustion, from mobilization of stored nitrogen such as from deforestation, and from increased nitrogen fixation (Vitousek et al. 1997a). Global rates of N fixation have increased due to increased cultivation of N-fixing crops and rice (which promotes fixation by microbes), and to industrial processes in the manufacture of fertilizer. The current rate of fixation due to all these anthropogenic processes approximately equals or slightly exceeds the background rate of fixation (Table 8.6).

Human impacts on the phosphorus cycle mostly revolve around mining of P-rich sedimentary deposits, refining and relocating this material, and applying it to the landscape for fertilizer. Globally, the rate at which P is mined is approximately four times the background level of chemical weathering (Table 8.6). This P runs off the landscape with varying efficiency and enters aquatic ecosystems as a major pollutant (Tiessen 1995a,b).

Although the preceding several paragraphs provide perspective on individual cycles, it is equally important to recognize that humans are simultaneously accelerating the global cycles of all the elements listed in Table 8.6. Therefore, let us ask the question, "What is the C:N:P stoichiometry of anthropogenic change in global cycles?" Table 8.6 makes the point that the percentage increase in the P cycle is much larger than in the N cycle, which in turn is much larger than in the C cycle. These percentage increases suggest that humans are having a disproportionate effect on the global P cycle (mainly due to perturbations associated with agriculture). A

TABLE 8.6
Estimates of human interventions in global biogeochemical cycles of several elements. Data are for the mid-1990s. From Falkowski et al. (2000)

Element	Flux	Magnitude of Flux (billions of metric tons per year)		% Change due to Human Activities
		Natural	Anthropogenic	
C	Terrestrial respiration and decay CO_2	61		+13
	Fossil fuel and land use CO_2		8	
N	Natural biological fixation	0.13		+108
	Fixation from rice cultivation, fossil fuel combustion, and fertilizer production		0.14	
P	Chemical weathering	0.003		+400
	Mining		0.012	
S	Natural emissions to atmosphere at Earth's surface	0.08		+113
	Fossil fuel and biomass burning emissions		0.012	
O and H	Precipitation over land	$111 \cdot 10^9$		+16
	Global water usage		$18 \cdot 10^9$	

different means of examining these data is to look at the C:N:P ratios of the "natural background" fluxes in comparison to the anthropogenic processes. One should note that these calculations are rather arbitrary in construction; they capture only certain of the fluxes (e.g., terrestrial respiration but not marine). However, they will give one means of comparing human actions to a background. The background "natural" proportions from this table are 52,527:96:1, a very C-rich and nutrient-poor composition. This global background says that a single atom of N fixed or P weathered is involved in multiple rounds of photosynthesis and respiration (taking the median of 250 from the NCEAS terrestrial C:P value in Figure 3.13, there would be 52,527/250 = 210 such rounds of C fixed per P weathered). In contrast, the anthropogenic stoichiometry is 1722:26:1, also C rich compared to the elemental composition of most living things, but much closer to biomass ratios than the background fluxes. Elsewhere (Houghton and Woodwell 1983), the summed release of combustion plus nutrient translocations due to mining and fertilizers has been calculated to have C:N:P of 1100:12:1, a similar value. By comparing these figures we see again that

humans are accelerating global cycles of the nutrients N and P at disproportionate rates compared to carbon.

Human impacts on global cycles can be divided into two broad classes:
biomass production (i.e., mainly agriculture) and energy consumption. Fertilizer runoff and human and animal wastes result from the former. The
C:N:P released as a product of fossil fuel combustion (including N fixed
during combustion) is very carbon rich and nutrient poor (24,000:80:1)
(Houghton and Woodwell 1983), a stoichiometry that bears little resemblance to living things. However, global anthropogenic C:N:P ratios are
tipped much more heavily toward higher N and P (see above). Thus, the
global human signal on biogeochemistry is much influenced by biomass
production. In other words, in terms of overall impact on global cycles,
biomass production to feed an increasing human population is an extremely important process and seems to outweigh, at least in the biogeochemical terms indexed by the stoichiometric fingerprint, the impact of
fossil fuel mobilization. We are enriching the planet with the nutrients N
and P, not just in absolute terms, but in relative (to carbon) ones as well.
This then opens up questions about the nutrient use efficiency of the
world's biota as a whole. In estimating impacts of increased N and P on
global C cycles, for example, it may be reasonable to expect a reduced
NUE due to this broad-scale eutrophication. Calculations such as these
make one wonder whether studies that seek to understand the fate and
impact of increased carbon without accounting for the simultaneously increased N and P are asking the right questions.

Stoichiometry will also come into play when we consider how the
Earth's biosphere will adjust to these relative changes. Downing et al.
(1999a) proposed an overall lowering of N:P balance in tropical aquatic
ecosystems as human uses in the surrounding land intensify. With pro-
gressive human-caused changes in the landscape, beginning with native
forest and progressing through deforestation, burning, agriculture, and finally urban and industrial development, N export from surrounding land
increases, but aquatic N:P generally decreases due to disproportionate effects on P loading during intensification of agriculture. This hypothesized
signal is similar to the one we discussed for large rivers in an earlier section ("Rivers of the world").

From what we know about the effects of stoichiometric balance on communities (Chapter 7), perturbing the relative availability of different nutrients worldwide will have broad and far-reaching consequences. A great
deal will depend on the stoichiometric homeostatic properties operating at
these large spatial and temporal scales. In several research areas, potential
effects of shifted stoichiometric balance have already received considerable attentions. These include effects on herbivores and litter decomposition as a result of shifted element ratios in plant matter.

"Direct" Carbon Dioxide Effects

Because we have discussed the effects of autotroph C:nutrient ratios else-where in this book, we will consider here only a few additional points that are relevant to global change issues, or that have come up within global change research. In Chapter 3, we considered effects of changed carbon supply on autotroph nutrient content. Often, although not always, auto-trophs respond to increased carbon availability with increased C:nutrient ratios. Consistent with this general expectation, Peñuelas and Matamala (1990) observed a decrease in nutrient content in leaves in historical her-barium specimens during the period of historical CO_2 increase. Nutrient balance theory (Chapter 3; Aerts and Chapin 2000) is a great help in un-derstanding these shifts. Chapter 5 considered numerous examples of how shifted autotroph elemental composition affects herbivores. A number of studies have put these processes together and evaluated herbivore dy-namics in CO_2-enriched environments.

Although responses of autotrophs to increased carbon are reasonably consistent (Cotrufo and Ineson 1996), response of consumers to changed foliar chemistry is somewhat less predictable. This probably should come as little surprise. The plant-herbivore system is a complex coupled non-linear biological system with a variety of indirect effects. Although in-creased carbon fixation (via direct CO_2 effects or increased light) may de-crease insect herbivore growth rates (Agrell et al. 2000), herbivores may alter their feeding rate or selection patterns in response to changed foliar chemistry (Lincoln 1993). Over the long term, shifts in plant species also may be very important. Changes in both herbivore growth and survivor-ship affect herbivore populations, and although the former may be fairly tightly coupled to food intake and hence to stoichiometry, the latter may not be. For example, different dynamics can be expected under a scenario where herbivores demonstrate compensatory feeding than where herbi-vore growth and reproduction are directly reduced under enhanced CO_2. Still, the simple stoichiometric prediction that under enhanced CO_2 supply autotroph C:nutrient ratios increase and herbivore performance decreases is often valid (Cotrufo et al. 1998).

Furthermore, it seems logical to expect that elevated autotroph C:nutri-ent ratios will produce litter with higher carbon and lower nutrients. Min-eralization patterns are dependent on litter nutrient concentration (Fig. 6.5), and in Chapter 7 we saw that (over at least some time scales) positive feedbacks may result. Mellilo (1996) reviewed the literature on CO_2 ef-fects on plant litter and concluded, "there is widespread experimental evi-dence for a reduction in nitrogen concentration of litter from plants grown in an enriched CO_2 environment." Fewer studies, however, went on to examine decay rates of that litter and results were mixed. In one study on trees (Melillo 1996), slower decomposition was observed for several species grown under high carbon, but not for others. Although elevated CO_2 may have fairly direct consequences for foliar nutrient levels, nutrient content of litter is also a product of nutrient translocation back into the plant before leaf abscission. In fact, Norby et al. (2001) concluded from a variety of studies that there was in fact no effect of elevated atmospheric carbon on litter decomposition. Another complication is that increased temperature itself should increase decay rate. Further, soil moisture affects decay rate and hydrological changes under future climate scenarios are still difficult to predict but may also play a large role. Finally, long-term changes in biomass types (wood vs. roots, etc.) and in species may produce results not predictable from shorter-term experiments. Any net effect of reduced litter quality against these changes of higher temperature and changed moisture may be difficult to predict. Certainly in this case stoichiometry is only one necessary piece of a complex puzzle.

The Oceanic Stoichiometric System

We now turn our attention to the stoichiometry of the oceans one last time. What occurs there is critical from a global perspective because of the importance of the marine environment in global cycles. The oceans account for about half of all primary production. They are the location of long-term burial of large quantities of organic matter. They both respond to and influence the climate. Some of the relevant processes are unrelated to ecological stoichiometry (e.g., the role of the "solubility pump" in the oceans, interactions among ocean temperature, cloud cover, and planetary albedo, etc.). However, many biological processes in the ocean contributing to variation in global dynamics have major stoichiometric aspects. One such process is the "biological pump."

The biological pump is the net downward movement of carbon from surface to deep waters resulting from assimilation of CO_2 by primary producers in the surface waters and net transport of fixed C to waters below the thermocline (Lalli and Parsons 1997). This promotes absorption of CO_2 from the atmosphere and its eventual burial in deep bottom waters and ocean sediments. Two aspects with strong stoichiometric overtones regulate the strength of the biological pump (Falkowski et al. 2000). One is the efficiency of utilization of dissolved nutrients (N, P) in upper ocean layers, a factor potentially affected by trace metal availability as well as other factors (such as light intensity). This will be discussed further below. The other is the C:N:P stoichiometry of organic matter formed during primary production, as the biological pump only produces a net transfer of carbon to deeper water if the C:nutrient ratio of settling materials is greater than the C:nutrient ratio of materials mixed to the surface from mesopelagic waters (Karl 1999). While we have previously emphasized the general uniformity of C:N:P in the marine environment, because of the large spatial scales and enormous amounts of biogenic matter involved, even modest deviations in C:N:P stoichiometry in the oceans can have global impacts. Thus, even relatively small variations in C:N:P stoichiometry of algal production in the oceans draw intense scrutiny (Sambrotto and Savidge 1993; Elser and Hassett 1994; Tyrell and Law 1997; Tyrell 1999; Pahlow and Riebesell 2000; Wu et al. 2000).

So, what might cause oceanic phytoplankton production to deviate significantly from the Redfield ratio (Falkowski et al. 2000)? Based on our discussion in Chapter 3, we hope it is clear that the relative importance of N versus P versus other factors (e.g., light, metals, pCO_2) as limiting factors has a major impact on the C:N:P stoichiometry of phytoplankton production. Several principles from Chapter 3 will help us think about these issues.

- Principle 1. Phytoplankton grow roughly at Redfield proportions when growing rapidly at nutrient-saturated rates but C:X increases as the severity of growth limitation by nutrient X increases.
- Principle 2. The N:P of phytoplankton biomass generally mirrors the N:P of the inorganic nutrient supply.
- Principle 3. Increases in light intensity (and possibly pCO_2) lead to increased algal C:X ratio under limitation by nutrient X. Addition of trace metals may have a similar effect, if trace metal limitation has its strongest effect on photosynthetic activity.

This suggests that oceanic primary production generally occurs at approximate Redfield proportions because of one or, more likely, a combination of the following factors:

1. Phytoplankton growth rates are uniformly high across the world's oceans, possibly kept at high rates in balance with high consumption rates coupled to rapid nutrient recycling (Goldman et al. 1979). (Derived from Principle 1, above.)

2. In most parts of the ocean, the overall nutrient supply is dominated by advective inputs of deeper water during seasonal mixing. These waters contain inorganic N and P in roughly Redfield proportions, thus resulting in algal growth not strongly divergent from the Redfield ratio, even if absolute rates of growth might be low. One possible mechanism maintaining balanced N:P in these huge water volumes is that most organic matter reaching such depths may be produced only by sporadic bloom events in which nutrients are not limiting and N and P are fixed in Redfield proportions (Anderson and Sarmiento 1994). During such rapid blooms, most production may escape consumption and is able to reach deep waters prior to mineralization. (Principle 2.)

3. Deficiencies in trace metals, especially iron or perhaps molybdenum, or limitation by light intensity, limit the activities of N-fixing microorganisms, preventing a strong imbalance in relative N and P supplies from becoming established (Rueter 1982; Howarth and Cole 1985; Howarth et al. 1988a; Wurtsbaugh 1988; Wilhelm 1995; Falkowski et al. 1998; Karl et al. 2001). Further complications arising from sulfate interference with molybdate assimilation may also have a role to play in reducing N fixation in the seas (Cole et al. 1993a). (Principle 2.)

4. Phytoplankton growth is limited by light, reflecting the intensity of hydrodynamic mixing processes in oceanic surface waters (Nelson and Smith 1991; Platt et al. 1991). The possibility of direct pCO_2 effects should at least be considered as well (Riebesell et al. 1993; Gervais and Riebesell 2001), as perturbation of pCO_2 levels may alter the efficiency of photosynthetic C uptake and thus alter the C:nutrient ratio of production, as is frequently the case for terrestrial autotrophs. (Principle 3.)

5. Limitation by other elements, especially iron (Martin et al. 1991), prevents strong limitation of growth by N or by P. If Fe limitation has its main effect on photosynthetic activity and energetic metabolism, then iron limitation of phytoplankton should manifest itself in C:N:P stoichiometry in the same way that light limitation does; that is, resulting in relatively balanced C:N:P ratios even under slow growth. This is equivocally supported by experimental investigations that were discussed in Chapter 3 (e.g., studies using the diatom *Phaeodactylum tricornatum*; Greene et al. 1991). As a reminder, in that study algal C:N:P ratios remained close to Redfield proportions even when growing under iron limitation at less than 20% of maximum growth rate. In fact, it is tempting to argue for a strong interaction between paragraphs 4 and 5, in that iron might have an effect on phytoplankton growth primarily by improving the energy harvesting abilities of phytoplankton experiencing chronically low illumination. (Principle 3.)

Consideration of these possibilities suggests that major large-scale perturbations of one or more of these factors might shift the oceans away from their current status. If such shifts are possible they may contribute to shifting the Earth between glacial and interglacial climatic regimes. Indeed, potential stoichiometric mechanisms underlying such shifts have been suggested by Broecker and Henderson (1998) and Archer and Winguth (2000). Support for a role of climatic and hydrodynamic factors in influencing ecosystem stoichiometry at large scales in the oceans is seen in long-term studies of biogeochemical cycling in the north Pacific subtropical gyre (the "world's largest ecosystem"; Karl 1999). Periods of contrasting climatic forcing due to El Niño southern oscillation (ENSO) events generate different biogeochemical patterns (Karl 1999; Karl et al. 2001) (Fig. 8.11). During the warm period of decadal-scale ENSO events, hydrodynamic conditions change: surface mixed layers become shallower and verti-

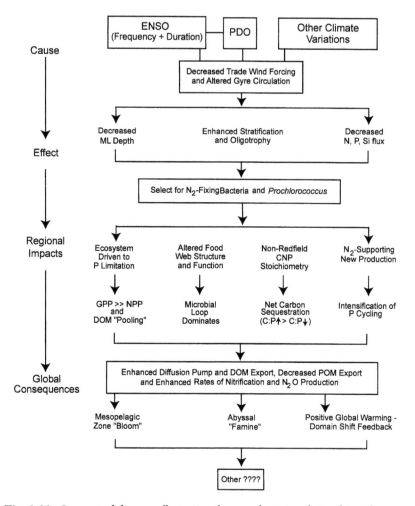

Fig. 8.11. Conceptual diagram illustrating the complex interrelationships of causes, effects, and consequences of climatic variation on the north Pacific subtropical gyre. During ENSO warm periods, less wind-induced mixing leads to a reduction in the depth of the mixed layer (ML), a more stable thermal stratification, and decreased supplies of nutrients from advective mixing. The increase in light intensity due to a shallower mixed layer appears to favor the proliferation of N-fixing cyanobacteria (*Trichodesmium* and other N fixers), generating unbalanced N:P supply ratios in the trophogenic zone. As a consequence, excess C accumulates in dissolved pools, trophic structures change (in favor of microbial processing), and particulate organic matter is generated with C:P and N:P in excess of Redfield proportions. Net effects of these changes also include a shift in the nature and rate of organic C reaching lower waters [from particulate (POM) to dissolved organic matter (DOM)] and increased rates of nitrification (of recycled NH_4) and subsequent production of N_2O. GPP = gross primary production; NPP = net primary production. Based on Karl (1999).

cal advective nutrient supplies diminish (Fig. 8.11). During these periods, an increase in the C:nutrient ratio of production is observed. One explanation for this shifted stoichiometry is that shallower mixing increases the light:nutrient ratio in the surface waters, which relieves light limitation and accentuates nutrient limitation. However, changed light and nutrient balance does not predict which nutrient (N, P, or other) should be limiting. Karl's work (Karl 1999; Karl et al. 2001) shows that ENSO conditions also favor the N-fixing alga *Trichodesmium*, resulting in increased N inputs and a shift toward P limitation in the ecosystem. Thus, it seems at least conceivable that one mechanism by which deviations from Redfield in the C:N:P stoichiometry of oceanic production might be generated is large-scale climatic change resembling extended warm ENSO events.

Alternatively, accentuation of P limitation might result when inputs of N and P become imbalanced due to stimulation of oceanic N fixation by major inputs of terrestrial iron. If the optimal N:P of *Trichodesmium* exceeds the Redfield ratio, large quantities of N fixation by this autotroph species will cause an overall excess of N relative to P. In such a scenario, N from Fe-stimulated N fixation will accumulate (Broecker and Henderson 1998; Archer and Winguth 2000), leading to an increase in the N:P of the overall nutrient supply experienced by phytoplankton and thus to P limitation of production and elevated C:P and N:P in biomass. Whether or not increased N from fixation could accumulate to a significant degree would depend on whether or not there was a compensatory increase in denitrification. Understanding these interactions requires adding yet another element (oxygen) to the mix. As discussed previously in this chapter, Van Cappelen and Ingall (1996) provide evidence that the availability of P in the oceans depends to a large extent on the development of anoxia (increased P release under anoxic conditions) as does the extent of denitrification. Thus, perturbations of organic matter production away from Redfield ratios due to P-limited growth conditions may result in increased anoxia (due to the elevated C:P of organic matter production; the greater amounts of organic C produced per unit nutrient can support more O_2 consumption when decomposed), leading to increased anoxia and subsequently decreased burial of P in sediments and increased denitrification. These feedbacks serve to shift the condition away from a P-limited state and a delicate balance is restored.

Many of the feedbacks just discussed operate on time scales of tens of thousands to hundreds of thousands of years (Van Cappelen and Ingall 1996). They point to the idea that the unavoidable coupling of C to N to P in biology is one of the key constraints at the organismal level that affords an autoregulatory capability to the biosphere (Lenton 1998; Lenton and Lovelock 2000). That is, to have a self-regulating system at the scale of the biosphere, one does not need Gaia to be a living entity, itself under evolution by natural selection. Instead, what is required is that the individual organisms driving the Earth's biogeochemical cycles have a set of functional constraints on their adaptation (Lenton and Lovelock 2000). In the case of global climate and C:N:P stoichiometry, we propose that these constraints are set by the core molecules of organismal function (carbohydrates, proteins, and nucleic acids; Chapter 2) and the biochemical mechanisms that produce them (photosynthesis, protein synthesis, nucleic acid replication). There is an unavoidable connection between biochemical machinery and planetary regulation. The face of Gaia vanishes, as in an optical illusion when one's attention is called to the underlying trick.

CATALYSTS FOR ECOLOGICAL STOICHIOMETRY

We are severely hampered by the lack of data on C, N, and P in different biomass components in ecosystems. Although certain components (leaves, zooplankton) may be reasonably well studied, others (belowground biomass, vertebrates) are seldom analyzed in the most useful way.

- At the ecosystem level, we need data sets of numerous of these components in single-habitat units. If we had a variety of such data sets, further stoichiometric patterns would be revealed, and they would teach us about new aspects of biogeochemistry. We would be able to test some of the contrasts we expect in stoichiometrically balanced and imbalanced ecosystems.
- One way to understand ecosystem diversity from a stoichiometric perspective would be to sample across major gradients. That is, instead of a haphazard approach where it may be difficult to identify the global target populations (all ecosystems? the grasslands of North America? small ponds in Iowa?), we could use major gradients of moisture, productivity, climate, etc., as organizing axes. The diversity of physical and chemical forcing ratios to be expected across such gradients should produce revealing patterns.
- Another major gap could be thought of as "geographic stoichiometry." We have suggested the existence of certain regional patterns. For example, "Canadian Shield lakes" relative to "Wisconsin lakes" have high C:P in seston (Hassett et al. 1997). Also, arctic versus temperate zooplankton differ in their stoichiometry (Dobberfuhl and Elser 2000; Elser et al. 2000a). However, we are hampered by lack of really good geographic coverage. The lessons of contrasting marine and terrestrial sites and progress in understanding N:P balance in aquatic systems argue for taking a landscape-level perspective on stoichiometry. But where really is the dominant stoichiometric variation at the landscape

level? Technical advances in remote sensing of N content of vegetation using infrared reflectance spectra measured from satellites (Martin and Aber 1997; Martin et al. 1998) open up the possibility of obtaining for the first time highly spatially resolved patterns of nutrient content in terrestrial vegetation (for an example, see Fig. 9 in Ollinger et al. 2002). An important point about basically all the survey work that we have mentioned in this book is that the results might be specific to the geographic region under investigation, but seldom have particular regions been sampled in a statistically rigorous way. Geographic differences need to be better examined. We are intrigued, for instance, by the observation of higher than expected zooplankton, and particularly *Daphnia*, abundance in saline prairie lakes of high P concentration (Evans et al. 1996). Stoichiometric surveys that include those sites as well as regions of known low seston P content would be extremely interesting and might capture the majority of the variation that can be seen in these variables.

Stoichiometric patterns across trophic levels are driven in part by patterns of biomass distribution (e.g., trophic pyramids). Although recent studies are revealing patterns in autotroph:heterotroph biomass ratios across productivity gradients, comparable information on a larger variety of ecosystems is still needed. At the ecosystem level, one or a small number of dominant species with distinct chemical signatures may sometimes drive strong stoichiometric signals.

- How does stoichiometry factor into the effects of "strong interactors" or "keystone species" at the ecosystem level? Are such species stoichiometrically distinct from less influential members of their functional group, trophic level, or guild?

It seems remarkable that we still lack a good understanding of the relative importance of the primary limiting chemical substances for most of the aquatic habitats on Earth. Early simplistic views are being replaced by a more sophisticated one that includes not just N and P but also micronutrients and light, and that recognizes not only temporal changes in limiting factors but also systematic differences in N:P balance across major habitat gradients and in upland-oceanic transects.

- A major challenge is to sort out these new views in the context of older classical ones and achieve a comprehensive understanding of the effects of N:P:micronutrient balance on primary production.

A new concept in this chapter is a formal theory describing the use of one substance to gain another. Here, we consider further cellular-physiological studies, such as *in vivo* measurements of protein synthesis rates of ribosomes in functionally significant biota, to be critical to further development. This theory provides a new way of linking mechanistic physiology to ecosystem ecology.

- When are efficiencies of interconversion constrained? Those physiological patterns will have ecosystem-level implications.
- Which constraints are avoidable evolutionarily and which are unavoidable due to physical and chemical constraints?

Several issues come to mind when thinking about challenges for ecological stoichiometry at the global scale. These include the following.

- How will anthropogenic shifts realign global cycles and the distribution of biota? We need to move toward an enhanced understanding of the homeostatic regulatory points at multiple temporal and spatial scales. To do this, we need more studies of the coupling of multiple chemical elements at large spatial scales in ecosystems. New analyses of coupled pools and fluxes of multiple elements in food webs of moderate resolution using methods of inverse analysis (Gaedke et al. 1996; Lyche et al. 1996a,b) may provide the necessary tools.
- Another research opportunity concerns the global homeostasis of biogeochemical functioning. Does global biogeochemistry revolve around production of homeostatic biota (and thus instabilities of abiotic cycles reign supreme), or do Gaia-like negative feedback processes dominate? In either case, stoichiometry is one set of tools to further our understanding.

SUMMARY AND SYNTHESIS

In this chapter we have seen that stoichiometric analysis has been widely used to good effect in the analysis of entire ecosystems. Nutrient ratios within organisms have helped in the understanding of biomass production, carbon burial, oxygen evolution, and other similar topics. As Reiners (1986) suggested years ago, stoichiometric analysis provides a "complementary model" (his term) of ecosystems (Chapter 1). So much of what Reiners pointed to in his paper, published some 15 years before this book was written, is consistent with today's data and today's views. However, there are new dimensions today as well. Linkages from evolution to ecosystems, for instance, are now better understood and new hypotheses are in the air (Chapter 4). Complex community dynamics with positive feedbacks are a relatively new addition to the stoichiometric picture (Chapter 7). In ecosystems, Reiners recognized that the concept of "balanced" chemical composition was a key point. The consequence of decoupled photosynthesis, nutrient uptake, and autotroph growth (Chapter 3) has a major imprint on ecosystem C flow and structure. In closing this chapter, let us look closely at this concept of stoichiometric balance by reviewing some of the points we have covered elsewhere in the book.

In many places, we have pointed to a major contrast between processes expected under balanced versus imbalanced stoichiometry. By "balanced" we mean here an ecosystem where major biotic pools all are similar and approximately at Redfield C:N:P composition. The Redfield ratio provides an absolute point of comparison both because this C:N:P balance perhaps resembles "protoplasmic life" (Reiners 1986) growing optimally and because it is close to the composition for most Metazoa (Chapter 4 and Fig. 8.2). Hence, when autotrophs and microbes also have near-Redfield composition, essentially the entire biotic portion of the ecosystem will as well. The drivers affecting ecosystem stoichiometric balance include the light-to-nutrient ratio (Chapter 3), major autotroph life form (e.g., aquatic vs. terrestrial, Fig. 3.6), autotroph size (Nielsen et al. 1996), and fertility of the habitat (this chapter). Some specific places we expect to find ecosystems that are stoichiometrically imbalanced are at low fertility and high light. They can be expected in slow-growing autotroph communities, and ones with relatively high amounts of structural carbon. Food-web structure and feedbacks too may be critical; "grazing lawns" are made up of high-quality, rapidly growing plants. In fact, due to the complex interactions we have described in Chapters 6 and 7, we might expect particular systems to shift continuously or abruptly along a gradient from balanced to imbalanced due to both extrinsic forcing and internal feedbacks.

Table 8.7 presents a collection of specific contrasts we have discussed in stoichiometrically balanced and imbalanced systems. This is not an exhaustive list, but it does include many of the major diverse contrasts that have been made, especially in this chapter and in Chapter 7. The balance of carbon to nutrients in autotrophs has consequences for herbivore nutrition as we discussed in Chapter 5. Stoichiometric imbalance is a nutritional challenge that is particularly a problem for high-nutrient (i.e., high-growth, Chapter 4) consumers. Hence, a major contrast that we can predict from these separate relationships is that high-growth-rate life histories ("r strategists") will be favored under stoichiometric balance and good competitors ("K strategists") will be favored under stoichiometric imbalance. This association implies a coupling between nutrient dynamics and time scales of population change. We also expect fundamental differences in autotroph-herbivore dynamics. Under stoichiometric balance, trophic cascades (favored by good conditions for high-growth-rate, generalist herbivores) should be prevalent. On the other hand, we expect dampening of these effects under stoichiometric imbalance. The nature of autotroph-herbivore equilibria differs under these two conditions as well, as we encountered in Chapter 7.

The balance of carbon and nutrients affects ecosystems in numerous

TABLE 8.7
Ecological contrasts in stoichiometrically balanced and unbalanced ecosystems

Scale, Level of Organization	Target	Stoichiometrically Balanced	Stoichiometrically Imbalanced
Defining Factor	Autotrophs	C:N:P ≈ Redfield	C:nutrient >> Redfield
Trophic Levels	Herbivores	r strategists favored	K strategists favored
	Trophic cascades	Favored	Disfavored
	Herbivore-plant dynamics	Classical models	Multiple equilibria
	Autotrophs' limiting factor	Light	Nutrients
	Herbivores' limiting factor	Carbon or energy	Nutrients
Ecosystems	NUE	Low	High
	CUE	High	Low
	Decomposition rate	Rapid	Slow
	Fate of primary production	Herbivory favored	Detritivory favored
	Thermodynamics	Paradox of enrichment	Paradox of energy enrichment
	Carbon burial	Disfavored	Favored
Example		Open ocean	Coniferous forests

ways. Stoichiometrically imbalanced ecosystems, by definition, are efficient users of nutrients for primary production (nutrient use efficiency is high), but this creates a different imbalance for herbivores, and they are inefficient at using carbon for their own growth (carbon use efficiency is low). Another major outcome of imbalance in ecosystems is to slow processing rates by herbivores and detritivores. Reduced heterotrophic growth (of both herbivores and detritivores) reduces the importance of herbivores as C conduits, leaving more C for detrital decomposition, which nevertheless occurs with reduced efficiency; hence, there are large standing stocks of detrital carbon in stoichiometrically imbalanced ecosystems. Some surprising outcomes result from stoichiometric imbalance. For example, under such conditions, production of higher trophic levels can be negatively affected by increased primary production (Figs. 7.12–7.14).

The contrasts in Table 8.7 range from true by definition (NUE, at least as we define it) to consistent with limited data (trophic cascades, paradox of energy enrichment) to largely speculative though theoretically reasonable (herbivore-autotroph equilibria). Our main purpose in constructing this table and considering it here in the book is to illustrate the great diversity of ecological phenomena that are amenable to stoichiometric analysis. We do believe that stoichiometric balance is a major key to understanding the variability among ecosystems in nature, even if not all of these specific contrasts hold up to the harsh reality of good empirical tests. This is a continuing story.

9

Recapitulation and Integration

They [atoms] move in the void and catching each other up jostle together, and some recoil in any direction that may chance, and others become entangled with one another in various degrees according to the symmetry of their shapes and positions and order, and they remain together and thus the coming into being of composite things is effected.—Simplicius of Cilicia (~530 A.D.)

So, how far have we come since the time of the ancient Greeks in understanding the "coming into being of composite things" in the living world? As we have now seen, we know quite a bit about the composite chemical nature of the living world. The preceding pages have taken us from the mundane (food ingestion, excretion, and egestion) to the esoteric (mathematical nullclines), from the staggeringly small (atom-to-atom interactions in biochemicals) to the numbingly large (material cycling at the global scale). Mass conservation applies at all these scales and all also are amenable to stoichiometric analysis. The fundamental nature of mass balance constraints suggests the possibility of moving even more ambitiously beyond ecology per se. Before discussing that possibility in a concluding synthesis, we first summarize and interconnect the major findings and ideas explored in this book in showing one last time that the composite nature of living things has vital implications for ecological systems.

RECAPITULATION

Chapter One

Ecological systems are tremendously complex. To approach such complexity, ecologists need firm foundations upon which to build. Key ideas from physical sciences and from other disciplines of biology are useful pieces. From outside biology these include the laws of thermodynamics and conservation (including the mass balance at the heart of chemical stoichiometry) and the principles of biological chemistry. From within biology, the principles of evolution by natural selection and the concept of physiological homeostasis are of central importance. Ecological stoichiometry seeks to take full advantage of these anchor points in constructing a mechanistic and predictive ecology. It uses the reality of chemical constraints to better understand factors limiting the abundance and distribution of organisms and to better incorporate key feedbacks among living and nonliving components of ecosystems. Ecological stoichiometry thus represents a simplification of natural complexity, distilling many of the fascinating details of species to a limited set of chemical parameters. The goal is to see how much variation in ecological dynamics can be successfully explained simply by acknowledging the fundamental chemical nature of life, thus highlighting those features of living systems that might require special, more detailed, explanations unique to biological inquiry. One key step in moving from chemical principles to biological dynamics is to determine which biological materials and organisms are constructed under tight homeostatic regulation and which have a more flexible composition. Chapter 1 provided a new, thorough, definition of stoichiometric homeostasis. We saw how it applied to several diverse organisms. Thus, understanding how stoichiometry affects ecological dynamics requires understanding the rules governing how living biomass is formed in growth. It also requires determining whether the ways that different organisms make a living in the natural world establish the mixtures of chemical elements they require and whether they can exhibit major variation in those proportions.

Chapter Two

The diversity of elements in the periodic table represents numerous chemical talents. Elements interact to differing degrees with each other by virtue of their atomic structure and, having come together in molecules, provide different biochemical functions at the molecular scale, such as roles in structure, electrochemical signaling, movement, or catalysis. We reviewed the ways that the elements carbon, nitrogen, and phosphorus are central to biology due to their functional dominance in biological polymers (carbohydrates, lipids, proteins, nucleic acids). Carbon, N, and P differ considerably in their relative proportions in these key molecules. Lipids and carbohydrates are C rich and low in nutrients, in contrast to proteins (rich in N, low in P) and nucleotides and nucleic acids (rich in both N and P). In turn, key cellular organelles differ in elemental composition because they differ in their biochemical composition. Specifically, cellular organelles differ most strongly in P content, with ribosomes, the organelle of protein synthesis and the engine of cellular growth, having particularly low C:P and N:P ratios. In sum, differences in the biochemical functions and the elemental composition of key molecules lead us to expect that the coupling of major elements in organisms will vary in understandable ways according to physiological condition or evolved life history. These rules link structure and function in stoichiometric terms.

Chapter Three

In Chapters 1 and 2 we saw the considerable variation in the relative abundance of key elements in the inorganic world and in the various biochemical constituents making up organisms. In Chapter 3 the range of variation began to narrow somewhat as C, N, and P were coupled to each other in the biomass of photoautotrophic organisms. This is because living cells require a full set of life-sustaining functions. Mixtures such as these are more bounded than the individual ingredients making them up. While variation in C:N:P stoichiometry in autotroph biomass is somewhat muted relative to the nonliving world, we noted that the elemental composition of autotrophs differs not only among taxa but also as a function of growing conditions because acquisition of C and nutrients are physiologically decoupled in autotroph metabolism. In both aquatic and terrestrial autotrophs, major increases in C:nutrient ratios are associated with more severe growth limitation by that nutrient, with higher light intensity, with elevated availability of CO_2 and perhaps trace metals, and with an increasingly imbalanced nutrient supply. We also saw that, due to the storage capabilities of autotrophs (especially in large vacuoles), autotroph N:P stoichiometry closely tracks the N:P stoichiometry of the environmental nutrient supply. Variation in autotroph C:N:P stoichiometry is also associated with size and with species-specific differences in competitive ability, as taxa that are good competitors for a particular nutrient will generally have low biomass requirements for that nutrient and thus will develop high C:nutrient ratios when it is in short supply. Variation in autotroph C:N:P stoichiometry lends itself to mathematical analysis and we saw how both the Droop quota equation and the Ågren productivity equation similarly accurately describe variation in autotroph stoichiometry with growth conditions. Consistent with wide variation in biomass C:N:P stoichiometry in physiological studies, autotroph biomass in nature exhibits considerable variation, both within and among habitat types. Marine pelagic food webs in particular are built on a relatively homogeneous and nutrient-rich autotroph base, while freshwater and terrestrial ecosystems have extremely variable C:N:P ratios in autotroph biomass. Terrestrial foliage has especially high C:nutrient ratios, reflecting a large allocation to C-rich, low-nutrient, structural materials necessary to attain large size and structural integrity in air. Intriguingly, terrestrial foliage, aquatic macrophytes, and freshwater seston have remarkably similar N:P, varying around an overall mean of ~30. Finally, we considered some correlative data suggesting that variability in C:nutrient ratios in autotrophs may reflect differences among habitats in light:nutrient balance, as high-light, low-nutrient conditions seem to produce a decoupling of C and nutrient acquisition in autotrophs, leading to unbalanced C:nutrient ratios at the base of the food web.

Chapter Four

In contrast to the relatively wide physiological variation in autotroph C:N:P stoichiometry, animal C:N:P stoichiometry is under tighter homeostatic control and falls within a narrower range. Variation in body C:N:P ratios in animals is largely inter- rather than intraspecific. Body P content varies fivefold across species in both crustaceans and insects. One key trait that seems associated with this variation in body C:N:P ratios is specific growth rate: rapidly growing organisms tend to have high concentrations of P-rich ribosomal RNA. Thus, body C:N:P stoichiometry in many invertebrate animals may have its proximate basis in the structure of ribosomal genes and its ultimate, evolutionary, basis in life history evolution. In vertebrates, a key determinant of body C:N:P stoichiometry is allocation to bones, as the P content of bones is extremely high. Allometric scaling patterns of growth rate and bone allocation suggest that biomass N:P may have a unimodal shape as a function of size across the full range of heterotroph size.

Chapter Five

Animal biomass is nutrient rich (Chapter 4) compared to autotroph biomass, particularly under nutrient-deficient conditions (Chapter 3). Thus, imbalance in elemental composition of autotrophs and herbivores is the rule. We examined the consequences of stoichiometric imbalance for consumer growth. One general consequence of the combination of consumer homeostasis and elemental imbalance with its food is that gross growth efficiencies (GGE_x) of different elements must differ. In particular, as food C:nutrient ratio increases relative to C:nutrient in the consumer's body, GGE_C declines and $GGE_{nutrient}$ increases. Thus, a major consequence of stoichiometric imbalance is in reducing growth and trophic transfer efficiencies between autotrophs and herbivores. The threshold elemental ratio (TER) is the C:nutrient ratio above which a consumer's growth is limited by the nutrient (rather than carbon or energy) intake in its diet. We saw that, in theory, TER's are a function of the consumer's ability to extract a potentially limiting nutrient from food and of the consumer's own nutrient requirements. Thus, one of the consequences of evolutionary processes that generate variation in a species' body C:nutrient ratio (Chapter 4) is an alteration in the quality of food it requires. In addition to simple formulations to estimate consumer TER's, homeostasis and mass balance facilitate the mathematical formalization of the stoichiometry of grazer growth. Even a minimal stoichiometric model of consumer growth yields correct predictions of consumer growth dynamics and nutrient processing efficiencies. We tested the most general predictions of stoichiometric models of secondary production by summarizing studies examining the effects of food nutrient content on dynamics of consumers as diverse as saprotrophic microorganisms, bacterivorous flagellates, crustacean zooplankton, termites, vertebrate herbivores, and detritivorous insects in streams. All show the expected patterns: as food C:nutrient ratio increases, the consumer's GGE_C declines and various mechanisms operate to increase its $GGE_{nutrient}$. These responses indicate that stoichiometric balance between food and consumers contributes substantially to variation in trophic transfer efficiency of food webs in nature. Thus, the mismatch between the nutritional physiology of autotrophs (Chapter 3) and Metazoa (Chapter 4) has major ramifications for herbivore performance and subsequent trophic dynamics.

Chapter Six

Since large elemental imbalance can occur between food and consumer (Chapter 5), mass conservation dictates that there should be major differences in the relative efficiency of nutrient recycling by consumers. In this chapter we reviewed evidence that the rates and ratios of consumer-driven nutrient recycling depend strongly on the stoichiometric balance of food and consumer. Phosphorus-rich, low-N consumers tend to retain P with high efficiency and recycle N liberally; in contrast, N-rich, low-P consumers do the opposite. Theory predicts that the strength of this skewing of recycled N:P depends not only on the imbalance between food and consumer biomass but also on the consumer's ability to extract limiting nutrient and on its overall gross growth efficiency. Since consumer-driven nutrient recycling can be a significant internal source of nutrients sustaining primary production in many ecosystems, this differential nutrient recycling can shift the identity of the limiting nutrient from N to P or from P to N and in doing so possibly alter the competitive arena for autotrophs. Here we can begin to see how imbalance in the C:N:P stoichiometry of food and consumers generates complex feedbacks in ecosystems.

Chapter Seven

Incorporating the balance of energy and multiple chemical elements into ecological theory leads us to expect that stoichiometry has not only quantitative but also qualitative effects on ecological interactions. In Chapter 7 we explored how stoichiometry can shift competition between herbivores to facilitation, alter the nature of nutrient limitation (N vs. P) of autotroph growth, tip the delicate balance maintaining symbioses, modulate the strength of the trophic cascade, and change the effects of solar energy input on secondary production from stimulatory to inhibitory. That such a broad set of outcomes can be predicted from a relatively simple set of assumptions is noteworthy. We saw that addition of stoichiometric constraints to theoretical models of autotroph-grazer interactions introduces dynamical complexity. With stoichiometrically explicit models, we saw the appearance of multiple stable states, deterministic extinction of grazers at high food abundance, and a new menagerie of cycles far beyond those that simple Lotka-Volterra expressions can generate. These mathematical systems seem much more like nature and represent a formal way to integrate population ecology and ecosystem ecology, as they express predator-prey and other ecological interactions in biogeochemically meaningful terms.

Chapter Eight

Here we examined how the coupling of elements in organisms creates patterns at the large scale, including global biogeochemical cycles and climate. Even at these largest spatial and temporal scales, stoichiometric variability generated ultimately at the biochemical scale is meaningful. Carbon:nutrient stoichiometry alters carbon fluxes and carbon burial. We were introduced to long-distance stoichiometric couplings when we considered how stoichiometric variability in organic matter burial in the deep ocean might determine atmospheric O_2. Humans seem to have a consistent stoichiometric footprint, making ecosystems more Redfield-like. In general, most studies show that aquatic and terrestrial ecosystems of increased fertility have lower autotroph C:nutrient ratios (i.e., they have lower nutrient use efficiency). Based on what we understand of consumer growth, those same conditions should lead to more efficient use of carbon to create secondary production of higher trophic levels. In contrast to the CNR models of Chapter 4, at these large scales, homeostatic biota balance biogeochemical cycles and make them more Redfield-like. This happens because different biogeochemical pathways (N fixation, denitrification) are favored under imbalanced conditions. In this chapter, we also identified how stoichiometric equilibrium, growth rate, and nutrient content must be related under certain conditions. The necessary and sufficient condition for a growth rate stoichiometry linkage (as in the growth rate hypothesis of Chapter 4) is that the efficiency of using one substance to control flux of another (e.g., N controls C, or ribosomes control mass) must be constrained. We considered present knowledge of human-induced changes to biogeochemical global cycles and drew some stoichiometric conclusions. Humans are having large impacts on the cycles of C, N, and P, but the overall C:N:P stoichiometry of human domination of Earth's ecosystems has a strong biomass-growth signal, not a strong energy-use signal.

INTEGRATION: TOWARD A BIOLOGICAL STOICHIOMETRY
OF LIVING SYSTEMS

In the latter part of the nineteenth and earlier part of the twentieth centuries, scientists such as Liebig and Lotka—unaware of so much of what we
consider modern biology—made great conceptual advances from which
we continue to draw inspiration. During the latter part of the twentieth
century, biology became highly sophisticated and increasingly predictive.
But with increasing sophistication came increasing specialization (Maienschein 1991), not only at particular levels of organization (e.g., ecosystem
science, molecular genetics) but also on particular model organisms (e.g.,
Arabidopsis, *Daphnia*, or *Drosophila*) or particular habitats (e.g., lakes,
temperate forests). As a result, knowledge has become increasingly frag-
mented, making it difficult to connect disparate areas of inquiry and appreciate living systems as functional wholes (Appel 1988; Allen and Hoekstra
1992; Pickett et al. 1994; Smocovitis 1996; Vogel 1998). As the twenty-first
century begins, biologists face the challenge of developing theories that
can reconnect our detailed understanding of individual levels of biological
organization and improve the compatibility of data and hypotheses generated in the study of diverse biota and habitats (Wilson 1998).

Reconciling our growing but fractured knowledge across multiple levels
of organization, diverse types of organisms, and contrasting habitats is a
thrilling challenge. Such reconciliation may be more feasible than is generally acknowledged. In ecology we are so often focused on the particulars of
our study system or set of interactions that we have difficulty imagining that
the phenomena we see there could somehow apply to other situations. Our
scope is often poorly defined at the outset, and so the safe route is to
assume the worst: that all phenomena are special cases. Part of this mind-
set comes from the necessary biological training in the myriad details of
morphology, taxonomy, and habitat-specific terminology. Given such a situation, why would anyone working on *Daphnia* from a boat imagine that
back on shore there might be an organism, an aphid, that shares a large
suite of similar ecological, life history, physiological, biochemical, and genetic features? (Here we note with some amusement that the letters in
"*Daphnia*" can be rearranged to spell "an aphid," changing one species into
another while conserving typographical mass balance.) It would seem that
an understanding of the various factors that determine the success of *Daphnia* in lakes would be useless in understanding aphids. After all, there are no
minnows on land preying on aphids, nor are there any phytoplankton to eat.

But such a mind-set betrays that the *Daphnia* specialist is not asking the
right kind of questions in wondering about ecological generalization and
fundamental rules of living systems. It would be akin to a planetary scien-

tist bemoaning the lack of generality of Newtonian theory because it cannot explain why the sixth (and not the fifth or seventh) planet from the Sun is the one with the big rings. The key idea in the burgeoning field of complexity theory is that simple rules can yield intricately complex outcomes (Lewin 1992). It is our contention that stoichiometry is one of those simple rules underlying ecological and biological complexity, and in this book we endeavored to show that stoichiometric mechanisms are at the heart of many ecological phenomena. In addition, stoichiometric analysis links many areas of biological investigation, from the biochemical composition of subcellular structures and the organization of genes to the dynamics of food webs in diverse ecosystems. We are confident that stoichiometric thinking will contribute to greater consilience (*sensu* Wilson 1998) within biology.

The ease with which stoichiometric analysis traverses disparate phenomena is illustrated in Figure 9.1, where we have plotted the stoichiometric signatures, in terms of N:P, of a variety of biotic and abiotic entities. Several interesting patterns are apparent in Figure 9.1. The first is that biological entities (especially tissues and whole organisms) are confined to the center of these diagrams while nonliving things (inorganic compounds, individual biochemicals, geological aspects) span the gamut of N:P. This emphasizes the winnowing of variation in elemental composition that results from the fact that only certain combinations of elements and biomolecules can constitute a functional living system. Also apparent is the nature of the shift in stoichiometric signature imposed by human actions: "human-free" signals, such as background global element cycling and runoff from unfertilized watersheds, occur at relatively high N:P, far from the P-rich biological signal associated with life (nucleic acids, ribosomes, bones). However, the signature of human impacts (fertilizer, sewage, the anthropogenic global fingerprint) is P rich. Thus, the net impact of humans on biogeochemical cycling appears to draw environmental nutrient ratios into closer balance with the demands of the human body itself. It also happens that this signature is the signature of rapidly proliferating biota, due to the P-rich demands of the ribosome. In this light, the intense eutrophication of both freshwater and coastal marine ecosystems accompanying human population expansion should not be surprising.

A brief retrospective highlighting some of the key findings from the book also illustrates these connections and the integrative power of stoichiometric analysis (Fig. 9.2). Differences in the nutritional strategies in autotrophs and heterotrophs are exemplified in the degree of C:N:P homeostasis in autotrophs and Metazoa. Intra- and interspecific variation in C:N:P stoichiometry is understood in terms of variations in relative allocation of biomass to key organelles and biochemicals that themselves differ significantly in elemental composition. Thus, the C:N:P stoichiometry of an

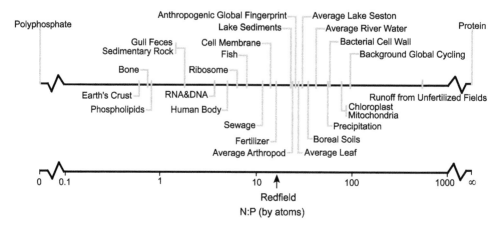

Fig. 9.1. A stoichiometric "number line" depicting the N:P ratios of certain abiotic and biotic entities we have discussed. The figure displays these ratios on an axis delineated by two fundamental molecules, proteins (N:P ~ infinity) and polyphosphate (N:P of 0). It is interesting to note, given current paradigms of life first evolving from autocatalytic RNA complexes, that fundamental signatures of biosynthetic capacity (RNA, ribosomes, genetic material) are found in the vicinity of a relatively P-rich signature of the Earth's crust. The overall "background" nutrient cycling in the modern world has high N:P but it appears that global nutrient cycling is being drawn back in the direction of the purported "primordial soup" under the influence of human action. The figure also illustrates how abiotic components (molecules, inorganic pools) are relatively unconstrained in their stoichiometry but living things, and the structures from which they are built, cluster in the middle of the number line.

organism, and its plasticity, is under genetic control, subject to evolution by natural selection, and thus connected to major features of the organism's evolved lifestyle. In turn, disparities in the nutritional strategies of autotrophs and heterotrophs have major ecological ramifications for trophic efficiencies, for consumer-driven nutrient recycling, and ultimately for population dynamics and community structure of herbivores in communities. Finally, C:N:P stoichiometry impinges on the global cycling of C and nutrients, as differences in the C:nutrient ratio of biomass production by dominant autotrophs modulate the fate of organic matter production in ecosystems, affecting the processes that dissipate or sequester organic C in the biosphere. In turn, human impacts have a distinct stoichiometric signature at the global scale and thus we might use our understanding of the coupling of C, N, and P in biota to better predict the impact of anthropogenic perturbations on both local ecosystems and the biosphere as a whole. Throughout, we have highlighted key unanswered questions that will help in strengthening our understanding of how stoichiometric mechanisms underpin various phenomena in ecological systems.

The evidence in this book convinces us that a stoichiometric perspective can contribute much to our understanding of diverse biota and diverse ecosystems. Might the stoichiometric approach be extended to achieve even greater coherence in biological science? In our view, a major task of integration in biology is to connect the central foci of ecosystem ecology (patterns of energy and material storage and flow in the environment) to major biological features all the way to the gene. The sketch in Figure 9.2 indicates a tentative outline of some of the possible connections identified in this book. To firmly establish these links we will need to determine how energy and material flows are coupled at all levels of biological organization and to understand the scaling rules among those levels (Elser et al. 2000d). This rich suite of principles, "biological stoichiometry," emerges from interweaving matter conservation, the principle of evolution by natural selection, physiological homeostasis, and the central dogma of molecular biology. In this approach, the ramifications of stoichiometric limitations would be recognized for all types of biological dynamics. Further, that recognition would then be used to better delineate the hard constraints that the laws of matter conservation and combination impose on the myriad of states a biological system can take. Better understanding of the communications among an organism's internal state, its genome, and its nutritional environment will emerge as modern techniques of molecular genetics are applied in physiological and ecological settings. We will begin to develop more realistic views of how the particulars of cellular machinery are connected to growth and other major features of the organism and to the organism's overall metabolic state. Further, understanding the stoichiometric implications of the whole complex of life history traits of an organism, as well as their genetic mechanisms, will allow us to incorporate evolutionary change and constraints into our models of ecological dynamics. Finally, by identifying the general underlying mechanisms that couple energy and nutrient elements in organisms, we will move beyond site-specific paradigms derived from analysis of special cases and make progress toward a robust, predictive, and internally consistent ecological theory. More broadly, in the elaboration of biological stoichiometry a more balanced relationship among the subdisciplines of biology will also be encouraged. Biological stoichiometry will itself be a "composite thing" and its coming into being will involve merging the most modern information from genomics to global change.

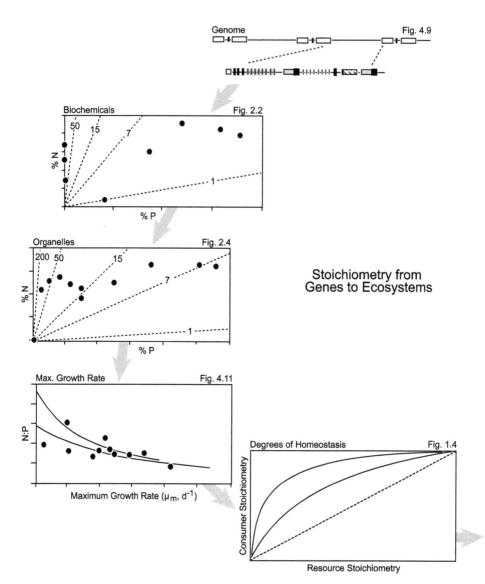

Fig. 9.2. A composite figure, which connects insights drawn from earlier chapters. Beginning at the molecular genetic scale represented by the rDNA at the top left, this mechanistic progression takes us to the biochemical and subcellular scales (stoichiometric diagrams of N and P in molecules and organelles), associated with ideas involved in the growth rate hypothesis (plot of N:P vs. specific growth rate). In the central part of the figure, we see the notion of differing degrees of homeostasis in autotrophs and consumers which, when coupled to the diversity of stoichiometric patterns in autotrophs and animals, has major ramifications for consumer growth and nutrient recycling. At the very top the connections lead to consequences at the

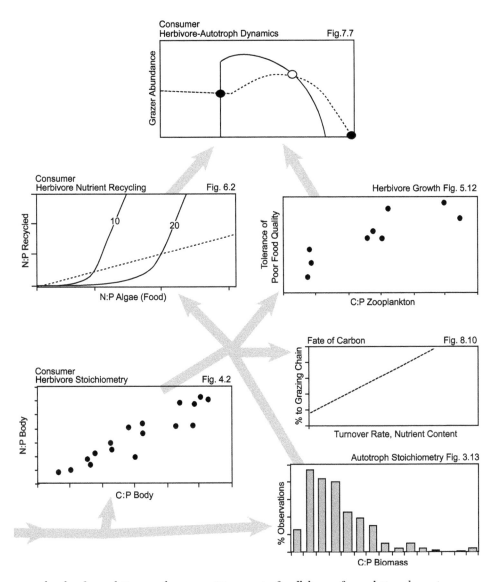

levels of populations and communities: a set of nullclines of population dynamics with equilibria that include a stable grazer extinction point and evidence for differential sensitivity of low-P and high-P consumers to autotroph C:P in pelagic communities. A branch to large-scale ecosystem and global-level patterns of C flux and anthropogenic impacts is shown on the right side of the diagram. This is but one set of causal pathways through the diverse web of information that makes up ecological stoichiometry. We hope you will find your own connections and bridges that are based on the laws of mass conservation and stoichiometric combination.

Appendix

LIST OF SYMBOLS AND THEIR DEFINITIONS

The mathematical symbols used in equations in this book often have been changed from the ones used in the original publication. Lower-case letters indicate rates. Upper-case letters indicate either (1) pools, standing stocks, or biomasses, or (2) efficiencies or dimensionless fractions. In general, definitions are based upon molar units, but when it is necessary to indicate mass units, a prime is used (e.g., Q vs. Q'). Nutrient ratios are always given using standard symbols for the elements (with X indicating any arbitrary element) and colons, e.g., (C:X) is the ratio of carbon to some element. To these nutrient ratios, subscripts are used to provide additional information, e.g., $(P:C)_{min}$ is the minimal ratio of P to C in some observational set. Other subscripts used frequently are as follows: t = time, x = element, and i = observational unit such as species or ecosystem component (e.g., soil). Several symbols used only once, or easily identified abbreviations (e.g., TP = total phosphorus or g = growth), do not appear here. We use "log" to indicate base 10 and "ln" to indicate base e logarithms. In general, our units are based on volumetric pools and processes. For specific applications, conversions to an areal basis would be necessary. In some cases, the text further clarifies specific units of parameters as used in specific contexts.

Symbol	Meaning	Example Units
a	Constant in composite parameter in error propagation	varies
A_X	Environmental concentration of substance X bound in prey (often autotroph, in one model autotroph plus inorganic) mass	moles L^{-1}
c	Integration constant in homeostasis model	varies
d	Grazer mortality loss rate	d^{-1}
e	Volumetric feeding rate	L consumer mass^{-1} time^{-1}

f'	Specific rate of food ingestion	mass carbon ingested (consumer carbon mass)$^{-1}$ time^{-1}, or simply time^{-1}
f'_{max}	Maximal specific rate of food ingestion at high food	time^{-1}
F	Incipient limiting food concentration	mass C L^{-1}
g_X	Absolute rate of consumer growth based on substance X (molar)	moles time^{-1}
g'	Absolute rate of consumer growth (mass)	mass time^{-1}
G_X	Environmental concentration of substance X bound in consumer (often grazer) biomass	moles L^{-1}
GGE_X	Gross growth efficiency of substance X	dimensionless fraction from one to zero
k	Rate coefficient of detrital decay	time^{-1}
K_μ	Half-saturation constant for growth	moles L^{-1}
K_v	Half-saturation constant for uptake	moles L^{-1}
L'	Total phosphorus concentration in loading	µg P L^{-1}
l_X	Specific loss rate of substance X	time^{-1}
$l_{X,\ min}$	Minimal specific loss for a limiting substance	time^{-1}
M	Biomass of an observation unit (individual, population, etc.)	g
m	Specific mortality or loss rate	time^{-1}
NUE	Nutrient use efficiency	mass mass^{-1} (many other units also used in literature)
P	Phosphorus concentration	moles L^{-1}
P_L	Phosphorus concentration in loading	moles L^{-1}
Q_{max}	Maximum achieved quota at growth μ_m	moles individual^{-1}
Q_{min}	Minimal cell quota at μ of zero	moles of substance individual^{-1}
Q	Quota (molar)	moles of substance individual^{-1}
Q'	Quota (mass)	grams of substance individual^{-1}
R	Concentration of limiting resource in external medium	moles L^{-1}
R^*	Equilibrium resource concentration in environment required by population in order for growth to balance mortality losses	moles L^{-1}
r	Consumer respiration rate	d^{-1}
RGR	Fraction of growth potential actually achieved (often μ:μ_m)	dimensionless fraction from one to zero
S_X	Assimilation efficiency; fraction of ingested substance X passing the gut wall and assimilated into the consumer's body	dimensionless fraction from one to zero

T'	Threshold algal P:C mass ratio where consumer phosphorus release is zero	mass mass^{-1}
TER	Threshold element ratio; element ratio in resources where consumer is equally limited by both elements	moles:moles (element ratio)
u	Numerator in composite parameter in error propagation	varies
v	Rate of nutrient uptake	moles time^{-1}
v_{max}	Maximum rate of nutrient uptake at high R	moles time^{-1}
w	Denominator in composite parameter in error propagation	varies
x	Any measure of resource stoichiometry	varies
X_X	Environmental abundance of an element X in some observational unit	moles L^{-1}
y	Any measure of consumer stoichiometry	varies
Y	Abundance of a nonlimiting element in a consumer	varies
z	Composite parameter in error propagation	varies
Γ_X	Concentration of nutrient X in biomass ($X{:}M$)	mass X (total mass)$^{-1}$
H	Homeostasis regulatory coefficient	varies
λ'	Rate of P release per unit consumer C biomass	P (C time)$^{-1}$
μ	Specific growth rate	time^{-1}
μ_m	Maximum specific growth rate	time^{-1}
μ'_m	Theoretical specific growth rate at infinite quota	time^{-1}
σ	Algal sinking loss rate (also, in statistical contexts, standard deviation)	d^{-1}
Φ	Nutrient productivity for carbon (rate of production of new plant carbon produced per unit plant nutrient)	moles carbon (moles nutrient time)$^{-1}$
Φ'_C	Nutrient productivity for carbon mass	mass carbon (nutrient mass time)$^{-1}$
Φ'_M	Nutrient productivity for mass	mass plant (nutrient mass time)$^{-1}$
Ψ	Generalized efficiency of obtaining element X_2 per unit investment in element X_1	moles X_2 (moles X_1 time)$^{-1}$

Literature Cited

Abrahamson, W. G., and H. Caswell. 1982. On the comparative allocation of biomass, energy, and nutrients in plants. *Ecology* 63: 982–991.

Adams, T. S., and R. W. Sterner. 2000. The effect of dietary nitrogen content on trophic level $\delta^{15}N$ enrichment. *Limnol. Oceanogr.* 45: 601–607.

Aerts, R. 1995. The advantage of being evergreen. *Trends Ecol. Evol.* 10: 402–407.

Aerts, R., R.G.A. Boot, and P.J.M. Van der Aart. 1991. The relation between above- and belowground biomass allocation patterns and competitive ability. *Oecologia* 87: 551–559.

Aerts, R., and F. S. Chapin III. 2000. The mineral nutrition of wild plants revisited: a re-evaluation of processes and patterns. *Adv. Ecol. Res.* 30: 1–55.

Agrell, J., E. P. McDonald, and R. L. Lindroth. 2000. Effects of CO_2 and light on tree phytochemistry and insect performance. *Oikos* 88: 259–272.

Ågren, G. I. 1988. Ideal nutrient productivities and nutrient proportions in plant growth. *Plant, Cell Environ.* 11: 613–620.

Ågren, G. I., and E. Bosatta. 1996. *Theoretical Ecosystem Ecology: Understanding Nutrient Cycles.* Cambridge University Press, Cambridge, U.K.

Aitkenhead, J. A., and W. H. McDowell. 2000. Soil C:N ratios as a predictor of annual riverine DOC flux at local and global scales. *Global Biogeochem. Cycles* 14: 127–138.

Akim, L. G., N. Cordeiro, C. P. Neto, and A. Gandini. 2000. Comparative analysis of the lignins of cork from *Quercus suber* L. and wood from *Eucalyptus globulus* L. by dry hydrogen iodide cleavage. Pages 291–302 in *Lignin: Historical, Biological, and Materials Perspectives*, edited by W. G. Gasser, R. A. Northey, and T. P. Schultz. American Chemical Society, Washington, D.C.

Alberts, B., D. Bray, J. Lewis, M. Raff, K. Roberts, and J. D. Watson. 1994. *Molecular Biology of the Cell.* Garland, New York.

Allan, J. D. 1976. Life history patterns in zooplankton. *Am. Nat.* 110: 165–180.

Allen, T.F.H., and T. W. Hoekstra. 1992. *Toward a Unified Ecology.* Columbia University Press, New York.

Andersen, O. K., J. C. Goldman, D. A. Caron, and M. R. Dennett. 1986. Nutrient cycling in a microflagellate food chain: III. Phosphorus dynamics. *Mar. Ecol. Prog. Ser.* 31: 47–55.

Andersen, T. 1997. *Pelagic Nutrient Cycles: Herbivores as Sources and Sinks for Nutrients.* Springer-Verlag, Berlin.

Andersen, T., and D. O. Hessen. 1991. Carbon, nitrogen, and phosphorus content of freshwater zooplankton. *Limnol. Oceanogr.* 36: 807–814.

Anderson, J. F., H. Rahn, and H. D. Prange. 1979. Scaling of supportive tissue mass. *Q. Rev. Biol.* 54: 139–148.

Anderson, L. A., and J. L. Sarmiento. 1994. Redfield ratios of remineralization determined by nutrient data analysis. *Global Biogeochem. Cycles* 8: 65–80.

Anderson, T. R. 1992. Modelling the influence of food C:N ratio and respiration on growth and nitrogen excretion in marine zooplankton and bacteria. *J. Plankton Res.* 14: 1645–1671.

Anderson, T. R. 1994. Relating C:N ratios in zooplankton food and fecal pellets using a biochemical model. *J. Exp. Mar. Biol. Ecol.* 184: 183–199.

Anderson, T. R. and D. O. Hessen. 1995. Carbon or nitrogen limitation in marine copepods? *J. Plankton Res.* 17: 317–331.

Anderson, T. R. and D. W. Pond. 2000. Stoichiometric theory extended to micronutrients: Comparison of the roles of essential fatty acids, carbon, and nitrogen in the nutrition of marine copepods. *Limnol. Oceanogr.* 45: 1162–1167.

Angel, M. V. 1989. Does mesopelagic biology affect the vertical flux? Pages 155–173 in *Productivity of the Ocean: Present and Past*, edited by W. H. Berger, V. S. Smetacek, and G. Wefer. Wiley, New York.

Appel, T. A. 1988. Organizing biology: The American Society of Naturalists and its "Affiliated Societies," 1883–1923. Pages 87–120 in *The American Development of Biology*, edited by R. Rainger, K. R. Benson, and J. Maienschein. Rutgers University Press, New Brunswick, N.J.

Archer, D., and A. Winguth. 2000. What caused the glacial/interglacial atmospheric pCO_2 cycles? *Rev. Geophys.* 38: 159–189.

Arendt, J. D. 1997. Adaptive intrinsic growth rates: an integration across taxa. *Q. Rev. Biol.* 72: 149–177.

Arnott, D. L., and M. J. Vanni. 1996. Nitrogen and phosphorus recycling by the zebra mussel (*Dreissena polymorpha*) in the western basin of Lake Erie. *Can. J. Fish. Aquat. Sci.* 53: 646–659.

Arsuffi, T. L., and K. Suberkropp. 1984. Leaf processing capabilities of aquatic hyphomycetes: Interspecific differences and influence on shredder feeding preferences. *Oikos* 42: 144–154.

Arsuffi, T. L., and K. Suberkropp. 1986. Growth of two stream caddisflies (Trichoptera) on leaves colonized by different fungal species. *J. N. Am. Benthol. Soc.* 5: 297–305.

Ascenzi, A., and G. H. Bell. 1972. Bone as a mechanical engineering problem. Pages 311–352 in *The Biochemistry and Physiology of Bone*, edited by G. H. Bourne Academic, New York.

Atchley, R., and D. Anderson. 1978. Ratios and the statistical analysis of biological data. *Syst. Zool.* 27: 78–83.

Atchley, W. R., C. T. Gaskins, and D. Anderson. 1976. Statistical properties of ratios. I. Empirical results. *Syst. Zool.* 25: 137–148.

Atkinson, M. J., and S. V. Smith. 1983. C:N:P ratios of benthic marine plants. *Limnol. Oceanogr.* 28: 568–574.

Ayres, M. P., R. T. Wilkens, J. J. Ruel, M. J. Lombardero, and E. Vallery. 2000. Nitrogen budgets of phloem-feeding bark beetles with and without symbiotic fungi. *Ecology* 81: 2198–2210.

Baines, S. B., and N. S. Fisher. 2001. Interspecific differences in the bioconcentration of selenite by phytoplankton and their ecological implications. *Mar. Ecol. Prog. Ser.* 213: 1–12.

Båmstedt, U. 1986. Chemical composition and energy content. Pages 1–58 in *The Biological Chemistry of Marine Copepods*, edited by E.D.S. Corner and S.C.M. O'Hara. Oxford University Press, Oxford.

Bannister, T. T. 1974a. A general theory of steady state phytoplankton growth in a nutrient saturated mixed layer. *Limnol. Oceanogr.* 19: 13–30.

Bannister, T. T. 1974b. Production equations in terms of chlorophyll concentration, quantum yield, and upper limit to production. *Limnol. Oceanogr.* 19: 1–12.

Barclay, R.M.R. 1994. Constraints on reproduction by flying vertebrates: energy and calcium. *Am. Nat.* 144: 1021–1031.

Bärlocher, F., and B. Kendrick. 1973a. Fungi and food preferences of *Gammarus pseudolimnaeus*. *Arch. Hydrobiol.* 72: 501–516.

Bärlocher, F., and B. Kendrick. 1973b. Fungi in the diet of *Gammarus pseudolimnaeus* (Amphipoda). *Oikos* 24: 295–300.

Bärlocher, F., and B. Kendrick. 1974. Dynamics of the fungal population on leaves in a stream. *J. Ecol.* 62: 761–791.

Baudouin, M. F., and O. Ravera. 1972. Weight, size and chemical composition of some freshwater zooplankton: *Daphnia hyalina* (Leydig). *Limnol. Oceanogr.* 17: 645–649.

Baudouin, M. F., and P. Scoppa. 1975. The determination of nucleic acids in freshwater zooplankton and its ecological implications. *Freshwater Biol.* 5: 115–120.

Baudouin-Cornu, P., Y. Surdin-Kerjan, P. Marliere, and D. Thomas. 2001. Molecular evolution of protein atomic composition. *Science* 293: 297–300.

Bazely, D. R. 1989. Carnivorous herbivores: mineral nutrition and the balanced diet. *Trends Ecol. Evol.* 4: 155–156.

Bazett-Jones, D. P., M. J. Hendzel, and M. J. Kruhlak. 1999. Stoichiometric analysis of protein- and nucleic acid-based structures in the cell nucleus. *Micron* 30: 151–157.

Becker, W. M. 1986. *The World of the Cell*. Benjamin Cummings, Menlo Park, Calif.

Bedford, B. L., M. R. Walbridge, and A. Aldous. 1999. Patterns in nutrient availability and plant diversity of temperate North American wetlands. *Ecology* 80: 2151–2169.

Beja, O., E. N. Spudich, J. L. Spudich, M. Leclerc, and E. F. DeLong. 2001. Proteorhodopsin phototrophy in the ocean. *Nature* 411: 786–789.

Bender, E. A., T. J. Case, and M. E. Gilpin. 1984. Perturbation experiments in community ecology: Theory and practice. *Ecology* 65: 1–13.

Beniac, D. R., G. J. Czarnota, B. J. Rutherford, F. P. Ottensmeyer, and G. Harauz. 1997. The in situ architecture of *Escherichia coli* ribosomal RNA derived by electron spectroscopic imaging and three-dimensional reconstruction. *J. Microsc.* 188: 24–35.

Benrey, B., and R. F. Denno. 1997. The slow-growth–high-mortality hypothesis: A test using the cabbage butterfly. *Ecology* 78: 987–999.

Berberovic, R. 1990. Biomass composition of two sympatric *Daphnia* species: impact of environmental factors and life history strategies. Ph.D. thesis, University of Konstanz, Konstanz.

Berendse, F., and R. Aerts. 1987. Nitrogen-use efficiency: a biologically meaningful definition? *Funct. Ecol.* 1: 293–296.

Berendse, F., and W. T. Elberse. 1989. Competition and nutrient losses from the plant. Pages 269–284 in *Causes and Consequences of Variation in Growth Rate and Productivity of Higher Plants*, edited by H. Lambers, M. L. Cambridge, H. Konings and T. L. Pons. SPB Academic, The Hague.

Berges, J. A. 1997. Ratios, regression statistics, and "spurious" correlations. *Limnol. Oceanogr.* 42: 1006–1007.

Berman-Frank, I., and Z. Dubinsky. 1999. Balanced growth in aquatic plants: Myth or reality? *BioScience* 49: 29–37.

Bernard, R.T.F., and A. Davison. 1996. Does calcium constrain reproductive activity in insectivorous bats—some empirical evidence for Schreibers long-fingered bat (*Miniopterus schreibersii*). *SAFR J. Zool.* 31: 218–220.

Bernays, E. A. 1982. The insect on the plant—a closer look. Pages 3–17 in *Proceedings of the 5th International Symposium on Insect-Plant Relationships*, edited by J. H. Viser and A. K. Minks. Pudoc, Washington, D.C.

Berner, Y. N., and M. Shike. 1988. Consequences of phosphate imbalance. *Annu. Rev. Nutr.* 8: 121–148.

Berry, J. P. 1996. The role of lysosomes in the selective concentration of mineral elements: A microanalytical study. *Cell. Mol. Biol.* 42: 395–411.

Bevington, P. R. 1969. *Data Reduction and Error Analysis for the Physical Sciences*. McGraw-Hill, New York.

Biddanda, B., M. L. Ogdahl, and J. B. Cotner. 2001. Dominance of bacterial metabolism in oligotrophic relative to eutrophic waters. *Limnol. Oceanogr.* 46: 730–739.

Bielefeld, M., A. J. Bale, Z. S. Kolber, J. Aiken, and P. G. Falkowski. 1996. Confirmation of iron limitation of phytoplankton photosynthesis in the Equatorial Pacific Ocean. *Nature* 383: 508–510.

Bieleski, R. L. 1973. Phosphate pools, phosphate transport, and phosphate availability. *Annu. Rev. Plant Physiol.* 24: 225–252.

Bieleski, R. L., and I. B. Ferguson. 1983. Physiology and metabolism of phosphate and its compounds. Pages 422–449 in *Encyclopedia of Plant Physiology, New Series*, edited by A. Lauchli and R. L. Bieleski. Springer-Verlag, Berlin.

Billick, I., and T. J. Case. 1994. Higher order interactions in ecological communities: What are they and how can they be detected? *Ecology* 75: 1529–1543.

Blackman, R. L., and V. F. Eastop. 1994. *Aphids on the World's Trees*. Oxford University Press, Oxford.

Bloom, A. J., F. S. Chapin III, and H. A. Mooney. 1985. Resource limitation in plants—an economic analogy. *Annu. Rev. Ecol. Syst.* 16: 363–392.

Boerner, E. J. 1984. Foliar nutrient dynamics and nutrient use efficiency of four deciduous tree species in relation to site fertility. *J. Appl. Ecol.* 11: 335–407.

Boersma, M., C. Schöps, and E. McCauley. 2001. Nutritional quality of seston for the freshwater herbivore *Daphnia galeata* X *hyalina*: Biochemical versus mineral limitations. *Oecologia* 129: 342–348.

Boseley, P., T. Moss, M. Machler, R. Portmann, and M. Birnstiel. 1979. Sequence organisation of the spacer DNA in a ribosomal gene unit of *Xenopus laevis*. Cell 17: 19–31.

Bowen, H.J.M. 1979. *Environmental Chemistry of the Elements*. Academic, London.

Bratbak, G., and T. F. Thingstad. 1985. Phytoplankton-bacteria interactions: an apparent paradox? Analysis of a model system with both competition and commensalism. *Mar. Ecol. Prog. Ser.* 25: 23–30.

Breman, H., and C. T. de Wit. 1983. Rangeland productivity and exploitation in the Sahel. *Science* 221: 1341–1347.

Brett, M. 1993. Comment on "Possibility of N or P limitation for planktonic cladocerans: An experimental test" (Urabe and Watanabe) and "Nutrient element limitation of zooplankton production" (Hessen). *Limnol. Oceanogr.* 38: 1333–1337.

Brett, M. T., and J. C. Goldman. 1996. A meta-analysis of the freshwater trophic cascade. *Proc. Natl. Acad. Sci. USA* 93: 7723–7726.

Brett, M. T., and D. Müller-Navarra. 1997. The role of essential fatty acids in aquatic food-web processes. *Freshwater Biol.* 38: 483–499.

Brett, M. T., K. Wiackowski, F. S. Lubnow, A. Mueller-Solger, J. J. Elser, and C. R. Goldman. 1995. *Diacyclops*, *Daphnia*, *Diaptomus*, and *Holopedium* effects on planktonic ecosystem structure in Castle Lake, California. *Ecology* 75: 2243–2254.

Bridgham, S. D., J. Pastor, C. A. McClaugherty, and C. J. Richardson. 1995. Nutrient-use efficiency: a litterfall index, a model, and a test along a nutrient-availability gradient in North Carolina peatlands. *Am. Nat.* 145: 1–21.

Brock, T. D., and M. T. Madigan. 1991. *The Biology of Microorganisms*, 6th ed. Prentice-Hall, Englewood Cliffs, N.J.

Broecker, W. S., and G. M. Henderson. 1998. The sequence of events surrounding Termination II and their implications for the cause of glacial-interglacial CO_2 changes. *Paleoceanography* 13: 352–364.

Brown, J. H., and G. B. West, eds., 2000. *Scaling in Biology*. Oxford University Press, Oxford.

Brown, R. D. 1990. Nutrition and antler development. Pages 426–441 in *Horns, Pronghorns, and Antlers*, edited by G. A. Bubenik and A. B. Bubenik. Springer, New York.

Buonaccorsi, J. P., and A. M. Leibhold. 1988. Statistical methods for estimating ratios and products in ecological studies. *Ecol. Entomol.* 17: 572–580.

Burkhardt, S. and U. Riebesell. 1997. CO_2 availability affects elemental composition (C:N:P) of the marine diatom *Skeletonema costatum*. *Mar. Ecol. Prog. Ser.* 155: 67–76.

Burkhardt, S., I. Zondervan, and U. Riebesell. 1999. Effect of CO_2 concentration on C:N:P ratio in marine phytoplankton: A species comparison. *Limnol. Oceanogr.* 44: 683–690.

Burmaster, D. 1979. The continuous culture of phytoplankton: mathematical equivalence among three steady state models. *Am. Nat.* 113: 123–134.

Buwalda, J. G., and K. M. Goh. 1982. Host-fungus competition for carbon as a cause of growth depressions in vesicular-arbuscular mycorrhizal ryegrass. *Soil Biol. Biochem.* 14: 103–106.

Calder, W. A. 1984. *Size, Function, and Life History*. Harvard University Press, Cambridge, Mass.

Calow, P. 1977a. Conversion efficiencies in heterotrophic organisms. *Biol. Rev.* 52: 385–409.

Calow, P. 1977b. Ecology, evolution and energetics: A study in metabolic adaptation. *Adv. Ecol. Res.* 10: 1–60.

Calow, P. 1982. Homeostasis and fitness. *Am. Nat.* 120: 416–419.

Campana, T., and L. M. Schwartz. 1981. RNA and associated enzymes. Pages 877–944 in *Advanced Cell Biology*, edited by L. M. Schwartz and M. M. Azar. Van Nostrand Reinhold, New York.

Caron, D. A., J. C. Goldman, O. K. Andersen, and M. R. Dennett. 1985. Nutrient cycling in a microflagellate food chain: II. Population dynamics and carbon cycling. *Mar. Ecol. Prog. Ser.* 24: 243–254.

Carpenter, S. R., and J. F. Kitchell. 1993. *The Trophic Cascade in Lakes*. Cambridge University Press, Cambridge, U.K.

Carpenter, S. R., J. F. Kitchell, and J. R. Hodgson. 1985. Cascading trophic interactions and lake productivity. *Bioscience* 35: 634–639.

Carpenter, S. R., J. F. Kitchell, J. R. Hodgson, P. A. Cochran, J. J. Elser, M. M. Elser, D. Lodge, D. Kretchmer, X. He, and C. N. von Ende. 1987. Regulation of lake primary productivity by food web structure. *Ecology* 68: 1863–1876.

Carpenter, S. R., C. E. Kraft, R. Wright, H. Xi, P. A. Soranno, and J. R. Hodgson. 1992. Resilience and resistance of a lake phosphorus cycle before and after food web manipulation. *Am. Nat.* 140: 781–798.

Carrillo, P., M. Villar-Argaiz, and J. M. Medina-Sánchez. 2001. Relationship between N:P ratio and growth rate during the life cycle of calanoid copepods: An *in situ* measurement. *J. Plankton Res.* 23: 537–547.

Case, T. J. 2000. *An Illustrated Guide to Theoretical Ecology*. Oxford University Press, New York.

Cassadevall, M., A. Casinos, C. Viladiu, and M. Ontañon. 1990. Scaling of skeletal mass and mineral content in teleosts. *Zool. Anz.* 225: 144–150.

Cavaletto, J. F., and W. S. Gardner. 1999. Seasonal dynamics of lipids in freshwater benthic invertebrates. Pages 109–131 in *Lipids in Freshwater Ecosystems*, edited by M. T. Arts and B. C. Wainman. Springer, New York.

Cavalier-Smith, T., ed. 1985. *The Evolution of Genome Size*. Wiley, New York.

Cebrián, J. 1999. Patterns in the fate of production in plant communities. *Am. Nat.* 154: 449.

Cebrián, J., and C. M. Duarte. 1994. The dependence of herbivory on growth rate in natural plant communities. *Funct. Ecol.* 8: 518–525.

Cebrián, J., and C. M. Duarte. 1995. Plant growth-rate dependence of detrital carbon storage in ecosystems. *Science* 268: 1606–1608.

Cebrián, J., and C. M. Duarte. 1998. Patterns in leaf herbivory on seagrasses. *Aquat. Bot.* 60: 67–82.

Cebrián, J., C. M. Duarte, N. Marba, S. Enriquez, M. Gallegos, and B. Olesen. 1996. Herbivory on *P. oceanica* (L.) Delile: magnitude and variability in the Spanish Mediteranean. *Mar. Ecol. Prog. Ser.* 130: 147–155.

Cebrián, J., M. Williams, J. McClelland, and I. Valiella. 1998. The dependence of heterotrophic consumption and C accumulation on autotrophic nutrient content in ecosystems. *Ecol. Lett.* 1: 165–170.

Chapin, F. S., III. 1980. The mineral nutrition of wild plants. *Annu. Rev. Ecol. Syst.* 11: 233–260.

Chapin, F. S., III. 1991a. Effects of multiple environmental stresses on nutrient availability and use. Pages 67–88 in *Responses of Plants to Multiple Stresses*, edited by H. A. Mooney, W. E. Winner, and E. J. Pell. Academic, San Diego.

Chapin, F. S., III. 1991b. Integrated responses of plants to stress. *BioScience* 41: 29–36.

Chapin, F. S., III. 1993. The functional role of growth forms in ecosystem and global processes. Pages 287–312 in *Scaling Processes Between Leaf and Landscape Levels*, edited by J. R. Ehleringer and C. B. Field. Academic, San Diego.

Chapin, F. S., III, A. J. Bloom, C. B. Field, and R. H. Waring. 1987. Plant responses to multiple environmental factors. *BioScience* 37: 49–57.

Chapin, F. S., III, P. M. Vitousek, and K. Van Cleve. 1986. The nature of nutrient limitation in plant communities. *Am. Nat.* 127: 48–58.

Chase, J. M. 2000. Are there real differences among aquatic and terrestrial food webs? *Trends Ecol. Evol.* 15: 408–412.

Chase, Z., and N. M. Price. 1997. Metabolic consequences of iron deficiency in heterotrophic marine protozoa. *Limnol. Oceanogr.* 42: 1673–1684.

Chisholm, S. W., and F.M.M. Morel, eds. 1991. *What Controls Phytoplankton Production in Nutrient-Rich Areas of the Open Sea?* Special issue of *Limnol. Oceanogr.* 36.

Chrzanowski, T. H., and M. Kyle. 1996. Ratios of carbon, nitrogen and phosphorus in *Pseudomonas fluorescens* as a model for bacterial element ratios and nutrient regeneration. *Aquat. Microb. Ecol.* 10: 115–122.

Chrzanowski, T. H., M. Kyle, J. J. Elser, and R. W. Sterner. 1997. Element ratios and growth dynamics of bacteria in an oligotrophic Canadian shield lake. *Aquat. Microb. Ecol.* 11: 119–125.

Church, R. G., and F. W. Robertson. 1966. A biochemical study of the growth of *Drosophila melanogaster. J. Exp. Biol.* 162: 337–352.

Cimbleris, A.C.P., and J. Kalff. 1998. Planktonic bacterial respiration as a function of C:N:P ratios across temperate lakes. *Hydrobiologia* 384: 89–100.

Clark, D. R. 2001. Growth rate relationships to physiological indices of nutrient status in marine diatoms. *J. Phycol.* 37: 249–256.

Cluster, P. D., D. Marinkovic, R. W. Allard, and F. J. Ayala. 1987. Correlations between development rates, enzyme activities, ribosomal DNA spacer-length phenotypes, and adaptation in *Drosophila melanogaster. Proc. Natl. Acad. Sci. USA* 84: 610–614.

Cole, J. J., J. M. Lane, R. Marino, and R. W. Howarth. 1993a. Molybdenum assimilation by cyanobacteria and phytoplankton in freshwater and salt water. *Limnol. Oceanogr.* 38: 25–35.

Cole, J. J., B. L. Peierls, N. F. Caraco, and M. L. Pace. 1993b. Nitrogen loading of rivers as a human-driven process. Pages 141–160 in *Humans as Components of Ecosystems: The Ecology of Subtle Human Effects and Populated Areas*, edited by M. J. McDonnell and S.T.A. Pickett. Springer, New York.

Colman, R. L., and A. Lazemby. 1970. Factors affecting the response of tropical and temperate grasses to fertilizer nitrogen. Pages 393–397 in *Proceedings of the 11th International Grassland Conference.* University of Queensland Press, St. Lucia.

Connell, J. H. 1961. The influence of interspecific competition and other factors on the distribution of the barnacle *Chthamalus stellatus. Ecology* 42: 710–723.

Connell, J. H., and W. P. Sousa. 1983. On the evidence needed to judge ecological stability or persistence. *Am. Nat.* 121: 789–821.

Copin-Montegut, C., and G. Copin-Montegut. 1983. Stoichiometry of carbon, nitrogen, and phosphorus in marine particulate matter. *Deep-Sea Res.* 30: 31–46.

Corner, E.D.S., C. B. Cowey, and S. M. Marshall. 1965. On the nutrition and metabolism of zooplankton. III. Nitrogen excretion by *Calanus. J. Mar. Biol. Ass. U.K.* 45: 429–442.

Corner, E.D.S., and A. G. Davies. 1971. Plankton as a factor in the nitrogen and phosphorus cycles in the sea. *Adv. in Mar. Biol.* 9: 101–204.

Corner, E.D.S., R. N. Head, and C. C. Kilvington. 1972. On the nutrition and metabolism of zooplankton VIII. The grazing of *Biddulphia* cells by *Calanus helgolandicus. J. Mar. Biol. Ass. U.K.* 52: 847–861.

Corner, E.D.S., R. N. Head, C. C. Kilvington, and L. Pennycuick. 1976. On the nutrition and metabolism of zooplankton. *J. Mar. Biol. Ass. U.K.* 56: 345–358.

Corner, E.D.S., and B. S. Newell. 1967. On the nutrition and metabolism of zooplankton. IV. The forms of nitrogen excreted by *Calanus. J. Mar. Biol. Ass. U.K.* 47: 113–120.

Cortadas, J., and M. C. Pavon. 1982. The organization of ribosomal genes in vertebrates. *EMBO J.* 1: 1075–1080.

Costanza, R., R. d'Arge, R. de Groot, S. Farber, M. Grasso, B. Hannon, K. Limburg, S. Naeem, R. V. O'Neill, J. Paruelo, R. G. Raskin, P. Sutton, and M. van den Belt. 1997. The value of the world's ecosystem services and natural capital. *Nature* 387: 253–260.

Cotner, J. B. 2001. Heterotrophic bacterial growth and nutrient limitation in large, oligotrophic lakes and oceans. *Verh. Int. Verein. Limnol.* 27: 1831–1835.

Cotner, J. B., and B. A. Biddanda. 2002. Small players, large role: microbial influence on auto-heterotrophic coupling and biogeochemical processes in aquatic ecosystems. *Ecosystems* 5: 105–121.

Cotrufo, M. F., M.J.I. Briones, and P. Ineson. 1998. Elevated CO_2 affects field decomposition rate and palatability of tree leaf litter: Importance of changes in substrate quality. *Soil Biol. Biochem.* 30: 1565–1571.

Cotrufo, M. F., and P. Ineson. 1996. Elevated CO_2 reduces field decomposition rates of *Betula pendula* (Roth.) leaf litter. *Oecologia* 106: 525–530.

Council, A. R. 1980. *The Nutrient Requirements of Ruminant Livestock*. Commonwealth Agricultural Bureau, Slough, U.K.

Cox, P. A. 1989. *The Elements: Their Origin, Abundance and Distribution*. Oxford University Press, Oxford.

Cox, P. A. 1995. *The Elements on Earth*. Oxford University Press, Oxford.

Cromack, K., Jr., and B. A. Caldwell. 1992. The role of fungi in litter decomposition and nutrient cycling. Pages 653–668 in *The Fungal Community: Its Organization and Role in the Ecosystem*, edited by G. C. Carroll and D. T. Wicklow. Marcel Dekker, New York.

Cruz-Rivera, E. and M. E. Hay. 2000. Can quantity replace quality? Food choice, compensatory feeding, and fitness of marine mesograzers. *Ecology* 81: 201–219.

Cullis, C. A., and L. Charlton. 1981. The induction of ribosomal DNA changes in flax. *Plant Sci. Lett.* 20: 213–217.

Curtis, P. S., and X. Wang. 1998. A meta-analysis of elevated CO_2 effects on woody plant mass, form, and physiology. *Oecologia* 113: 299–313.

Cyr, H., and M. L. Pace. 1992. Grazing by zooplankton and its relationship to community structure. *Can. J. Fish. Aquat. Sci.* 49: 1455–1465.

Cyr, H., and M. L. Pace. 1993. Magnitude and patterns of herbivory in aquatic and terrestrial ecosystems. *Nature* 361: 148–150.

Dadd, R. H. 1970. Arthropod nutrition. *Chem. Zool.* 5: 35–95.

Dagg, M. J., and J. L. Littlepage. 1972. Relationships between growth rate and RNA, DNA, protein, and dry weight in *Artemia salina* and *Eucheata elongata*. *Mar. Biol.* 17: 162–170.

Darbyshire, J. F., M. S. Davidson, S. J. Chapman, and S. Ritchie. 1994. Excretion of nitrogen and phosphorus by the soil ciliate *Colpoda steinii* when fed the soil bacterium *Arthrobacter* sp. *Soil Biol. Biochem.* 26: 1193–1199.

Davis, J. A., and C. E. Boyd. 1978. Concentrations of selected elements and ash in bluegill (*Lepomis macrochirus*) and certain other freshwater fish. *Trans. Am. Fish. Soc.* 107: 862–867.

De La Rocha, C. L., D. A. Hutchins, M. A. Brzezinski, and Y. H. Zhang. 2000. Effects of iron and zinc deficiency on elemental composition and silica production by diatoms. *Mar. Ecol. Prog. Ser.* 195: 71–79.

de Mazencourt, C., and M. Loreau. 2000. Effect of herbivory and plant species replacement on primary production. *Am. Nat.* 155: 735–754.

de Mazancourt, C., M. Loreau, and L. Abbadie. 1999. Grazing optimization and nutrient cycling: Potential impact of large herbivores in a savannah system. *Ecol. Applic.* 9: 784–797.

De Rosa, M., A. Trincone, B. Nicolaus, and A. Gambacorta. 1991. Achaebacteria: lipids, membrane structures, and adaptation to environmental stress. Pages 61–87 in *Life under Extreme Conditions*, edited by G. Di Prisco. Springer-Verlag, Berlin.

DeAngelis, D. L. 1992. *Dynamics of Nutrient Cycling and Food Webs*. Chapman and Hall, New York.

DeAngelis, D. L., W. M. Post, and C. C. Travis. 1986. *Positive Feedback in Natural Systems*. Springer-Verlag, Berlin.

Deevey, E. S. 1970. Mineral cycles. *Sci. Am.* 223: 148–158.

del Giorgio, P. A., and J. J. Cole. 1998. Bacterial growth efficiencies in natural aquatic systems. *Annu. Rev. Ecol. Syst.* 29: 503–541.

del Giorgio, P. A., and J. M. Gasol. 1995. Biomass distribution in freshwater plankton communities. *Am. Nat.* 146: 135–152.

del Giorgio, P. A., and R. H. Peters. 1994. Patterns in planktonic P:R ratios in lakes: influence of lake trophy and dissolved organic matter. *Limnol. Oceanogr.* 39: 772–787.

Delany, M. E., A. Emsley, M. B. Smiley, J. R. Putnam, and S. E. Bloom. 1994a. Nucleolar size polymorphisms in commercial layer chickens—determination of incidence, inheritance, and nucleolar sizes within and among chicken lines selected for enhanced growth. *Poultry Sci.* 73: 1211–1217.

Delany, M. E., and A. B. Krupkin. 1999. Molecular characterization of ribosomal gene variation within and among NORs segregating in specialized populations of chicken. *Genome* 42: 60–71.

Delany, M. E., D. E. Muscarella, and S. E. Bloom. 1994b. Effects of rRNA gene copy number and nucleolar variation on early development: inhibition of gastrulation in rDNA-deficient chick embryos. *J. Hered.* 85: 211–217.

Delgenes, J. P., E. Rustrian, N. Bernet, and R. Moletta. 1998. Combined degradation of carbon, nitrogen and phosphorus from wastewaters. *J. Mol. Catal. B—Enzym.* 5: 429–433.

DeMelo, R., R. France, and D. J. McQueen. 1992. Biomanipulation: Hit or myth? *Limnol. Oceanogr.* 37: 192–207.

Demment, M. W., and P. J. Van Soest. 1983. *Body Size, Digestive Capacity, and Feeding Strategies of Herbivores.* Winrock International, Morrilton, Ark.

DeMott, W. R., and R. D. Gulati. 1999. Phosphorus limitation in *Daphnia*: Evidence from a long term sudy of three hypereutrophic Dutch lakes. *Limnol. Oceanogr.* 44: 1557.

DeMott, W. R., R. D. Gulati, and K. Siewertsen. 1998. Effects of phosphorus-deficient diets on the carbon and phosphorus balance of *Daphnia magna*. *Limnol. Oceanogr.* 43: 1147–1161.

DeMott, W. R., and D. C. Müller-Navarra. 1997. The importance of highly unsaturated fatty acids in zooplankton nutrition: Evidence from experiments with *Daphnia*, a cyanobacterium and lipid emulsions. *Freshwater Biol.* 38: 649–664.

DeRuiter, P., A.-M. Neutel, and J. C. Moore. 1995. Energetics, patterns of interaction strengths, and stability in real ecosystems. *Science* 269: 1257–1260.

Desalle, R., J. Slightom, and E. Zimmer. 1986. The molecular through ecological genetics of abnormal abdomen: II. Ribosomal DNA polymorphism is associated with the abnormal abdomen syndrome in *Drosophila mercatorum*. *Genetics* 112: 861–876.

DeZwann, A., and G. vd Thillart. 1985. Low and high power output modes of anaerobic metabolism: invertebrate and vertebrate strategies. Pages 166–192 in *Circulation, Respiration, and Metabolism*, edited by R. Gilles. Springer-Verlag, Berlin.

DiBacco, C. and L. A. Levin. 2000. Development and application of elemental fingerprinting to track the dispersal of marine larvae. *Limnol. Oceanogr.* 45: 871–880.

Dixon, A.F.G. and R. Kundu. 1998. Resource tracking in aphids: Programmed reproductive strategies anticipate seasonal trends in habitat quality. *Oecologia* 114: 73.

Dobberfuhl, D. R. 1999. N:P stoichiometry in crustacean zooplankton: Phylogenetic patterns, physiological mechanisms, and ecological consequences. Ph.D. thesis, Arizona State University, Tempe, Ariz.

Dobberfuhl, D. R. and J. J. Elser. 2000. Elemental stoichiometry of lower food web components in arctic and temperate lakes. *J. Plankton Res.* 22: 1341–1354.

Dolan, J. R. 1997. Phosphorus and ammonia excretion by planktonic protists. *Mar. Geol.* 139: 109–122.

Dover, G. A. 1982. Molecular drive: a cohesive mode of species evolution. *Nature* 295: 564–568.

Downing, J. A. 1997. Marine nitrogen:phosphorus stoichiometry and the global N:P cycle. *Biogeochemistry* 37: 237–252.

Downing, J. A., and E. McCauley. 1992. The nitrogen:phosphorus relationship in lakes. *Limnol. Oceanogr.* 37: 936–945.

Downing, J. A., M. McClain, R. Twilley, J. M. Melack, J. Elser, N. N. Rabalais, W. M. Lewis, Jr., R. E. Turner, J. Corredor, D. Soto, A. Yanez-Arancibia, J. A.

Kopaska, and R. A. Howarth. 1999a. The impact of accelerating land-use change on the N-cycle of tropical aquatic ecosystems: Current conditions and projected changes. *Biogeochemistry* 46: 109–148.

Downing, J. A., C. W. Osenberg, and O. Sarnelle. 1999b. Meta-analysis of marine nutrient-enrichment experiments: Variation in the magnitude of nutrient limitation. *Ecology* 80: 1157–1167.

Downing, J. A., C. Plante, and S. Lalonde. 1990. Fish production correlated with primary productivity, not the morphoedaphic index. *Can. J. Fish. Aquat. Sci.* 47: 1929–1936.

Droop, M. R. 1970. Vitamin B_{12} and marine ecology V. Continuous culture as an approach to nutritional kinetics. *Helgoländer Meeresuntersuchungen* 20: 629–636.

Droop, M. R. 1973. Some thoughts on nutrient limitation in algae. *J. Phycol.* 9: 264–272.

Droop, M. R. 1974. The nutrient status of algal cells in continuous culture. *J. Mar. Biol. Assoc. U.K.* 54: 825–855.

Droop, M. R. 1983. 25 years of algal growth kinetics: A personal view. *Bot. Mar.* 26: 99–112.

Duarte, C. M. 1992. Nutrient concentration of aquatic plants: patterns across species. *Limnol. Oceanogr.* 37: 882–889.

Duarte, C. M., and J. Cebrián. 1996. The fate of marine autotrophic production. *Limnol. Oceanogr.* 41: 1758–1766.

Eisele, K. A., D. S. Schimel, L. A. Kapustka, and W. J. Parton. 1989. Effects of available P and N:P ratios on non-symbiotic dinitrogen fixation in tallgrass prairie soils. *Oecologia* 79: 471–474.

Elliott, K. J., and A. S. White. 1994. Effects of light, nitrogen, and phosphorus on red pine seedling growth and nutrient use efficiency. *For. Sci.* 40: 47–58.

Elrifi, I. R., and D. H. Turpin. 1985. Steady-state luxury consumption and the concept of optimum nutrient ratios: A study with phosphate and nitrate limited *Selenastrum minutum* (Chlorophyta). *J. Phycol.* 21: 592–602.

Elser, J. J., T. H. Chrzanowski, R. W. Sterner, and K. H. Mills. 1998. Stoichiometric constraints on food-web dynamics: A whole-lake experiment on the Canadian Shield. *Ecosystems* 1: 120–136.

Elser, J. J., T. H. Chrzanowski, R. W. Sterner, J. H. Schampel, and D. K. Foster. 1995a. Elemental ratios and the uptake and release of nutrients by phytoplankton and bacteria in three lakes of the Canadian Shield. *Microb. Ecol.* 29: 145–162.

Elser, J. J., D. Dobberfuhl, N. A. MacKay, and J. H. Schampel. 1996. Organism size, life history, and N:P stoichiometry: Towards a unified view of cellular and ecosystem processes. *BioScience* 46: 674–684.

Elser, J. J., T. Dowling, D. R. Dobberfuhl, and J. O'Brien. 2000a. The evolution of ecosystem processes: Ecological stoichiometry of a key herbivore in temperate and arctic habitats. *J. Evol. Biol.* 13: 845–853.

Elser, J. J., M. M. Elser, N. A. MacKay, and S. R. Carpenter. 1988. Zooplankton-mediated transitions between N and P limited algal growth. *Limnol. Oceanogr.* 33: 1–14.

Elser, J. J., W. F. Fagan, R. F. Denno, D. R. Dobberfuhl, A. Folarin, A. Huberty,

S. Interlandi, S. S. Kilham, E. McCauley, K. L. Schulz, E. H. Siemann, and R. W. Sterner. 2000b. Nutritional constraints in terrestrial and freshwater food webs. *Nature* 408: 578–580.

Elser, J. J., and D. K. Foster. 1998. N:P stoichiometry of sedimentation in lakes of the Canadian Shield: Relationships with seston and zooplankton elemental composition. *EcoScience* 5: 56–63.

Elser, J. J., D. K. Foster, and R. E. Hecky. 1995b. Effects of zooplankton on sedimentation in pelagic ecosystems: theory and test in two lakes of the Canadian Shield. *Biogeochemistry* 30: 143–170.

Elser, J. J., and N. B. George. 1993. The stoichiometry of N and P in the pelagic zone of Castle Lake, California. *J. Plankton Res.* 15: 977–992.

Elser, J. J., and R. P. Hassett. 1994. A stoichiometric analysis of the zooplankton-phytoplankton interaction in marine and freshwater ecosystems. *Nature* 370: 211–213.

Elser, J. J., H. Hayakawa, and J. Urabe. 2001. Nutrient limitation reduces food quality for zooplankton: *Daphnia* response to seston phosphorus enrichment. *Ecology* 82: 898–903.

Elser, J. J., F. S. Lubnow, M. T. Brett, E. R. Marzolf, G. Dion, and C. R. Goldman. 1995c. Abiotic and biotic factors associated with inter- and intra-annual variation of nutrient limitation on phytoplankton growth in Castle Lake, California. *Can. J. Fish. Aquat. Sci.* 52: 93–104.

Elser, J. J., C. Luecke, M. T. Brett, and C. R. Goldman. 1995d. Effects of food web compensation after manipulation of rainbow trout in an oligotrophic lake. *Ecology* 76: 52–69.

Elser, J. J., E. R. Marzolf, and C. R. Goldman. 1990. Phosphorus and nitrogen limitation of phytoplankton growth in the freshwaters of North America: A review and critique of experimental enrichments. *Can. J. Fish. Aquat. Sci.* 47: 1468–1477.

Elser, J. J., R. W. Sterner, A. E. Galford, T. H. Chrzanowski, D. L. Findlay, K. H. Mills, M. J. Paterson, M. P. Stainton, and D. W. Schindler. 2000c. Pelagic C:N:P stoichiometry in a eutrophied lake: Responses to a whole-lake food-web manipulation. *Ecosystems* 3: 293–307.

Elser, J. J., R. W. Sterner, E. Gorokhova, W. F. Fagan, T. A. Markow, J. B. Cotner, J. F. Harrison, S. E. Hobbie, G. M. Odell, and L. J. Weider. 2000d. Biological stoichiometry from genes to ecosystems. *Ecol. Lett.* 3: 540–550.

Elser, J. J., and J. Urabe. 1999. The stoichiometry of consumer-driven nutrient recycling: Theory, observations, and consequences. *Ecology* 80: 735–751.

Elwood, H. J., G. J. Olsen, and M. L. Sogin. 1985. The small-subunit ribosomal RNA gene sequences from the hypotrichous ciliates *Oxytricha nova* and *Stylonychia pustulata*. *Mol. Biol. Evol.* 2: 399–410.

Enquist, B. J., and K. J. Niklas. 2001. Invariant scaling relations across tree-dominated communities. *Nature* 410: 655–660.

Enriquez, S., C. M. Duarte, and K. Sand-Jensen. 1993. Patterns in decomposition rates among photosynthetic organisms: The importance of detritus C:N:P content. *Oecologia* 94: 457–471.

Epstein, E. 1972. *Mineral Nutrition of Plants: Principles and Perspectives*. Wiley, New York.

Evans, M. S., M. T. Arts, and R. D. Robarts. 1996. Algal productivity, algal biomass, and zooplankton biomass in a phosphorus-rich, saline lake: Deviations from regression model predictions. *Can. J. Fish. Aquat. Sci.* 53: 1048–1060.

Evans, W. H. 1989. *Membrane Structure and Function.* Oxford University Press, Oxford.

Evers, R., and I. Grummt. 1995. Molecular evolution of the mammalian ribosomal gene terminator sequences and the transcription termination factor TTF-I. *Proc. Natl. Acad. Sci. USA* 92: 5827–5831.

Fagan, W. F., E. H. Siemann, R. F. Denno, C. Mitter, A. Huberty, H. A. Woods, and J. J. Elser. 2002. Nitrogen in insects: Implications for trophic complexity and species diversification. *Am. Nat.* in press.

Falkowski, P. G. 1997. Evolution of the nitrogen cycle and its influence on the biological sequestration of CO_2 in the ocean. *Nature* 387: 272–275.

Falkowski, P. G. 2000. Rationalizing elemental ratios in unicellular algae. *J. Phycol.* 36: 3–6.

Falkowski, P. G., R. T. Barber, and V. Smetacek. 1998. Biogeochemical controls and feedbacks on ocean primary productivity. *Science* 281: 200–206.

Falkowski, P. G., and J. A. Raven. 1997. *Aquatic Photosynthesis.* Blackwell, Malden, Mass.

Falkowski, P. G., R. J. Scholes, E. Boyle, J. Canadell, D. Canfield, J. Elser, N. Gruber, K. Hibbard, P. Högberg, S. Linder, F. T. Mackenzie, B. I. Moore, T. Pedersen, Y. Rosenthal, S. Seitzinger, V. Smetacek, and W. Steffen. 2000. The global carbon cycle: a test of our knowlege of Earth as a system. *Science* 290: 291–296.

Fee, E. J., R. E. Hecky, S.E.M. Kasian, and D. R. Cruikshank. 1996. Effects of lake size, water clarity, and climatic variability on mixing depths in Canadian Shield lakes. *Limnol. Oceanogr.* 41: 912–920.

Fee, E. J., R. E. Hecky, G. W. Regehr, L. L. Hendzel, and P. Wilkinson. 1994. Effects of lake size on nutrient availability in the mixed layer during summer stratification. *Can. J. Fish. Aquat. Sci.* 51: 2756–2768.

Feeny, P. P. 1976. Plant apparency and chemical defense. *Rec. Adv. Phytochem.* 10: 1–40.

Fenchel, T., G. M. King, and T. H. Blackburn. 1998. *Bacterial Biogeochemistry: The Ecophysiology of Mineral Cycling.* Academic, San Diego.

Fernández-Aláez, M., C. Fernández-Aláez, and E. Bécares. 1999. Nutrient content in macrophytes in Spanish shallow lakes. *Hydrobiologia* 408/409: 317–326.

Ferris, T. 2001. *The Whole Shebang: A State of the Universe(s) Report.* Simon and Schuster, New York.

Field, C., and H. A. Mooney. 1986. The photosynthesis-nitrogen relationship in wild plants. Pages 25–55 in *On the Economy of Plant Form and Function,* edited by T. Givnish. Cambridge University Press, Cambridge, U.K.

Findlay, D. L., R. E. Hecky, L. L. Hendzel, M. P. Stainton, and G. W. Regehr. 1994. Relationship between N_2-fixation and heterocyst abundance and its relevance to the nitrogen budget of Lake 227. *Can. J. Fish. Aquat. Sci.* 51: 2254–2266.

Fitter, A. H. 1991. Costs and benefits of mycorrhizas: implications for functioning under natural conditions. *Experientia* 47: 350–355.

Flavell, R. B., M. O'Dell, P. Sharp, and E. Nevo. 1986. Variation in the intergenic spacer of ribosomal DNA of wild wheat, *Triticum dicoccoides*, in Israel. *Mol. Biol. Evol.* 3: 547–558.

Fogg, G. E. 1975. *Algal Cultures and Phytoplankton Ecology*, 2nd ed. The University of Wisconsin Press, Madison, Wis.

Fourqurean, J. W., J. C. Zieman, and G.V.N. Powell. 1992. Phosphorus limitation of primary production in Florida Bay: Evidence from C:N:P ratios of the dominant seagrass *Thalassia testudinum*. *Limnol. Oceanogr.* 37: 162–171.

Fox, L. R., and B. J. Macauley. 1977. Insect grazing on *Eucalyptus* in response to variation in leaf tannins and nitrogen. *Oecologia* 29: 145–162.

Franz, G., and W. Kunz. 1981. Intervening sequences in ribosomal RNA genes and bobbed phenotype in *Drosophila hydei*. *Nature* 292: 638–640.

Fraústo da Silva, J.J.R., and R.J.P. Williams. 1991. *The Biological Chemistry of the Elements: The Inorganic Chemistry of Life*. Clarendon, Oxford.

Gaedke, U., D. Straile, and C. Pahl-Wostl. 1996. Trophic structure and carbon flow dynamics in the pelagic community of a large lake. Pages 60–71 in *Food Webs: Integration of Pattern and Dynamics*, edited by G. A. Polis and K. O. Winemiller. Chapman and Hall, New York.

Gall, J. G. 1968. Differential synthesis of the genes for ribosomal RNA during amphibian oogenesis. *Proc. Natl. Acad Sci. USA* 60: 553–560.

Ganley, A.R.D., and B. Scott. 1998. Extraordinary ribosomal spacer length heterogeneity in a neotyphodium endophyte hybrid: Implications for concerted evolution. *Genetics* 150: 1625–1637.

Gannes, L. Z., D. M. O'Brien, and C. M. Del Rio. 1997. Stable isotopes in animal ecology: Assumptions, caveats, and a call for more laboratory experiments. *Ecology* 78: 1271–1276.

Garten, C. T., Jr. 1978. Multivariate perspectives on the ecology of plant mineral element composition. *Am. Nat.* 112: 533–544.

Gauthier, M. J., G. N. Flatau, and R. L. Clément. 1990. Influence of phosphate ions and alkaline phosphatase activity of cells on survival of *Escherichia coli* in seawater. *Microb. Ecol.* 20: 245–251.

Geider, R. J., H. L. MacIntyre, and T. M. Kana. 1998. A dynamic regulatory model of phytoplanktonic acclimation to light, nutrients, and temperature. *Limnol. Oceanogr.* 43: 679–694.

Gervais, F., and U. Riebesell. 2001. Effect of phosphorus-limitation on elemental composition and stable carbon isotope fractionation in a marine diatom growing under different CO_2 concentrations. *Limnol. Oceanogr.* 46: 497–504.

Gibbs, R. J. 1970. Mechanisms controlling world water chemistry. *Science* 170: 1088–1090.

Giese, A. C. 1979. *Cell Physiology*. Saunders, Philadelphia.

Gilbert, L. I. 1967. Lipid metabolism and function in insects. Pages 70–212 in *Advances in Insect Physiology*, edited by J.W.L. Beament, J. E. Treherne and V. B. Wigglesworth. Academic, London.

Gisselson, L.-A., and E. Granéli. 2001. Variation in cellular nutrient status within a population of *Dinophysis norvigica* (Dinophyceae) growing in situ: Single-cell elemental analysis by use of nuclear microprobe. *Limnol. Oceanogr.* 46: 1237–1242.

Gniazdowska, A., B. Szal, and A. M. Rychter. 1999. The effect of phosphate deficiency on membrane phospholipid composition of bean (*Phaseolus vulgaris* L.) roots. *Acta Physiol. Plant.* 21: 263–269.

Goldman, J. C. 1984. Oceanic nutrient cycles. Pages 137–170 in *Flows of Energy and Materials in Marine Ecosystems Theory*, edited by M. J. Fasham. Plenum, New York.

Goldman, J. C. 1986. On phytoplankton growth rates and particulate C:N:P ratios at low light. *Limnol. Oceanogr.* 31: 1358–1363.

Goldman, J. C., D. A. Caron, O. K. Andersen, and M. R. Dennett. 1985. Nutrient cycling in a microflagellate food chain: I. Nitrogen dynamics. *Mar. Ecol. Prog. Ser.* 24: 231–242.

Goldman, J. C., D. A. Caron, and M. R. Dennett. 1987a. Nutrient cycling in a microflagellate food chain: IV. Phytoplankton-microflagellate interactions. *Mar. Ecol. Prog. Ser.* 38: 75–87.

Goldman, J. C., D. A. Caron, and M. R. Dennett. 1987b. Regulation of gross growth efficiency and ammonium regeneration in bacteria by substrate C:N ratio. *Limnol. Oceanogr.* 32: 1239–1252.

Goldman, J. C., J. J. McCarthy, and D. G. Peavey. 1979. Growth rate influence on the chemical composition of phytoplankton in oceanic waters. *Nature* 279: 210–215.

Goldman, W. E., G. Goldberg, L. H. Bowman, D. Steinmetz, and D. Schlessinger. 1983. Mouse rDNA: Sequences and evolutionary analysis of spacer and mature RNA regions. *Mol. Cell. Biol.* 3: 1488–1500.

Goodwin, T. W., and E. I. Mercer. 1972. *Introduction to Plant Biochemistry*. Pergamon, Oxford.

Gordon, W. S., and R. B. Jackson. 2000. Nutrient concentrations in fine roots. *Ecology* 81: 275–280.

Gorham, E., and F. M. Boyce. 1989. Influence of lake surface area and depth upon thermal stratification and the depth of the summer thermocline. *J. Gt. Lakes Res.* 15: 233–245.

Gorham, E., W. E. Dean, and J. E. Sanger. 1983. The chemical composition of lakes in the North-Central United States. *Limnol. Oceanogr.* 28: 287–391.

Gorham, E., P. M. Vitousek, and W. A. Reiners. 1979. The regulation of chemical budgets over the course of terrestrial ecosystem succession. *Annu. Rev. Ecol. Syst.* 10: 53–84.

Gragnani, A., M. Scheffer, and S. Rinaldi. 1999. Top-down control of cyanobacteria: A theoretical analysis. *Am. Nat.* 153: 59–72.

Grasman, B. T., and E. C. Hellgren. 1993. Phosphorus nutrition in white-tailed deer—nutrient balance, physiological responses, and antler growth. *Ecology* 74: 2279–2296.

Graveland, J., and T. Vangijzen. 1994. Arthropods and seeds are not sufficient as calcium sources for shell formation and skeletal growth in passerines. *Ardea* 82: 299–314.

Gray, M. W., and M. N. Schnare. 1996. Evolution of rRNA gene organization. Pages 49–70 in *Ribosomal RNA: Structure, Evolution, Processing, and Function in Protein Biosynthesis*, edited by R. A. Zimmerman and A. E. Dahlberg. CRC, Boca Raton, Fla.

Greene, R. M., R. J. Geider, and P. Falkowski. 1991. Effect of iron limitation on photosynthesis in a marine diatom. *Limnol. Oceanogr.* 36: 1772–1782.

Greenwood, E.A.N. 1976. Nitrogen stress in plants. *Adv. Agron.* 28: 1–35.

Grimaldi, G., and P. O. Di Nocera. 1988. Multiple repeated units in *Drosophila melanogaster* ribosomal DNA spacer stimulate rRNA precursor transcription. *Proc. Natl. Acad. Sci. USA* 85: 5502–5506.

Grime, J. P. 1977. Evidence for the existence of three primary strategies in plants and its relevance to ecological and evolutionary theory. *Am. Nat.* 111: 1169–1194.

Grime, J. P., and R. Hunt. 1975. Relative growth rate: Its range and adaptive significance in a local flora. *J. Ecol.* 63: 393–422.

Groot, J.J.R., and J.H.J. Spiertz. 1991. The role of nitrogen in yield formation and achievement of quality standards in cereals. Pages 227–247 in *Plant Growth: Interactions with Nutrition and Environment*, edited by J. R. Porter and D. W. Lawlor. Cambridge University Press, Cambridge, U.K.

Grover, J. P. 1989. Influence of cell shape and size on algal competitive ability. *J. Phycol.* 25: 402–405.

Grover, J. P. 1991a. Algae grown in non-steady state continuous culture: Population dynamics and phosphorus uptake. *Verh. Int. Verein. Limnol.* 24: 2661–2664.

Grover, J. P. 1991b. Non-steady state dynamics of algal population growth: Experiments with two chlorophytes. *J. Phycol.* 27: 70–79.

Grover, J. P. 1991c. Resource competition in a variable environment: phytoplankton growing according to the variable-internal-stores model. *Am. Nat.* 138: 811–835.

Grover, J. P. 1997. *Resource Competition*. Chapman and Hall, London.

Grover, J. P., and J. H. Lawton. 1994. Experimental studies on community convergence and alternative stable states: Comments on a paper by Drake et al. *J. Anim. Ecol.* 63: 484–487.

Guerrini, F., A. Mazzotti, L. Boni, and R. Pistocchi. 1998. Bacterial-algal interactions in polysaccharide production. *Aquat. Microb. Ecol.* 15: 247–253.

Guildford, S. J., and R. E. Hecky. 2000. Total nitrogen, total phosphorus, and nutrient limitation in lakes and oceans: Is there a common relationship? *Limnol. Oceanogr.* 45: 1213–1223.

Gulati, R. D., and W. R. DeMott. 1997. The role of food quality for zooplankton: Remarks on the state-of-the-art, perspectives and priorities. *Freshwater Biol.* 38: 753–768.

Gulati, R. D., K. Siewetsen, and L. Van Liere. 1991. Carbon and phosphorus relationships of zooplankton and its seston food in Loosdrecht lakes. *Mem. Ist. Ital. Idrobiol.* 48: 279–298.

Gunderson, L. H. 2000. Ecological resilience—in theory and application. *Annu. Rev. Ecol. Syst.* 31: 425–439.

Gurung, T. B., J. Urabe, and M. Nakanishi. 1999. Regulation of the relationship between phytoplankton *Scenedesmus acutus* and heterotrophic bacteria by the balance of light and nutrients. *Aquat. Microb. Ecol.* 17: 27–35.

Guthrie, R. D. 1984a. Alaskan megabucks, megabulls, and megarams: The issue of Pleistocene gigantism. Pages 482–510 in *Contributions in Quaternary Vertebrate Paleontology: A Volume in Memorial to John E. Guilday*, edited by H. H. Genoways and M. R. Dawson. Special Publication 8. Carnegie Museum of Natural History, Pittsburgh, Pa.

Guthrie, R. D. 1984b. Mosaics, allelochemicals, and nutrients: An ecological theory of late Pleistocene gigantism. Pages 259–298 in *Quaternary Extinctions, Prehistoric Revolution*, edited by P. S. Martin and R. G. Klein. University of Arizona Press, Tucson, Ariz.

Haack, R. A. and F. Slansky, Jr. 1987. Nutritional ecology of wood-feeding Coleoptera, Lepidoptera, and Hymenoptera. Pages 449–486 in *Nutritional Ecology of Insects, Mites, Spiders, and Related Invertebrates*, edited by F. Slansky, Jr., and J. G. Rodriguez. Wiley, New York.

Hairston, N. G., Sr. 1989. *Ecological Experiments: Purpose, Design, and Execution.* Cambridge University Press, New York.

Hairston, N. G., F. E. Smith, and L. B. Slobodkin. 1960. Community structure, population control and competition. *Am. Nat.* 45: 421–425.

Halfpenny, J., and M. Heffernan. 1991. Nutrient transport to an alpine tundra: an aeolian insect component. *Southwest. Nat.* 37: 247–251.

Harmsen, G. W., and D. A. van Schreven. 1955. Mineralization of organic nitrogen in soil. *Adv. Agron.* 7: 299.

Harris, G. P. 1986. *Phytoplankton Ecology: Structure, Function, and Fluctuation.* Chapman and Hall, London.

Harris, G. P. 1999. Comparison of the biogeochemistry of lakes and estuaries: Ecosystem processes, functional groups, hysteresis effects and interactions between macro- and microbiology. *Mar. Freshwater Res.* 50: 791–811.

Harte, J., and A. P. Kinzig. 1993. Mutualism and competition between plants and decomposers: Implications for nutrient allocation in ecosystems. *Am. Nat.* 141: 829–846.

Hassett, R. P., B. Cardinale, L. B. Stabler, and J. J. Elser. 1997. Ecological stoichiometry of N and P in pelagic ecosystems: Comparison of lakes and oceans with emphasis on the zooplankton-phytoplankton interaction. *Limnol. Oceanogr.* 42: 648–662.

Hastings, A. 2001. Transient dynamics and persistence of ecological systems. *Ecol. Lett.* 4: 215–220.

Hatcher, A. 1991a. Effect of temperature on carbon, nitrogen and phosphorus turnover by the solitary ascidian *Herdmania momus* (Savigny). *J. Exp. Mar. Biol. Ecol.* 152: 15–31.

Hatcher, A. 1991b. The use of metabolic ratios for determining the catabolic substrates of a solitary ascidian. *Mar. Biol.* 108: 433–440.

Hatcher, A. 1994. Nitrogen and phosphorus turnover in some benthic marine invertebrates: Implications for the use of C:N ratios to assess food quality. *Mar. Biol.* 121: 161–166.

Healey, F. P. 1985. Interacting effects of light and nutrient limitation on the growth rate of *Synechococcus linearis* (Cyanophyceae). *J. Phycol.* 21: 134–146.

Healey, F. P., and L. L. Hendzel. 1979. Indicators of phosphorus and nitrogen deficiency in five algae in culture. *J. Fish. Res. Board Can.* 36: 1364–1369.

Healey, F. P., and L. L. Hendzel. 1980. Physiological indicators of nutrient deficiency in lake phytoplankton. *Can. J. Fish. Aquat. Sci.* 37: 442–543.

Hecky, R. E. 1984. African lakes and their trophic efficiencies: A temporal perspective. Pages 405–448 in *Trophic Interactions within Aquatic Ecosystems*, edited by D. G. Meyers and J. R. Strickler. Westview, Boulder, Colo.

Hecky, R. E., P. Campbell, and L. L. Hendzel. 1993. The stoichiometry of carbon, nitrogen, and phosphorus in particulate matter of lakes and oceans. *Limnol. Oceanogr.* 38: 709–724.

Hecky, R. E., and R. H. Hesslein. 1995. Contributions of benthic algae to lake food webs as revealed by stable isotope analysis. *J. N. Am. Benthol. Soc.* 14: 631–653.

Hecky, R. E., and P. Kilham. 1988. Nutrient limitation of phytoplankton in freshwater and marine environments: A review of recent evidence on the effects of enrichment. *Limnol. Oceanogr.* 33: 796–822.

Heldal, M., S. Norland, K. M. Fagerbakke, F. Thingstad, and G. Bratbak. 1996. The elemental composition of bacteria: A signature of growth conditions? *Mar. Pollut. Bull.* 33: 1–6.

Hendzel, L. L., R. E. Hecky, and D. L. Findlay. 1994. Recent changes of N_2-fixation in lake 227 in response to reduction of the N:P loading ratio. *Can. J. Fish. Aquat. Sci.* 51: 2247–2253.

Hendzel, M. J., and D. P. Bazett-Jones. 1996. Probing nuclear ultrastructure by electron spectroscopic imaging. *J. Microsc.* 182: 1–14.

Hendzel, M. J., F.-M. Boisvert, and D. P. Bazett-Jones. 1999. Direct visualization of a protein nuclear architecture. *Mol. Biol. Cell* 10: 2051–2062.

Herbert, D. 1961. The chemical composition of micro-organisms as a function of their environment. Pages 391–416 in *Symposium of the Society for General Microbiology, Vol. 11: Microbial Reaction to the Environment*, edited by G. G. Meynell and H. Goodet. Cambridge University Press, Cambridge, U.K.

Herbert, D. 1976. Stoichiometric aspects of microbial growth. Pages 1–27 in *Continuous Culture 6: Applications and New Fields*, edited by A.C.R. Dean, D. C. Ellwood, C.G.T. Evans and J. Melling. Horwood, Chichester, U.K.

Herbert, D., P. J. Phipps and R. E. Strange. 1971. Chemical analysis of microbial cells. Pages 209–343 in *Methods in Microbiology*, edited by J. R. Norris and D. W. Ribbons. Academic, New York.

Hessen, D. O. 1990. Carbon, nitrogen and phosphorus status in *Daphnia* at varying food conditions. *J. Plankton Res.* 12: 1239–1249.

Hessen, D. O. 1992. Nutrient element limitation of zooplankton production. *Am. Nat.* 140: 799–814.

Hessen, D. O. 1993. The role of mineral nutrients for zooplankton nutrition: Reply to the comment by Brett. *Limnol. Oceanogr.* 38: 1340–1343.

Hessen, D. O. 1997. Stoichiometry in food webs—Lotka revisited. *Oikos* 79: 195–200.

Hessen, D. O., and T. Andersen. 1990. Bacteria as a source of phosphorus for zooplankton. *Hydrobiologia* 206: 217–223.

Hessen, D. O., and T. Andersen. 1992. The algae-grazer interface: Feedback mechanisms linked to elemental ratios and nutrient cycling. *Arch. Hydrobiol. Beih. Ergebn. Limnol.* 35: 111–120.

Hessen, D. O., T. Andersen, and B. Faafeng. 1992. Zooplankton contribution to particulate phosphorus and nitrogen in lakes. *J. Plankton Res.* 14: 937–947.

Hessen, D. O., and B. Bjerkeng. 1997. A model approach to planktonic stoichiometry and consumer-resource stability. *Freshwater Biol.* 38: 447–472.

Hessen, D. O., and A. Lyche. 1991. Inter- and intraspecific variations in zooplankton element composition. *Arch. Hydrobiol.* 121: 355–363.

Hessen, D. O., K. Nygaard, K. Salonen, and A. Vähätalo. 1994. The effect of substrate stoichiometry on microbial activity and carbon degradation in humic lakes. *Environ. Int.* 20: 67–76.

Hewitt, S. W., and B. L. Johnson. 1987. A generalized bioenergetics model of fish growth for microcomputers, University of Wisconsin Sea Grant Institute, Madison, Wis.

Heymsfield, S. B., M. Waki, J. Kehayias, S. Lichtman, F. A. Dilmanian, Y. Kamen, J. Wang, and R. N. Pierson, Jr. 1991. Chemical and elemental analysis of humans *in vivo* using improved body composition models. *Am. J. Physiol.* 261: E190–E198.

Higashi, M., and T. Abe. 1997. Global diversification of termites driven by the evolution of symbiosis and sociality. Pages 83–111 in *Biodiversity: An Ecological Perspective*, edited by T. Abe, S. A. Levin, and M. Higashi. Springer, New York.

Higashi, M., T. Abe, and T. Burns. 1992. Carbon-nitrogen balance and termite ecology. *Proc. R. Soc. London B* 249: 303–308.

Hill, W. R., M. G. Ryon, and E. M. Schilling. 1995. Light limitation in a stream ecosystem: responses by primary producers and consumers. *Ecology* 76: 1297–1309.

Hillebrand, H., and U. Sommer. 1999. The nutrient stoichiometry of benthic microalgal growth: Redfield proportions are optimal. *Limnol. Oceanogr.* 44: 440–446.

Hirose, T. 1975. Relations between turnover rate, resource utility and structure of some plant populations: a study in the matter budgets. *J. Fac. Sci. Univ. Tokyo* 11: 335–407.

Hobbie, S. E. 1992. Effects of plant species on nutrient cycling. *Trends Ecol. Evol.* 7: 336–339.

Hobbie, S. E., D. B. Jensen, and F. W. Chapin. 1993. Resource supply and disturbance as controls over present and future plant diversity. Pages 385–408 in *Biodiversity and Ecosystem Function*, edited by E.-D. Schultze and H. A. Mooney. Springer-Verlag, Berlin.

Hobbie, S. E., and P. M. Vitousek. 2000. Nutrient limitation of decomposition in Hawaiian forests. *Ecology* 81: 1867–1877.

Holmes, F. L. 1964. Introduction. Pages vii–cxvi in *Animal Chemistry or Organic Chemistry in its Application to Physiology and Pathology by Justus Liebig*, edited by H. Woolf. Johnson Reprint Corporation, New York.

Holm-Hansen, D. 1969. Algae: amounts of DNA and organic carbon in single cells. *Science* 163: 87–88.

Hoppema, M., and L. Goeyens. 1999. Redfield behavior of carbon, nitrogen, and phosphorus depletions in Antarctic surface water. *Limnol. Oceanogr.* 44: 220–224.

Houghton, R. A., and G. M. Woodwell. 1983. Effect of increased C, N, P and S on the global storage of C. Pages 327–343 in *The Major Biogeochemical Cycles and their Interactions*, edited by B. Bolin and R. B. Cook. Wiley, New York.

Howarth, R. W. 1988. Nutrient limitation of net primary production in marine ecosystems. *Annu. Rev. Ecol. Syst.* 19: 89–110.

Howarth, R. W., and J. J. Cole. 1985. Molybdenum availability, nitrogen limitation, and phytoplankton growth in natural waters. *Science* 229: 653–655.

Howarth, R. W., R. Marino, and J. J. Cole. 1988a. Nitrogen fixation in freshwater, estuarine, and marine ecosystems. 2. Biogeochemical controls. *Limnol. Oceanogr.* 33: 688–701.

Howarth, R. W., R. Marino, J. Lane, and J. J. Cole. 1988b. Nitrogen fixation in freshwater, estuarine, and marine ecosystems. 1. Rates and importance. *Limnol. Oceanogr.* 33: 669–687.

Hsu, S. B., S. P. Hubbell, and P. Waltman. 1977. Mathematical theory for single-nutrient competiton in continuous cultures of microorganisms. *Siam J. Appl. Math.* 32: 366–383.

Huisman, J., J. P. Grover, R. van der Wal, and J. van Andel. 1999. Competition for light, plant-species replacement and herbivore abundance along productivity gradients. Pages 239–269 in *Herbivores, Plants and Predators*, edited by V. K. Brown, R. Drent and H. Olff. Blackwell, Oxford.

Hunt, H. W., J.W.B. Stewart, and C. V. Cole. 1983. A conceptual model for interactions among carbon, nitrogen, sulphur, and phosphorus in grasslands. Pages 303–325 in *The Major Biogeochemical Cycles and their Interactions*, edited by B. Bolin and R. B. Cook. Wiley, New York.

Hunter, M. D., and J. N. McNeil. 1997. Host-plant quality influences diapause and voltinism in a polyphagous insect herbivore. *Ecology* 78: 977–986.

Hynes, H.B.N. 1970. *The Ecology of Running Waters*. University of Toronto Press, Toronto.

Ingall, E. D., R. M. Bustin, and P. Van Cappellen. 1993. Influence of water column anoxia on the burial and preservation of carbon and phosphorus in marine shales. *Geochem. Cosmochim. Acta* 57: 303–313.

Ingestad, T. 1979a. Mineral nutrient requirements of *Pinus sylvestris* and *Picea abies* seedlings. *Physiol. Plant.* 45: 373–380.

Ingestad, T. 1979b. Nitrogen stress in birch seedlings. II. N, P, Ca and Mg nutrition. *Physiol. Plant.* 45: 149–157.

Interlandi, S. and S. S. Kilham. 2001. Limiting resources and the regulation of diversity in phytoplankton communities. *Ecology* 82: 1270–1282.

Ittekot, V., and S. Zhang. 1989. Pattern of particulate nitrogen transport in world rivers. *Global Biogeochem. Cycles* 3: 283–391.

Iversen, T. M. 1974. Ingestion and growth in *Sericostoma personatum* (Trichoptera) in relation to the nitrogen content of ingested leaves. *Oikos* 25: 278–282.

Jackson, D. A., H. H. Harvey, and K. M. Somers. 1990. Ratios in aquatic sciences: Statistical shortcomings with mean depth and the morphoedaphic index. *Can. J. Fish. Aquat. Sci.* 47: 1788–1795.

Jackson, P. J. 1980. Characterization of the ribosomal DNA of soybean cells. *Fed. Proc.* 39: 1878.

Jaeger, K. M., C. Johansson, U. Kunz, and H. Lehmann. 1997. Sub-cellular element analysis of a cyanobacterium (*Nostoc* sp.) in symbiosis with *Gunnera manicata* by ESI and EELS. *Bot. Acta* 110: 151–157.

James, S. W. 1991. Soil, nitrogen, phosphorus and organic matter processing by earthworms in tallgrass prarie. *Ecology* 72: 2101–2109.

Jansen, M.P.T., and N. E. Stamp. 1997. Effects of light availability on host plant chemistry and the consequences for behavior and growth of an insect herbivore. *Entomol. Exp. Appl.* 82: 319–333.

Janssen, J.A.M. 1994. Impact of the mineral composition and water content of excised maize leaf sections on fitness of the African armyworm, *Spodoptera exempta* (Lepidoptera: Noctuidae). *Bull. Entomol. Res.* 84: 233–245.

Jobling, M. 1994. *Fish Bioenergetics*. Chapman and Hall, London.

John, B., and G. Miklos. 1988. *The Eukaryote Genome in Development and Evolution*. Allen and Unwin, London.

Johnson, L. S., and R.M.R. Barclay. 1996. Effects of supplemental calcium on the reproductive output of a small passerine bird, the house wren (*Troglodytes aedon*). *Can. J. Zool.* 74: 278–282.

Johnstone, J. 1908. *Conditions of Life in the Sea; A Short Account of Quantitative Marine Biological Research*. Cambridge University Press, Cambridge, U.K.

Jones, C. G., and J. H. Lawton. 1995. *Linking Species and Ecosystems*. Chapman and Hall, New York.

Jones, M. D., D. M. Durall, and P. B. Tinker. 1990. Phosphorus relationships and production of extramatrical hyphae by two types of willow ecotomycorrhizas at different soil phosphorus levels. *New Phytol.* 115: 259–267.

Jones, M. D., D. M. Durall, and P. B. Tinker. 1991. Fluxes of carbon and phosphorus between symbionts in willow ectomycorrhizas and their changes with time. *New Phytol.* 108: 99–106.

Justic, D., N. N. Rabalais, and R. E. Turner. 1995. Stoichiometric nutrient balance and origin of coastal eutrophication. *Mar. Pollut. Bull.* 30: 41–46.

Kafatos, F. C., W. Orr, and K. Delidakis. 1985. Developmentally regulated gene amplification. *Trends Genet.* 1: 301–306.

Kahlert, M. 1998. C:N:P ratios of freshwater benthic algae. *Arch. Hydrobiol. Beih. Ergebn. Limnol.* 51: 105–114.

Karl, D., R. Letellier, L. Tupas, J. Dore, J. Christian, and D. Hebel. 1997. The role of nitrogen fixation in the biogeochemical cycling in the subtropical North Pacific Ocean. *Nature* 388: 533–538.

Karl, D. M. 1999. A sea of change: biogeochemical variability in the North Pacific Subtropical Gyre. *Ecosystems* 2: 181–214.

Karl, D. M., K. M. Björkman, J. E. Dore, L. Fujieki, D. V. Hebel, T. Houlihan, R. M. Letelier, and L. M. Tupas. 2001. Ecological nitrogen-to-phosphorus stoichiometry at Station ALOHA. *Deep-Sea Res. Part II* 48: 1529–1566.

Karl, D. M., R. Letelier, D. Hebel, L. Tupas, J. Dore, J. Christian, and C. Winn. 1995. Ecosystem changes in the North Pacific subtropical gyre attributed to the 1991–92 El Niño. *Nature* 373: 230–234.

Karl, D. M., G. Tien, J. Dore, and C. D. Winn. 1993. Total dissolved nitrogen and phosphorus concentrations at United-States-JGOFS station ALOHA—Redfield reconciliation. *Mar. Chem.* 41: 203–208.

Karner, M. B., E. F. DeLong, and D. M. Karl. 2001. Archaeal dominance in the mesopelagic zone of the Pacific Ocean. *Nature* 409: 507–510.

Keefe, A. D., and S. L. Miller. 1995. Are polyphosphates or phosphate esters prebiotic reagents? *J. Mol. Evol.* 41: 693–702.

Ketchum, B. H. 1962. Regeneration of nutrients by zooplankton. *Rapp. P. V. Reun. Cons. Inst. Explor. Mer* 152: 142–146.

Kihlberg, R. 1972. The microbe as a source of food. *Annu. Rev. Microbiol.* 26: 427–466.

Kilham, P. 1990. Mechanisms controlling the chemical composition of lakes and rivers: Data from Africa. *Limnol. Oceanogr.* 35: 80–83.

Killingbeck, K. T. 1993. Nutrient resorption in desert shrubs. *Rev. Chil. Hist. Nat.* 66: 345–355.

Kiørboe, T. 1989. Phytoplankton growth rate and nitrogen content: Implications for feeding and fecundity in a herbivorous copepod. *Mar. Ecol. Prog. Ser.* 55: 229–234.

Kirk, J.T.O. 1983. *Light and Photosynthesis in Aquatic Ecosystems.* Cambridge University Press, Cambridge, U.K.

Kirk, J.T.O., and R.A.E. Tilney-Bassett. 1978. *The Plastids*, 2nd. ed. Elsevier, Amsterdam.

Kitchell, J. F., J. F. Koonce, and P. S. Tennis. 1975. Phosphorus flux through fishes. *Verh. Int. Verein. Limnol.* 19: 2478–2484.

Kitchell, J. F., R. V. O'Neill, D. Webb, G. W. Gallepp, S. M. Bartell, J. F. Koonce, and B. S. Ausmus. 1979. Consumer regulation of nutrient cycling. *BioScience* 29: 28–34.

Knauer, G. A., J. H. Martin, and K. W. Bruland. 1979. Fluxes of particulate carbon, nitrogen, and phosphorus in the upper water column of the northeast Pacific. *Deep-Sea Res.* 26A: 97–108.

Knops, J.M.H., W. D. Koenig, and T. H. Nash III. 1997. On the relationship between nutrient use efficiency and fertility in forest ecosystems. *Oecologia* 110: 550–556.

Koch, G. W., and H. A. Mooney, eds. 1996. *Carbon Dioxide and Terrestrial Ecosystems.* Academic. New York.

Koerselman, W., and A.F.M. Meuleman. 1996. The vegetation N:P ratio: A new tool to detect the nature of nutrient limitation. *J. Appl. Ecol.* 33: 1441–1450.

Koide, R., and G. Elliott. 1989. Cost, benefit and efficiency of the vesicular-arbuscular mycorrhizal symbiosis. *Funct. Ecol.* 3: 252–255.

Kolowith, L. C., E. D. Ingall, and R. B. Ingall. 2000. Composition and cycling of marine organic phosphorus. *Limnol. Oceanogr.* 46: 309–320.

Kooijman, S.A.L.M. 1993. *Dynamic Energy Budgets in Biological Systems.* Cambridge University Press, New York.

Kooijman, S.A.L.M. 1995. The stoichiometry of animal energetics. *J. Theor. Biol.* 177: 139–149.

Körner, C. 1989. The nutritional status of plants from high altitudes. *Oecologia* 81: 379–391.

Körner, C. 1993. Scaling from species to vegetation: the usefulness of functional groups. Pages 117–140 in *Biodiversity and Ecosystem Function*, edited by E. D. Schulze and H. Mooney. Springer-Verlag, Berlin.

Kraft, C. E. 1992. Estimates of phosphorus and nitrogen cycling by fish using a bioenergetics approach. *Can. J. Fish. Aquat.* Sci. 49: 2596–2604.

Kramer, P. J., and T. T. Kozlowski. 1979. *Physiology of Woody Plants.* Academic, New York.

Krause-Jensen, D., and K. Sand-Jensen. 1998. Light attenuation and photosynthesis of aquatic plant communities. *Limnol. Oceanogr.* 43: 396–407.

Kraut, S., U. Feudel, and C. Grebogi. 1999. Preference of attractors in noisy multistable systems. *Phys. Rev. E* 59: 5253–5260.

Krivtsov, V., E. G. Bellinger, and D. C. Sigee. 1999. Modelling of elemental associations in *Anabaena*. *Hydrobiologia* 414: 77–83.

Kump, L. R., and F. T. Mackenzie. 1996. Regulation of atmospheric O_2: Feedback in the microbial feedbag. *Science* 271: 459–460.

Labandeira, C. C., and J. J. Sepkoski, Jr. 1993. Insect diversity in the fossil record. *Science* 261: 310–315.

Lahooti, M., and C. R. Harwood. 1999. Transcriptional analysis of the *Bacillis subtilis* teichuronic acid operon. *Microbiology–UK* 145: 3409–3417.

Lalli, C. M., and T. R. Parsons. 1997. *Biological Oceanography: An Introduction*, 2nd ed. Butterworth-Heinemann, Boston.

Lambers, H., M. L. Cambridge, H. Konings, and T. L. Pons. 1989. *Causes and Consequences of Variation in Growth Rate and Productivity of Higher Plants.* SPB Academic, The Hague.

Lampert, W. 1999. Forum: Nutrient ratios. Editor's comment. *Arch. Hydrobiol.* 146: 1.

Lampert, W., and C. J. Loose. 1992. Plankton towers—bridging the gap between laboratory and field experiments. *Arch. Hydrobiol.* 126: 53–66.

Lampert, W., and U. Schober. 1980. The importance of "threshold" food concentrations. Pages 264–267 in *Evolution and Ecology of Zooplankton Communities*, edited by W. C. Kerfoot. University Press of New England, Hanover, N.H.

Langworthy, T. A. 1985. Lipids of Archaebacteria. Pages 459–480 in *The Bacteria*, edited by C. R. Woese and R. S. Wolfe. Academic, Orlando, Fla.

Laska, M. S., and J. T. Wootton. 1998. Theoretical concepts and empirical approaches to measuring interaction strength. *Ecology* 79: 461.

Lawlor, D. W., W. Day, A. E. Johnston, B. J. Legg, and K. T. Parkinson. 1981. Growth of spring barley under drought: crop development, photosynthesis, dry matter accumulation, and nutrient content. *J. Agric. Sci., Cambridge* 96: 167–186.

Lawlor, D. W., M. Konturri, and A. T. Young. 1989. Photosynthesis by flax leaves of wheat in relation to protein, ribulose bisphosphate carboxylase, and nitrogen supply. *J. Exp. Bot.* 40: 43–52.

Lawlor, L. R. 1979. Direct and indirect effects of *n*-species competition. *Oecologia* 43: 355–364.

Laws, E. A., D. R. Jones, K. L. Terry, and J. A. Hirata. 1985. Modifications in recent models of phytoplankton growth: Theoretical developments and experimental examination of predictions. *J. Theor. Biol.* 114: 323–341.

Laws, E. A., D. G. Redalje, D. M. Karl, and M. S. Chalup. 1983. A theoretical and experimental examination of the predictions of two recent models of phytoplankton growth. *J. Theor. Biol.* 105: 469–491.

Lawson, D. L., M. J. Klug, and R. W. Merritt. 1984. The influence of the physical, chemical, and microbiological characteristics of decomposing leaves on the growth of the detritivore *Tipula abdominalis* (Diptera: Tipulidae). *Can. J. Zool.* 62: 2339–2343.

Le Borgne, R. P. 1982. Zooplankton production in the eastern tropical Atlantic Ocean: Net growth efficiency and P:B in terms of carbon, nitrogen and phosphorus. *Limnol. Oceanogr.* 27: 681–698.

Lee, R. B., and R. G. Ratcliffe. 1983. Phosphorus nutrition and the intracellular distribution of inorganic phosphate in pea root tips: A quantitative study using ^{31}P-NMR. *J. Exp. Bot.* 34: 1222–1244.

Legendre, L., and J. Le Fèvre. 1989. Hydrodynamical singularities as controls of recycled versus export production in oceans. Pages 49–63 in *Productivity of the*

Ocean: Present and Past, edited by W. H. Berger, V. S. Smetacek, and G. Wefer. Wiley, New York.

Lehman, J. T. 1980. Release and cycling of nutrients between planktonic algae and herbivores. *Limnol. Oceanogr.* 25: 620–632.

Lehman, J. T. 1984. Grazing, nutrient release, and their impacts on the structure of phytoplankton communities. Pages 49–72 in *Trophic Dynamics within Aquatic Ecosystems*, edited by D. G. Meyers and J. R. Strickler. AAAS Selected Symposium 85, Westview, Boulder, Colo.

Lehninger, A. L. 1971. *Bioenergetics*. W.A. Benjamin, Menlo Park, Calif.

Lehninger, A. L., D. L. Nelson, and M. M. Cox. 1993. *Principles of Biochemistry*. Worth, New York.

Leigh, R. A., and R. G. Wyn-Jones. 1985. Cellular compartmentation in plant nutrition: the selective cytoplasm and the promiscuous vacuole. Pages 249–277 in *Advances in Plant Nutrition*, edited by B. Tinker and A. Läuchli. Praeger Scientific, New York.

Lenton, T. M. 1998. Gaia and natural selection. *Nature* 394: 439–447.

Lenton, T. M., and J. Lovelock. 2000. Daiseyworld is Darwinian: Constraints on adaptation are important for planetary self-regulation. *J. Theor. Biol.* 206: 109–114.

Levi, M. P., and E. B. Cowling. 1969. Role of nitrogen in wood deterioration. VII. Physiological adaptation of wood-destroying and other fungi to substrates deficient in nitrogen. *Phytopathology* 59: 460–468.

Levins, R., and R. Lewontin. 1980. Dialectics and reductionism in ecology. *Synthese* 43: 47–78.

Lewin, B. 1980. *Gene Expression, Vol. 2 Eucaryotic Chromosomes*. Wiley, New York.

Lewin, R. 1992. *Complexity: Life at the Edge of Chaos*. Collier, New York.

Lewis, W. M. 1985. Nutrient scarcity as an evolutionary cause of haploidy. *Am. Nat.* 125: 692–701.

Lincoln, D. E. 1993. The influence of plant carbon dioxide and nutrient supply on susceptibility to insect herbivores. *Vegetatio* 104/105: 273–280.

Linley, E. A., and R. C. Newell. 1984. Estimates of bacterial growth yields based on plant detritus. *Bull. Mar. Sci.* 35: 409–425.

Lockaby, B. G., and W. H. Conner. 1999. N:P balance in wetland forests: productivity across a biogeochemical continuum. *Bot. Rev.* 65: 171–185.

Loladze, Y. K., Y. Kuang, and J. J. Elser. 2000. Stoichiometry in producer-grazer systems: linking energy flow with element cycling. *Bull. Math. Biol.* 62: 1137–1162.

Loneragan, J. F. 1997. Plant nutrition in the 20th and perspectives for the 21st century. *Plant Soil* 196: 163–174.

Long, E. O., and I. B. Dawid. 1980. Repeated genes in eukaryotes. *Annu. Rev. Biochem.* 49: 727–764.

Lotka, A. J. 1925. *Elements of Physical Biology*. Williams and Wilkins, Baltimore, Md.

Lovelock, J. 1988. *The Ages of Gaia*. Norton, New York.

Lyche, A., T. Andersen, K. Christoffersen, D. O. Hessen, P. H. Berger Hansen, and A. Klysner. 1996a. Mesocosm tracer studies. 1. Zooplankton as sources and sinks in the pelagic phosphorus cycles of a mesotrophic lake. *Limnol. Oceanogr.* 41: 460–474.

Lyche, A., T. Andersen, K. Christoffersen, D. O. Hessen, P. H. Berger Hansen, and A. Klysner. 1996b. Mesocosm tracer studies. 2. The fate of primary production and the role of consumers in the pelagic carbon cycle of a mesotrophic lake. *Limnol. Oceanogr.* 41: 475–487.

Maaløe, O., and N. O. Kjeldgaard. 1966. *Control of Macromolecular Synthesis: A Study of DNA, RNA, and Protein Synthesis in Bacteria.* Benjamin, New York.

MacArthur, R. H., and E. O. Wilson. 1967. *The Theory of Island Biogeography.* Vol. 1. Princeton University Press, Princeton, N.J.

MacKay, N. A., and J. J. Elser. 1998a. Factors potentially preventing trophic cascades: food quality, invertebrate predation, and their interaction. *Limnol. Oceanogr.* 43: 339–347.

MacKay, N. A., and J. J. Elser. 1998b. Nutrient recycling by *Daphnia* reduces N_2 fixation by cyanobacteria. *Limnol. Oceanogr.* 43: 347–354.

Mahony, N., E. Nol, and T. Hutchinson. 1997. Food-chain chemistry, reproductive success, and foraging behaviour of songbirds in acidified maple forests of central Ontario. *Can. J. Zool.* 75: 509–517.

Maienschein, J. 1991. *Transforming Traditions in American Biology, 1880–1915.* Johns Hopkins University Press, Baltimore, Md.

Main, T., D. R. Dobberfuhl, and J. J. Elser. 1997. N:P stoichiometry and ontogeny in crustacean zooplankton: a test of the growth rate hypothesis. *Limnol. Oceanogr.* 42: 1474–1478.

Makino, W., J. Urabe, J. J. Elser, and C. Yoshimizu. 2002. Evidence of phosphorus-limited individual and population growth of *Daphnia* in a Canadian shield lake. *Oikos* 96: 197–205.

Markow, T. A., B. Raphael, D. Dobberfuhl, C. M. Breitmeyer, J. J. Elser, and E. Pfeiler. 1999. Elemental stoichiometry of *Drosophila* and their hosts. *Funct. Ecol.* 13: 78–84.

Marr, A. G. 1991. Growth rate of *Escherichia coli. Microbiol. Rev.* 55: 316–333.

Marschner, H. 1995. *Mineral Nutrition of Higher Plants.* Academic, London.

Marshall, B., and J. R. Porter. 1991. Concepts of nutritional and environmental interactions determining plant productivity. Pages 99–124 in *Plant Growth: Interactions with Nutrition and Environment*, edited by J. R. Porter and D. W. Lawlor. Cambridge University, Cambridge, U.K.

Martin, J. H., R. M. Gordon, and S. E. Fitzwater. 1991. The case for iron. *Limnol. Oceanogr.* 36: 1793–1802.

Martin, M. E., and J. D. Aber. 1997. High spectral resolution remote sensing of forest canopy lignin, nitrogen and ecosystem processes. *Ecol. Applic.* 7: 431–443.

Martin, M. E., S. D. Newman, J. D. Aber, and R. G. Congalton. 1998. Determining forest species composition using high spectral resolution remote sensing data. *Remote Sensing Environ.* 65: 249–254.

Mathers, E. M., D. F. Houlihan, I. D. McCarthy, and L. J. Burren. 1993. Rates of growth and protein synthesis correlated with nucleic acid content in fry of rainbow trout, *Oncorhynchus mykiss*: effects of age and temperature. *J. Fish Biol.* 43: 245–263.

Matsumoto, T. 1976. The role of termites in an equatorial rain forest ecosystem of

West Malaysia. I. Population density, biomass, carbon, nitrogen and calorific content and respiration rate. *Oecologia* 22: 153–178.

Matsuzaki, H., M. Uehara, K. Suzuki, Q. L. Liu, S. Sato, Y. Kanke, and S. Goto. 1997. High phosphorus diet rapidly induces nephrocalcinosis and proximal tubular injury in rats. *J. Nutr. Sci. Vitaminol.* 43: 627–641.

Mattson, W. J., Jr. 1980. Herbivory in relation to plant nitrogen content. *Annu. Rev. Ecol. Syst.* 11: 119–161.

Mattson, W. J., and J. N. Scriber. 1987. Nutritional ecology of insect folivores of woody plants: Nitrogen, water, fiber, and mineral considerations. Pages 105–146 in *Nutritional Ecology of Insects, Mites, Spiders, and Related Invertebrates*, edited by F. Slansky and J. G. Rodriguez. Wiley, New York.

May, R. M. 1976. Models for two interacting populations. Pages 49–70 in *Theoretical Ecology: Principles and Applications*, edited by R. M. May. Saunders, Philadelphia.

Mazumder, A., and D.R.S. Lean. 1994. Consumer-dependent responses of lake ecosystems to nutrient loading. *J. Plankton Res.* 16: 1567–1580.

Mazumder, A., and W. D. Taylor. 1994. Thermal structure of lakes varying in size and water clarity. *Limnol. Oceanogr.* 39: 968–979.

Mazumder, A., W. D. Taylor, D. J. McQueen, and D.R.S. Lean. 1989. Effects of fertilization and planktivorous fish on epilimnetic phosphorus and phosphorus sedimentation in large enclosures. *Can. J. Fish. Aquat. Sci.* 46: 1735–1742.

McCarthy, J. L., and A. Islam. 2000. Lignin chemistry, technology, and utlization: A brief history. Pages 2–99 in *Lignin: Historical, Biological, and Materials Perspectives*, edited by W. G. Gasser, R. A. Northey, and T. P. Schultz. American Chemical Society, Washington, D.C.

McCauley, E., J. A. Downing, and S. Watson. 1989. Sigmoid relationships between nutrients and chlorophyll among lakes. *Can. J. Fish. Aquat. Sci.* 46: 1171–1175.

McCauley, E., and W. W. Murdoch. 1987. Cyclic and stable populations: Plankton as paradigm. *Am. Nat.* 129: 97–121.

McCauley, E., W. W. Murdoch, and S. Watson. 1988. Simple models and variation in plankton densities among lakes. *Am. Nat.* 132: 383–403.

McCauley, E., R. M. Nisbet, W. W. Murdoch, A. M. de Roos, and W.S.C. Gurney. 1999. Large-amplitude cycles of *Daphnia* and its algal prey in enriched environments. *Nature* 402: 653–656.

McDowell, L. R. 1992. *Minerals in Animal and Human Nutrition*. Academic, San Diego.

McGuire, D., J. M. Melillo, and L. A. Joyce. 1995. The role of nitrogen in the response of forest net primary production to elevated atmospheric carbon dioxide. *Annu. Rev. Ecol. Syst.* 26: 473–503.

McKee, M., and C. O. Knowles. 1987. Levels of protein, RNA, DNA, glycogen and lipids during growth and development of *Daphnia magna* Straus (Crustacea: Cladocera). *Freshwater Biol.* 18: 341–351.

McMahon, J. W., and F. H. Rigler. 1965. Feeding rate of *Daphnia magna* Straus in different foods labeled with radioactive phosphorus. *Limnol. Oceanogr.* 10: 105–114.

McNaughton, S. J., M. Oesterheld, D. A. Frank, and K. J. Williams. 1989. Ecosystem-level patterns of primary productivity and herbivory in terrestrial habitats. *Nature* 341: 142–144.

McNaughton, S. J., M. Oesterheld, D. A. Frank, and K. J. Williams. 1991. Primary and secondary production in terrestrial ecosystems. Pages 120–139 in *Comparative Analyses of Ecosystems: Patterns, Mechanisms, and Theories*, edited by J. Cole, G. Lovett, and S. Findlay. Springer, New York.

Megard, R. O. 1972. Phytoplankton, photosynthesis, and phosphorus in Lake Minnetonka, Minnesota. *Limnol. Oceanogr.* 17: 68–87.

Megard, R. O., W. S. Combs, P. D. Smith, and A. S. Knoll. 1979. Attentuation of light and daily integral rates of photosysnthesis attained by planktonic algae. *Limnol. Oceanogr.* 24: 1038–1050.

Melillo, J. M. 1996. Elevated carbon dioxide, litter quality and decomposition. Pages 199–206 in *Global Change: Effects on Coniferous Forests and Grasslands*, edited by A. I. Breymeyer, D. O. Hall, J. M. Melillo, and G. I. Ågren. Wiley, New York.

Menge, J. A., D. Steirle, J. Babyaraj, E.L.V. Jonson, and R. T. Leonard. 1978. Phosphorus concentrations in plants responsible for inhibition of mycorrhizal infection. *New Phytol.* 80: 575–578.

Merrill, W., and E. B. Cowling. 1966. Role of nitrogen in wood deterioration: Amounts and distribution of nitrogen in tree stems. *Can. J. Bot.* 44: 1555–1580.

Meybeck, M. 1982. Carbon, nitrogen and phosphorus transport by world rivers. *Am. J. Sci.* 282: 401–450.

Meybeck, M. 1993. C, N, P and S in rivers: from sources to global inputs. Pages 163–193 in *Interactions of C, N, P and S Biogeochemical Cycles and Global Change*, edited by R. Wollast, F. T. Mackenzie, and L. Chou. Springer-Verlag, Berlin.

Miceli, M. V., T. O. Henderson, and T. C. Myers. 1980. 2-aminoethylphosphonic acid metabolism during embryonic development of the planoborid snail *Helisoma*. *Science* 209: 1245–1247.

Mieyal, J. J., and J. L. Blumer. 1981. The endoplasmic reticulum. Pages 641–690 in *Advanced Cell Biology*, edited by L. M. Schwartz and M. M. Azar. Van Nostrand Reinhold, New York.

Mimura, T., K. Sakano, and T. Shimmen. 1996. Studies on the distribution, retranslocation, and homeostasis in inorganic phosphate in barley leaves. *Plant, Cell, Environ.* 19: 311–320.

Minagawa, M., and E. Wada. 1984. Stepwise enrichment of ^{15}N along food chains: Further evidence and the relation between δ^{15}N and animal age. *Geochem. Cosmochim. Acta* 48: 1135–1140.

Minnich, J. 1979. *The Rodale Guide to Composting*. Rodale, Emmaus, Pa.

Mitchell, S. F., F. R. Trainor, P. H. Rich, and C. E. Goulden. 1992. Growth of *Daphnia magna* in the laboratory in relation to the nutritional state of its food species, *Chlamydomonas reinhardtii*. *J. Plankton Res.* 14: 379–391.

Miyashita, S., and T. Miyazaki. 1992. Seasonal changes in neutral sugars and amino acids of particulate matter in Lake Nakanuma, Japan. *Hydrobiologia* 245: 95–104.

Moegenburg, S. M., and M. Vanni. 1991. Nutrient regeneration by zooplankton:

Effects on nutrient limitation of phytoplankton in a eutrophic lake. *J. Plankton Res.* 13: 573–588.

Moen, R., Y. Cohen, and J. Pastor. 1998. Linking moose population and plant growth models with a moose energetics model. *Ecosystems* 1: 52–63.

Moen, R., and J. Pastor. 1998. A model to predict nutritional requirements for antler growth in moose. *Alces* 34: 59–74.

Moen, R., J. Pastor, and Y. Cohen. 1997. A spatially explicit model of moose foraging and energetics. *Ecology* 78: 505–521.

Moen, R. A., J. Pastor, and Y. Cohen. 1999. Antler growth and extinction of Irish Elk. *Evol. Ecol. Res.* 1: 235–249.

Moorhead, D. L., and J. F. Reynolds. 1992. Modeling the contributions of decomposer fungi in nutrient cycling. Pages 691–714 in *The Fungal Community: Its Organization and Role in the Ecosystem*, edited by G. C. Carroll and D. T. Wicklow. Marcel Dekker, New York.

Morel, F.M.M. 1987. Kinetics of nutrient uptake and growth in phytoplankton. *J. Phycol.* 23: 137–150.

Morowitz, H. J. 1968. *Energy Flow in Biology*. Academic, New York.

Morowitz, H. J. 1992. *Beginnings of Cellular Life: Metabolism Recapitulates Biogenesis*. Yale University Press, New Haven, Conn.

Mortimer, C. H. 1941. The exchange of dissolved substances between mud and water in lakes. I. *J. Ecol.* 29: 280–329.

Mortimer, C. H. 1942. The exchange of dissolved substances between mud and water in lakes. II. *J. Ecol.* 30: 147–201.

Moss, B. 1998. *Ecology of Fresh Waters: Man and Medium, Past to Future*. Blackwell, Malden, Mass.

Mosse, B. 1973. Plant growth responses to vesicular-arbuscular mycorrhize. IV. In soil given additional phosphate. *New Phytol.* 72: 127–136.

Muggli, D. L. and P. J. Harrison. 1996. Effects of nitrogen source on the physiology and metal nutrition of *Emiliania huxleyi* grown under different iron and light conditions. *Mar. Ecol. Prog. Ser.* 130: 255–267.

Muggli, D. L., M. Lecourt, and P. J. Harrison. 1996. Effects of iron and nitrogen source on the sinking rate, physiology and metal composition of an oceanic diatom from the subarctic Pacific. *Mar. Ecol. Prog. Ser.* 132: 215–227.

Müller-Navarra, D. 1995a. Biochemical versus mineral limitation in *Daphnia*. *Limnol. Oceanogr.* 40: 1209–1214.

Müller-Navarra, D. 1995b. Evidence that a highly unsaturated fatty acid limits *Daphnia* growth in nature. *Arch. Hydrobiol.* 132: 297–307.

Müller-Navarra, D., M. T. Brett, A. M. Liston, and C. R. Goldman. 2000. A highly unsaturated fatty acid predicts carbon transfer between primary producers and consumers. *Nature* 403: 74–77.

Müller-Navarra, D., and W. Lampert. 1996. Seasonal patterns of food limitation in *Daphnia galeata*: Separating food quantity and food quality effects. *J. Plankton Res.* 18: 1137–1158.

Munro, H. N. 1969. Evolution of protein metabolism in mammals. Pages 133–182 in *Mammalian Protein Metabolism*, edited by H. N. Munro. Academic, New York.

Murneek, A. E. 1942. Quantitative distribution of nitrogen and carbohydrates in apple trees. *Mo. Agric. Exp. Stn. Res. Bull.* 348.

Murphy, A. E., B. B. Sageman, and D. J. Hollander. 2000. Eutrophication by de-coupling of the marine biogeochemical cycles of C, N and P: A mechanism for the Late Devonian mass extinction. *Geology* 28: 427–430.

Nakano, S. 1994a. Carbon:nitrogen:phosphorus ratios and nutrient regeneration of a heterotrophic flagellate fed on baceria with different elemental ratios. *Arch. Hydrobiol.* 129: 257–271.

Nakano, S. 1994b. Estimation of phosphorus release rate by bacterivorous flagel-lates in Lake Biwa. *Jpn. J. Limnol.* 55: 201–211.

Nakano, S. 1994c. Rates and ratios of nitrogen and phosphorus released by a bacte-rivorous flagellate. *Jpn. J. Limnol.* 55: 115–123.

Nelson, D. M., and W. O. Smith, Jr. 1991. Sverdrup revisited: Critical depths, maxi-mum chlorophyll levels, and the control of Southern Ocean productivity by the irradiance-mixing regime. *Limnol. Oceanogr.* 36: 1650–1661.

Nelson, W. A., E. McCauley, and F. J. Wrona. 2001. Multiple dynamics in a single predator-prey system: experimental effects of food quality. *Proc. R. Soc. London B Biol. Sci.* 268: 1223–1230.

Newbold, J. D., J. W. Elwood, R.V. O'Neill, and A. L. Sheldon. 1983. Phosphorus dynamics in a woodland stream ecosystem: a study of nutrient spiralling. *Ecology* 64: 1249–1265.

Newman, E. I., and P. Reddell. 1987. The distribution of mycorrhizas among fami-lies of vascular plants. *New Phytol.* 106: 745–751.

Nielsen, S. L., S. Enríquez, C. M. Duarte, and K. Sand-Jensen. 1996. Scaling maxi-mum growth rates across photosynthetic organisms. *Funct. Ecol.* 10: 167–175.

Nisbet, I.C.T. 1997. Female common terns *Sterna hirundo* eating mollusc shells—evidence for calcium deficits during egg laying. *Ibis* 139: 400–401.

Nisbet, R. M., A. H. Ross, and A. J. Brooks. 1996. Empirically-based dynamic energy budget models: theory and an application to ecotoxicology. *Nonlin. World* 3: 85–106.

Nomura, M., R. Gourse, and G. Baughman. 1984. Regulation of the synthesis of ribosomes and ribosomal components. *Annu. Rev. Biochem.* 53: 75–117.

Norby, R. J., M. F. Cotrufo, P. Ineson, E. G. O'Neill, and J. G. Canadell. 2001. Elevated CO_2 litter chemistry, and decomposition: A synthesis. *Oecologia* 127: 153–165.

Norby, R. J., E. G. O'Neill, and R. J. Luxmoore. 1986. Effects of atmospheric CO_2 enrichment on the growth and mineral nutrition of *Quercus alba* seedlings in nutrient poor soils. *Plant Physiol.* 82: 83–89.

Norby, R. J., S. D. Wullschleger and C. A. Guderson. 1996. Tree responses to elevated CO_2 and implications for forests. Pages 1–21 in *Carbon Dioxide and Terrestrial Ecosystems*, edited by H. A. Mooney and G. W. Koch. Academic, New York.

Northcote, T. G. 1988. Fish in the structure and function of freshwater ecosystems: A "top-down" view. *Can. J. Fish. Aquat. Sci.* 45: 361–379.

Obernosterer, I., and G. J. Herndl. 1995. Phytoplankton extracellular release and bacterial growth: Dependence on the inorganic N:P ratio. *Mar. Ecol. Prog. Ser.* 116: 247–257.

Odum, H. T., and R. C. Pinkerton. 1955. Time's speed regulator: The optimum efficiency for maximum power output in physical and biological systems. *Am. Sci.* 43: 331–343.

Oedekoven, M. A., and A. Joern. 2000. Plant quality and spider predation affects grasshoppers (Acrididae): Food-quality-dependent compensatory mortality. *Ecology* 81: 66–77.

Oksanen, L., and T. Oksanen. 2000. The logic and realism of the hypothesis of exploitation ecosystems. *Am. Nat.* 155: 703–723.

Olins, A. L., D. E. Olins, and D. P. Bazett-Jones. 1996. Osmium ammine-B and electron spectroscopic imagine of ribonucleoproteins: Correlations of stain and phosphorus. *Biol. Cell* 87: 143–147.

Ollinger, S. V., M. L. Smith, M. E. Martin, P. A. Hallett, C. L. Goodale, and J. D. Aber. 2002. Regional variation in foliar chemistry and N cycling among forests of diverse history and composition. *Ecology* 83: 339–355.

Olsen, G. J. 1994. Microbial ecology—archaea, archaea, everywhere. *Nature* 371: 657–658.

Olsen, Y., A. Jensen, H. Reinertsen, K. Y. Børsheim, M. Heldal, and A. Langeland. 1986. Dependence of the rate of release of phosphorus by zooplankton on the P:C ratio in the food supply, as calculated by a recycling model. *Limnol. Oceanogr.* 31: 34–44.

Olsen, Y., and K. Østgaard. 1985. Estimating release rates of phosphorus from zooplankton: Model and experimental verification. *Limnol. Oceanogr.* 30: 844–852.

O'Neill, E. G., R. J. Luxmoore, and R. J. Norby. 1987. Elevated atmospheric CO_2 effects on seedling growth, nutrient uptake, and rhizosphere bacterial populations of *Liriodendron tulipifera* L. *Plant Soil* 104: 3–11.

Ovarzun, S. E., G. J. Crawshaw, and E. V. Valdes. 1996. Nutrition of the *Tamandua*. 1. Nutrient composition of termites (*Nasutitermes* spp.) and stomach contents from wild tamanduas (*Tamandua tetradactyla*). *Zoo Biol.* 15: 509–524.

Ovington, J. D. 1957. Dry matter production in *Pinus sylvestris* L. *Ann. Bot. (London)* 21: 287–314.

Pace, M. L., J. J. Cole, S. R. Carpenter, and J. F. Kitchell. 1999. Trophic cascades revealed in diverse ecosystems. *Trends Ecol. Evol.* 14: 483–488.

Pahlow, M., and U. Riebesell. 2000. Temporal trends in deep ocean Redfield ratios. *Science* 287: 831–833.

Paine, R. T. 1966. Food web complexity and species diversity. *Am. Nat.* 100: 65–75.

Pandian, T. J., and F. J. Vernberg, eds. 1987. *Animal Energetics*. Academic. San Diego.

Pastor, J., J. D. Aber, and C. A. McClaugherty. 1984. Aboveground production and N and P cycling along a nitrogen mineralization gradient on Blackhawk Island, Wisconsin. *Ecology* 65: 256–268.

Pastor, J. and S. D. Bridgham. 1998. Nutrient efficiency along nutrient availability gradients. *Oecologia* 118: 50–58.

Pastor, J. and Y. Cohen. 1997. Herbivores, the functional diversity of plant species, and the cycling of nutrients in ecosystems. *Theor. Popul. Biol.* 51: 165–179.

Pastor, J., and R. J. Naiman. 1992. Selective foraging and ecosystem processes in boreal forests. *Am. Nat.* 139: 690–705.

Paul, E. A., and F. E. Clark. 1989. *Soil Microbiology and Biochemistry*. Academic, San Diego.

Paule, M. R. 1994. Transcription of ribosomal RNA by eukaryotic RNA polymerase

I. Pages 83–106 in *Transcription: Mechanisms and Regulation*, edited by R. C. Conaway and J. W. Conaway. Raven, New York.

Paule, M. R., and A. K. Lofquist. 1996. Organization and expression of eukaryotic ribosomal RNA genes. Pages 395–420 in *Ribosomal RNA: Structure, Evolution, Processing, and Function in Protein Biosynthesis*, edited by R. A. Zimmerman and A. E. Dahlberg. CRC, New York.

Pautard, F.G.E. 1978. Phosphorus and bone. Pages 261–354 in *New Trends in Bio-Inorganic Chemistry*, edited by R.J.P. Williams and J.R.R.F. Da Silva. Academic, London.

Pendleton, B. F., I. Newman, and R. S. Marshall. 1983. A Monte Carlo approach to correlational spuriousness and ratio variables. *J. Stat. Comput. Simul.* 18: 93–124.

Peñuelas, J., and R. Matamala. 1990. Changes in N and S leaf content, stomatal density, and specific leaf areas of 14 plant species during the last three centuries of CO_2 increase. *J. Exp. Bot.* 4: 1119–1124.

Persson, L., J. Bengtsson, B. A. Menge, and M. A. Power. 1996. Productivity and consumer regulation—concepts, patterns, and mechanisms. Pages 396–434 in *Food Webs: Integration of Pattern and Process*, edited by G. A. Polis and K. O. Winemiller. Chapman and Hall, New York.

Peters, R. H. 1983. *The Ecological Implications of Body Size.* Cambridge University Press, Cambridge, U.K.

Peters, R. H. 1991. *A Critique for Ecology.* Cambridge University Press, New York.

Pickett, S.T.A., J. Kolasa, and C. G. Jones. 1994. *Ecological Understanding.* Academic, San Diego.

Pierotti, R., and C. A. Annett. 1991. Diet choice in the herring gull: Constraints imposed by reproductive and ecological factors. *Ecology* 72: 319–328.

Pirozynski, K. A., and D. W. Malloch. 1975. The origin of land plants: A matter of mycotrophism. *Biosystems* 6: 153–164.

Plath, K., and M. Boersma. 2001. Mineral limitation of zooplankton: Stoichiometric constraints and optimal foraging. *Ecology* 82: 1260–1269.

Platt, T., D. F. Bird, and S. Sathyendranath. 1991. Critical depth and marine primary production. *Proc. R. Soc. London B Biol. Sci.* 246: 205–217.

Polanco, C., and M. Perez de la Vega. 1997. Intragenic ribosomal spacer variability in hexaploid oat cultivars and landraces. *Heredity* 78: 115–123.

Polis, G. A. 1981. The evolution and dynamics of intraspecific predation. *Annu. Rev. Ecol. Syst.* 12: 225–251.

Polis, G. A. 1999. Why are parts of the world green? Multiple factors control productivity and the distribution of biomass. *Oikos* 86: 3–15.

Polis, G. A., A.L.W. Sears, G. R. Huxel, D. R. Strong, and J. Maron. 2000. When is a trophic cascade a trophic cascade? *Trends Ecol. Evol.* 15: 473–475.

Poorter, H., and J. R. Evans. 1998. Photosynthetic nitrogen-use efficiency of species that differ inherently in specific leaf area. *Oecologia* 116: 26–37.

Porter, C. A. 1965. Biosynthesis of chitin during various stages in the metamorphosis of *Prodenia eridania*. *J. Insect Physiol.* 11: 1151–1160.

Potrikus, C. J., and J. A. Breznak. 1981. Gut bacteria recycle uric acid nitrogen in termites: A strategy for nutrient conservation. *Proc. Natl. Acad. Sci. USA* 78: 4601–4605.

Prairie, Y. T., and D. F. Bird. 1989. Some misconceptions about the spurious correlation problem in the ecological literature. *Oecologia* 81: 285–288.

Prairie, Y. T., C. M. Duarte, and J. Kalff. 1989. Unifying nutrient chlorophyll relationships in lakes. *Can. J. Fish. Aquat. Sci.* 46: 1176–1182.

Prange, H. D., J. F. Anderson, and H. Rahn. 1979. Scaling of skeletal mass to body mass in birds and mammals. *Am. Nat.* 113: 103–122.

Prestwich, G. D., and B. L. Bentley. 1981. Nitrogen fixation by intact colonies of the termite *Nasutitermes conrniger*. *Oecologia* 49: 249–251.

Price, P. W., C. E. Bouton, P. Gross, B. A. McPheron, J. N. Thompson, and A. E. Weis. 1980. Interactions among three trophic levels: Influence of plants on interactions between insect herbivores and natural enemies. *Annu. Rev. Ecol. Syst.* 11: 41–65.

Quintana, C., S. Marco, N. Bonnet, C. Risco, M. L. Gutierreza, A. Gueero, and J. L. Carrascosa. 1998. Optimization of phosphorus localization by EFTEM of nucleid acid containing structures. *Micron* 29: 297–307.

Rao, Y. K. 1985. *Stoichiometry and Thermodynamics of Metallurgical Processes*. Cambridge University Press, Cambridge, U.K.

Rastetter, E. B., and G. G. Shaver. 1992. A model of multiple-element limitation for acclimating vegetation. *Ecology* 73: 1157–1174.

Raubenheimer, D., and S. J. Simpson. 1994. The analysis of nutrient budgets. *Funct. Ecol.* 8: 783–791.

Redfield, A. C. 1934. On the proportions of organic derivatives in sea water and their relation to the composition of plankton. Pages 176–192 in *James Johnstone Memorial Volume*, edited by R.J. Daniel. Liverpool University Press, Liverpool, U.K.

Redfield, A. C. 1942. The processes determining the concentration of oxygen, phosphate and other organic derivatives within the depths of the Atlantic Ocean. *Pap. Phys. Oceanogr. Meteorol.* 9: 1–22.

Redfield, A. C. 1958. The biological control of chemical factors in the environment. *Am. Sci.* 46: 205–221.

Redfield, A. C., B. H. Ketchum, and F. A. Richards. 1963. The influence of organisms on the composition of seawater. Pages 26–77 in *Comparative and Descriptive Oceanography*, edited by M. N. Hill. Wiley, New York.

Reeder, R. H., J. G. Roan, and M. Dunaway. 1983. Spacer regulation of *Xenopus* ribosomal gene transcription: Competition in oocytes. *Cell* 35: 449–456.

Reich, P. B., D. S. Ellsworth, M. B. Walters, J. M. Vose, C. Gresham, J. C. Volin, and W. D. Bowman. 1999. Generality of leaf trait relationships: A test across six biomes. *Ecology* 80: 1955–1969.

Reiners, W. A. 1986. Complementary models for ecosystems. *Am. Nat.* 127: 59–73.

Reiners, W. A. 1992. Twenty years of ecosystem reorganization following experimental deforestation and regrowth supression. *Ecol. Monogr.* 62: 505–523.

Reinhardt, S. B., and E. S. Van Vleet. 1986. Lipid composition of twenty-two species of Antarctic midwater species and fish. *Mar. Biol.* 91: 149–159.

Reynolds, C. S. 1984. *The Ecology of Freshwater Phytoplankton*. Cambridge University Press, Cambridge, U.K.

Rhee, G.-Y. 1973. Continuous culture study of phosphate uptake, growth rate and polyphosphate in *Scenedesmus* sp. *J. Phycol.* 9: 495–506.

Rhee, G.-Y. 1978. Effects of N:P atomic ratios and nitrate limitation on algal growth, cell composition and nitrate uptake. *Limnol. Oceanogr.* 23: 10–25.

Rhee, G.-Y., and I. J. Gotham. 1981. The effect of environmental factors on phytoplankton growth: Light and the interactions of light with nitrate limitation. *Limnol. Oceanogr.* 26: 649–659.

Richards, A. G. 1978. The chemistry of the insect cuticle. Pages 205–232 in *Biochemistry of Insects*, edited by M. Rockstein. Academic, New York.

Riebesell, U., D. A. Wolfgladrow, and V. Smetacek. 1993. Carbon dioxide limitation of marine phytoplankton growth rates. *Nature* 361: 249–251.

Rillig, M. C., G. Y. Hernández, and P.C.D. Newton. 2000. Arbuscular mycorrhizae respond to elevated atmospheric CO_2 after long-term exposure: Evidence from a CO_2 spring in New Zealand supports the resource balance model. *Ecol. Lett.* 3: 475–478.

Risebrow, A., and A.F.G. Dixon. 1987. Nutritional ecology of phloem-feeding insects. Pages 421–448 in *Nutritional Ecology of Insects, Mites, Spiders, and Related Invertebrates*, edited by F. Slansky, Jr., and J. G. Rodriguez. Wiley, New York.

Ritchie, M. E., and H. Olff. 1999. Herbivore diversity and plant dynamics: compensatory and additive effects. Pages 175–204 in *Herbivores, Plants and Predators*, edited by V. K. Brown, R. Drent, and H. Olff. Blackwell, Oxford.

Ritossa, F. 1976. The bobbed locus (*Drosophila*). *Genet. Biol. Drosophila* 1b: 801–846.

Rocheford, T. R., J. C. Osterman, and C. O. Gardner. 1990. Variation in the ribosomal DNA intergenic spacer of a maize population mass-selected for high grain yield. *Theor. Appl. Genet.* 79: 793–800.

Rodhe, W. 1948. Environmental requirements of freshwater plankton algae. Experimental studies in the ecology of phytoplankton. *Symb. Bot. Ups.* 10: 1–149.

Rodhe, W., R. A. Vollenweider, and A. Nauwerck. 1958. The primary production and standing crop of phytoplankton. Pages 299–322 in *Perspectives in Marine Biology*, edited by A. A. Buzzati-Traverso. University of California Press, Berkeley.

Roff, D. A. 1992. *The Evolution of Life Histories: Theory and Analysis*. Chapman and Hall, New York.

Rogers, S. O., and A. J. Bendich. 1987. Ribosomal RNA genes in plants: Variability in copy number and in the intergenic spacer. *Plant Mol. Biol.* 9: 509–520.

Rooney, J. M. 1994. The carbon and nitrogen dependence of plant development. Pages 217–228 in *A Whole-Plant Perspective on Carbon-Nitrogen Interactions*, edited by J. Roy and E. Garnier. SPB Academic, The Hague.

Rosenzweig, M. L. 1969. Why the prey curve has a hump. *Am. Nat.* 103: 81–87.

Rothhaupt, K.-O. 1995. Algal nutrient limitation affects rotifer growth rate but not ingestion rate. *Limnol. Oceanogr.* 40: 1201–1208.

Rothhaupt, K.-O. 1996. Laboratory experiments with a mixotrophic chrysophyte and obligately phagotrophic and phototrophic competitors. *Ecology* 77: 716–724.

Rothhaupt, K.-O. 1997. Grazing and nutrient influences of *Daphnia* and *Eudiaptomus* on phytoplankton in laboratory microcosms. *J. Plankton Res.* 19: 125–139.

Roy, J., and E. Garnier, eds. 1994. *A Whole Plant Perspective on Carbon-Nitrogen Interactions*. SPB Academic, The Hague.

Ruesink, J. L. 1998. Variation in per capita interaction strength: Thresholds due to nonlinear dynamics and nonequilibrium conditions. *Proc. Natl. Acad. Sci. USA* 95: 6843–6847.

Ruess, R. W., and S. J. McNaughton. 1987. Grazing and the dynamics of nutrient and energy regulated microbial processes in the Serengeti grasslands. *Oikos* 49: 101–110.

Rueter, J. G. 1982. Theoretical Fe limitations of microbial N_2 fixation in the oceans. *Eos* 63: 260–262.

Rueter, J. G., and D. R. Ades. 1987. The role of iron nutrition in photosynthesis and nitrogen assimilation in *Scenedesmus quadricauda* (Chlorophyceae). *J. Phycol.* 23: 452–457.

Rusak, J. A., N. D. Yan, and D. J. McQueen. 1999. The temporal coherence of zooplankton population abundances in neighboring north-temperate lakes. *Am. Nat.* 153: 46.

Russell-Hunter, M. D. 1970. *Aquatic Productivity: An Introduction to Some Basic Aspects of Biological Oceanography and Limnology.* Macmillan, New York.

Ryser, P., and H. Lambers. 1995. Root and leaf attributes accounting for the performance of fast- and slow-growing grasses at different nutrient supply. *Plant Soil* 170: 251–265.

Ryther, J. H., and W. M. Dunstan. 1971. Nitrogen, phosphorus, and eutrophication in the coastal marine environment. *Science* 171: 1008–1013.

Sadava, D. E. 1993. *Cell Biology: Organelles, Structure, and Function.* Jones and Bartlett, Boston.

Sambrotto, R. N. and G. Savidge. 1993. Elevated consumption of carbon relative to nitrogen in the surface ocean. *Nature* 363: 248–250.

Sang, J. H. 1978. The nutritional requirements of *Drosophila*. Pages 159–192 in *The Genetics and Biology of Drosophila*, edited by M. Ashburner and T.R.F. Wright. Academic, New York.

Sañudo-Wilhelmy, S. A., A. B. Kustka, C. J. Gobler, D. A. Hutchins, M. Yang, K. Lwiza, J. Burns, D. G. Capone, J. A. Raven, and E. J. Carpenter. 2001. Phosphorus limitation of nitrogen fixation by *Trichodesmium* in the central Atlantic Ocean. *Nature* 411: 66–69.

Sargent, J. R., and S. Falk-Petersen. 1988. The lipid biochemistry of calanoid copepods. *Hydrobiologia* 167/168: 101–114.

Sargent, J. R., and R. J. Henderson. 1986. Lipids. Pages 59–108 in *The Biological Chemistry of Marine Copepods*, edited by E.D.S. Corner and S.C.M. O'Hara. Oxford, Oxford University Press, Oxford.

Sarnelle, O. 1992. Nutrient enrichment and grazer effects on phytoplankton in lakes. *Ecology* 73: 551–560.

Sarnelle, O. 1999. Zooplankton effects on vertical particulate flux: Testable models and experimental results. *Limnol. Oceanogr.* 44: 357–370.

Schaus, M. H., M. J. Vanni, and R. A. Stein. 1997. Nitrogen and phosphorus excretion by detritivorous gizzard shad in a reservoir ecosystem. *Limnol. Oceanogr.* 42: 1386–1397.

Scheffer, M., S. Rinaldi, A. Gragnani, L. R. Mur, and E. H. van Nes. 1997. On the dominance of filamentous cyanobacteria in shallow, turbid lakes. *Ecology* 78: 272–282.

Schimel, D. S. 1995. Terrestrial ecosystems and the global carbon cycle. *Global Change Biol.* 1: 77–91.

Schindler, D. E., S. R. Carpenter, K. L. Cottingham, X. He, J. R. Hodgson, J. F. Kitchell, and P. A. Soranno. 1996. Food web structure and littoral zone coupling to pelagic trophic cascades. Pages 96–105 in *Food Webs: Integration of Pattern and Dynamics*, edited by G. A. Polis and K. O. Winemiller. Chapman and Hall, New York.

Schindler, D. E., and L. A. Eby. 1997. Stoichiometry of fishes and their prey: implications for nutrient recycling. *Ecology* 78: 1816–1831.

Schindler, D. E., J. F. Kitchell, X. He, S. R. Carpenter, J. R. Hodgson, and K. L. Cottingham. 1993. Food web structure and phosphorus cycling in lakes. *Trans. Am. Fish. Soc.* 122: 756–772.

Schindler, D. W. 1977. Evolution of phosphorus limitation in lakes. *Science* 195: 260–262.

Schindler, D. W. 1978. Factors regulating phytoplankton production and standing crop in the world's freshwaters. *Limnol. Oceanogr.* 23: 478–486.

Schindler, D. W. 1991. Whole lake experiments in the Experimental Lakes Area. Pages 108–122 in *Ecosystem Experiments*, edited by H. A. Mooney, E. Medina, D. Schindler, E.-D. Schulze, and B. H. Walker. Wiley, New York.

Schindler, D. W. 1998. The mysterious missing sink. *Nature* 398: 105–107.

Schlesinger, W. H. 1997. *Biogeochemistry: An Analysis of Global Change*, 2nd ed. Academic, San Diego.

Schlötterer, C., and D. Tautz. 1994. Chromosomal homogeneity of *Drosophila* ribosomal DNA arrays suggests intrachromosomal exchanges drive concerted evolution. *Curr. Biol.* 4: 777–783.

Schmidt, M. A., Y. H. Zhang, and D. A. Hutchins. 1999. Assimilation of Fe and carbon by marine copepods from Fe-limited and Fe-replete diatom prey. *J. Plankton Res.* 21: 1753–1764.

Schmidt-Nielsen, K. 1984. *Scaling: Why Is Animal Size so Important?* Cambridge University Press, Cambridge, U.K.

Schmitz, O. J., P. A. Hamback, and A. P. Beckerman. 2000. Trophic cascades in terrestrial systems: A review of the effects of carnivore removals on plants. *Am. Nat.* 155: 141–153.

Schopf, J. W., ed. 1982. *Earth's Earliest Biosphere: Its Origin and Evolution.* Princeton University Press. Princeton, N.J.

Schroeder, L. A. 1981. Consumer growth efficiencies: Their limits and relationships to ecological energetics. *J. Theor. Biol.* 93: 805–828.

Schulz, K. L. 1996. The nutrition of two cladocerans, the predacious *Bythotrephes cederstromei* and the herbivorous *Daphnia pulicaria*. PhD dissertation, The University of Michigan, Ann Arbor, Mich.

Schulz, K. L., and R. W. Sterner. 1999. Phytoplankton phosphorus limitation and food quality for *Bosmina. Limnol. Oceanogr.* 44: 1549–1556.

Schulze, E. D., and H. A. Mooney, eds. 1994. Design and execution of experiments on CO_2 enrichment. Commission of the European Communities, Luxembourg.

Schwab, S. M., J. A. Menge, and P. B. Tinker. 1991. Regulation of nutrient transfer between host and fungus in vesicular-arbuscular mycorrhizas. *New Phytol.* 117: 387–398.

Schwartz, L. M., and M. M. Lazar. 1981. *Advanced Cell Biology*. Van Nostrand Reinhold, New York.

Schwartz, M. W., and J. D. Hoeksema. 1998. Specialization and resource trade: Biological markets as a model of mutualisms. *Ecology* 79: 1029–1038.

Seitzinger, S. P. 1988. Denitrification in freshwater and coastal marine ecosystems: Ecological and geochemical significance. *Limnol. Oceanogr.* 33: 702–724.

Shapiro, J. 1980. The importance of trophic-level interactions to the abundance and species composition of algae in lakes. Pages 105–116 in *Hypertrophic Ecosystems*, edited by J. Barica and L. R. Mur. W. Junk, The Hague.

Shapiro, J. 1990. Biomanipulation: making it stable. *Hydrobiologia* 200/201: 13–27.

Shaver, G. R., and J. D. Aber. 1996. Carbon and nutrient allocation in terrestrial ecosystems. Pages 183–198 in *Global Change: Effects on Coniferous Forests and Grasslands*, edited by J. Mellilo and A. Bremeyer. SCOPE Synthesis Series. Wiley, New York.

Shaver, G. R., and J. M. Melillo. 1984. Nutrient budgets of marsh plants: Efficiency concepts and relation to availability. *Ecology* 65: 1491–1510.

Shermoen, A. W., and B. I. Kiefer. 1975. Regulation in rDNA-deficient *Drosophila melanogaster*. *Cell* 4: 275–280.

Shuter, B. 1979. A model of physiological adaptation in unicellular algae. *J. Theor. Biol.* 78: 519–552.

Siddiqi, M. Y., and A.D.M. Glass. 1981. Utilization index: a modified approach to the estimation and comparison of nutrient utilization efficiency in plants. *J. Plant Nutr.* 4: 289–302.

Siderius, M., A. Musgrave, H. Van den Ende, H. Koerten, P. Cambier, and P. Van der Meer. 1996. *Chlamydomonas eugametos* (Chlorophyta) stores phosphate in polyphosphate bodies together with calcium. *J. Phycol.* 32: 402–409.

Simard, S. W., D. A. Perry, M. D. Jones, D. D. Myrold, D. M. Durall, and M. Randy. 1997. Net transfer of carbon between ectomycorrhizal tree species in the field. *Nature* 388: 579–582.

Simkiss, K., and K. M. Wilbur. 1989. *Biomineralization: Cell Biology and Mineral Deposition*. Academic, San Diego.

Sims, D. A., J. R. Seemann, and Y. Luo. 1998. The significance of differences in the mechanisms of photosynthetic acclimation to light, nitrogen and CO_2 for return on investment in leaves. *Funct. Ecol.* 12: 185–194.

Sinclair, A.R.E., and R. P. Pech. 1996. Density dependence, stochasticity, compensation and predator regulation. *Oikos* 75: 164–173.

Slansky, F. J., and P. Feeny. 1977. Stabilization of the rate of nitrogen accumulation by larvae of the cabbage butterfly on wild and cultivated plants. *Ecol. Monogr.* 47: 209–228.

Slobodkin, L. B. 1988. Intellectual problems in applied ecology. *BioScience* 38: 337–342.

Smith, C. C. 1976. When and how much to reproduce: the trade-off between power and efficiency. *Am. Zool.* 16: 763–774.

Smith, V. H. 1979. Nutrient dependence of primary productivity in lakes. *Limnol. Oceanogr.* 24: 1051–1064.

Smith, V. H. 1982. The nitrogen and phosphorus dependence of algal biomass in lakes: An empirical and theoretical analysis. *Limnol. Oceanogr.* 27: 1101–1112.

Smith, V. H. 1983a. Light and nutrient dependence of photosynthesis by algae. *J. Phycol.* 19: 306–313.

Smith, V. H. 1983b. Low nitrogen to phosphorus ratios favor dominance by blue-green algae in lake phytoplankton. *Science* 221: 669–671.

Smith, V. H. 1993a. Applicability of resource-ratio theory to microbial ecology. *Limnol. Oceanogr.* 38: 239–249.

Smith, V. H. 1993b. Implications of resource-ratio theory for microbial ecology. *Adv. Microb. Ecol.* 13: 1–37.

Smith, V. H. 1993c. Resource competition between host and pathogen. *BioScience* 43: 21–30.

Smith, V. H. 1996. Resource competition and within-host dynamics. *Trends Ecol. Evol.* 11: 386–389.

Smith, V. H. 1998. Cultural eutrophication of inland, estuarine, and coastal waters. Pages 7–49 in *Successes, Limitations and Frontiers in Ecosystem Ecology*, edited by M. L. Pace, and P. M. Groffman. Springer, New York.

Smith, V. H. and S. J. Bennett. 1999. Forum: Nitrogen:Phosphorus supply ratios and phytoplankton community structure in lakes. *Arch. Hydrobiol.* 146: 37.

Smith, V. H., D. W. Graham, and D. D. Cleland. 1998. Application of resource-ratio theory to hydrocarbon biodegradation. *Environ. Sci. Technol.* 32: 3386–3395.

Smocovitis, V. B. 1996. *Unifying Biology*. Princeton University Press, Ewing, N.J.

Sollner-Webb, B., and J. Tower. 1986. Transcription of cloned eukaryotic ribosomal RNA genes. *Annu. Rev. Biochem.* 55: 801–803.

Sommer, U. 1984. The paradox of the plankton: Fluctuations of the phosphorus availability maintain diversity of phytoplankton in flow-through cultures. *Limnol. Oceanogr.* 29: 633–636.

Sommer, U. 1985. Comparison between steady state and nonsteady state competition: Experiments with natural phytoplankton. *Limnol. Oceanogr.* 30: 335–346.

Sommer, U. 1986. Nitrate- and silicate-competition among antarctic phytoplankton. *Mar. Biol.* 91: 345–351.

Sommer, U. 1988. Does nutrient competition among phytoplankton occur *in situ*? *Verh. Int. Verein. Limnol.* 23: 707–712.

Sommer, U. 1989. The role of competition for resources in phytoplankton succession. Pages 57–106 in *Plankton Ecology: Succession in Plankton Communities*, edited by U. Sommer. Springer-Verlag, Berlin.

Sommer, U. 1992. Phosphorus-limited *Daphnia*: Intraspecific facilitation instead of competition. *Limnol. Oceanogr.* 37: 966–973.

Stadler, B., B. Michalzik, and T. Müller. 1998. Linking aphid ecology with nutrient fluxes in a coniferous forest. *Ecology* 79: 1514–1525.

Stadler, B., and T. Müller. 1996. Aphid honeydew and its effect on the phyllosphere microflora of *Picea abies* (L.) Karst. *Oecologia* 108: 771–776.

Standiford, D. M. 1988. The development of a large nucleolus during oogenesis in *Acanthocyclops vernalis* (Crustacea, Copepoda) and its possible relationship to chromatin diminution. *Biol. Cell* 63: 35–40.

Stearns, S. C. 1976. Life-history tactics: A review of the ideas. *Q. Rev. Biol.* 51: 3–47.

Stearns, S. C. 1992. *The Evolution of Life Histories*. Oxford University Press, New York.

Stein, R. A., S. T. Threlkeld, C. D. Sandgren, W. G. Sprules, L. Persson, E. E. Werner, W. E. Neill, and S. I. Dodson. 1988. Size-structured interactions in lake communities. Pages 161–180 in *Complex Interactions in Lake Communities*, edited by S. R. Carpenter. Springer, New York.

Stephanopoulos, G., and J. J. Vallino. 1991. Network rigidity and metabolic engineering in metabolite overproduction. *Science* 252: 1675–1681.

Sterner, R. W. 1986. Herbivores' direct and indirect effects on algal populations. *Science* 231: 605–607.

Sterner, R. W. 1990a. Lake morphometry and light in the surface layer. *Can. J. Fish. Aquat. Sci.* 47: 687–692.

Sterner, R. W. 1990b. The ratio of nitrogen to phosphorus resupplied by herbivores: Zooplankton and the algal competitive arena. *Am. Nat.* 136: 209–229.

Sterner, R. W. 1993. *Daphnia* growth on varying quality of *Scenedesmus*: Mineral limitation of zooplankton. *Ecology* 74: 2351–2360.

Sterner, R. W. 1995. Elemental stoichiometry of species in ecosystems. Pages 240–252 in *Linking Species and Ecosystems*, edited by C. Jones and J. Lawton. Chapman and Hall, New York.

Sterner, R. W. 1997. Modelling interactions of food quality and quantity in homeostatic consumers. *Freshwater Biol.* 38: 473–481.

Sterner, R. W. 1998. Demography of a natural population of *Daphnia retrocurva* in a lake with low food quality. *J. Plankton Res.* 20: 471–490.

Sterner, R. W., J. Clasen, W. Lampert, and T. Weisse. 1998. Carbon:phosphorus stoichiometry and food chain production. *Ecol. Lett.* 1: 146–150.

Sterner, R. W., J. J. Elser, T. H. Chrzanowski, J. H. Schampel, and N. B. George. 1996. Biogeochemistry and trophic ecology: a new food web diagram. Pages 72–80 in *Food Webs: Integration of Patterns and Dynamics*, edited by G. A. Polis and K. O. Winemiller. Chapman and Hall, New York.

Sterner, R. W., J. J. Elser, E. J. Fee, S. J. Guildford, and T. H. Chrzanowski. 1997. The light:nutrient ratio in lakes: The balance of energy and materials affects ecosystem structure and process. *Am. Nat.* 150: 663–684.

Sterner, R. W., J. J. Elser, and D. O. Hessen. 1992. Stoichiometric relationships among producers, consumers, and nutrient cycling in pelagic ecosystems. *Biogeochemistry* 17: 49–67.

Sterner, R. W., and N. B. George. 2000. Carbon, nitrogen and phosphorus stoichiometry of cyprinid fishes. *Ecology* 81: 127–140.

Sterner, R. W., D. D. Hagemeier, W. L. Smith, and R. F. Smith. 1993. Phytoplankton nutrient limitation and food quality for *Daphnia*. *Limnol. Oceanogr.* 38: 857–871.

Sterner, R. W., and D. O. Hessen. 1994. Algal nutrient limitation and the nutrition of aquatic herbivores. *Annu. Rev. Ecol. Syst.* 25: 1–29.

Sterner, R. W., J. H. Schampel, K. L. Schulz, A. E. Galford, and J. J. Elser. 2001. Joint variation of zooplankton and seston stoichiometry in lakes and reservoirs. *Verh. Int. Verein. Limnol.* 27: 3009–3014.

Sterner, R. W., and K. L. Schulz. 1998. Zooplankton nutrition: Recent progress and a reality check. *Aquat. Ecol.* 32: 261–279.

Sterner, R. W., and M. Schwalbach. 2001. Diel integration of food quality by

Daphnia: Luxury consumption by a freshwater planktonic herbivore. *Limnol. Oceanogr.* 46: 410–416.

Stevens, C. E. and I. D. Hume. 1995. *Comparative Physiology of the Vertebrate Digestive System*, 2nd ed. Cambridge University Press, Cambridge, U.K.

Stocking, C. R., and A. Ongun. 1962. The intracellular distribution of some metallic elements in leaves. *Am. J. Bot.* 49: 284–289.

Stockner, J. G. 1987. Lake fertilization: the enrichment cycle and lake sockeye salmon (*Oncorhynchus nerka*) production. Pages 198–215 in *Sockeye Salmon (Oncorhynchus nerka) Population Biology and Future Management*, edited by H. D. Smith, L. Margolis, and C. C. Wood. *Can. Spec. Publ. Fish. Aquat. Sci.* 96.

Straile, D. 1997. Gross growth efficiencies of protozoan and metazoan zooplankton and their dependence on food concentration, predator-prey weight ratio, and taxonomic group. *Limnol. Oceanogr.* 42: 1375.

Strain, B. R., and J. D. Cure. 1994. Direct effects of atmospheric CO_2 enrichment on plants and ecosystems: An updated bibliography. Oak Ridge National Laboratory, Oak Ridge, Tenn.

Stribley, D. P., P. B. Tinker, and J. H. Rayner. 1980. Relation of internal phosphorus concentration and plant weight in plants infected by vesicular-arbuscular mycorrhizas. *New Phytol.* 86: 261–266.

Strong, D. R. 1992. Are trophic cascades all wet? Differentiation and donor-control in speciose ecosystems. *Ecology* 73: 747–754.

Strong, D. R., J. L. Maron, and P. G. Connors. 1996. Top down from underground? The underappreciated influences of subterranean food webs on aboveground ecology. Pages 170–178 in *Food Webs: Integration of Patterns and Dynamics*, edited by G. A. Polis and K. O. Winemiller. Chapman and Hall, New York.

Stumm, W., and J. J. Morgan. 1981. *Aquatic Chemistry: An Introduction Emphasizing Chemical Equilibria in Natural Waters*. Wiley, New York.

Suberkropp, K., and E. Chauvet. 1995. Regulation of leaf breakdown by fungi in streams: Influences of water chemistry. *Ecology* 76: 1433–1445.

Suberkropp, K., G. L. Godshalk, and M. J. Klug. 1976. Changes in the chemical composition of leaves during processing in a woodland stream. *Ecology* 57: 720–727.

Suberkropp, K., and M. J. Klug. 1976. Fungi and bacteria associated with leaves during processing in a woodland stream. *Ecology* 57: 707–719.

Sutcliffe, W.H.J. 1970. Relationship between growth rate and ribonucleic acid concentration in some invertebrates. *J. Fish. Res. Board Can.* 27: 606–609.

Sutherland, J. P. 1974. Multiple stable points in natural communities. *Am. Nat.* 108: 859–873.

Taghon, G. L. 1981. Beyond selection: Optimal ingestion rate as a function of food value. *Am. Nat.* 118: 202–214.

Takeda, S. 1998. Influence of iron availability on nutrient consumption ratio of diatoms in oceanic waters. *Nature* 393: 774–777.

Talling, J. F. 1957. The phytoplankton population as a compound photosynthetic system. *New Phytol.* 56: 133–149.

Tang, K. W., and H. G. Dam. 1999. Limitation of zooplankton production: Beyond stoichiometry. *Oikos* 84: 537–542.

Tanner, D. K., J. C. Brazner, and V. J. Brady. 2000. Factors influencing carbon, nitrogen, and phosphorus content of fish from a Lake Superior coastal wetland. *Can. J. Fish. Aquat. Sci.* 57: 1243–1251.

Tanner, E.V.J., P. M. Vitousek, and E. Cuevas. 1998. Experimental investigation of nutrient limitation of forest growth on wet tropical mountains. *Ecology* 79: 10–22.

Tateno, M., and F. S. Chapin III. 1997. The logic of carbon and nitrogen interactions in terrestrial ecosystems. *Am. Nat.* 149: 723–744.

Tautz, D., C. Tautz, D. Webb, and G. Dover. 1987. Evolutionary divergence of promoters and spacers in the rDNA family of four *Drosophila* species: Implications for molecular coevolution in multigene families. *J. Mol. Biol.* 195: 525–542.

Tayasu, I., A. Sugimoto, E. Wada, and T. Abe. 1994. Xylophagous termites depending on atmospheric nitrogen. *Naturwissenschaften* 81: 229–231.

Templeton, A. R., T. J. Crease, and F. Shah. 1985. The molecular through ecological genetics of abnormal abdomen in *Drosophila mercatorum*. I. Basic genetics. *Genetics* 111: 805–818.

Terry, K. L., E. A. Laws, and D. J. Burns. 1985. Growth rate variation in the N:P requirement ratio of phytoplankton. *J. Phycol.* 21: 323–329.

Tessier, A. J., M. A. Leibold, and J. Tsao. 2000. A fundamental trade-off in resource exploitation by *Daphnia* and consequences to plankton communities. *Ecology* 81: 826–841.

Tett, P., S. I. Heaney, and M. R. Droop. 1985. The Redfield ratio and phytoplankton growth rate. *J. Mar. Biol. Assoc. U.K.* 65: 487–504.

Tezuka, Y. 1985. The C:N:P ratios of seston in Lake Biwa as indicators of nutrient deficiency in phytoplankton and decomposition process of hypolimnetic particulate matter. *Jpn. J. Limnol.* 46: 239–246.

Tezuka, Y. 1989. The C:N:P ratio of phytoplankton determines the relative amounts of dissolved inorganic nitrogen and phosphorus released during aerobic decomposition. *Hydrobiologia* 173: 55–62.

Thiebaud, C. H. 1979. Quantitative determination of amplified rDNA and its distribution during oogenesis in *Xenopus laevis*. *Chromosoma* 73: 37–44.

Thingstad, T. F. 1987. Utilization of N, P, and organic C by heterotrophic bacteria. I. Outline of a chemostat theory with a consistent concept of "maintenance" metabolism. *Mar. Ecol. Prog. Ser.* 35: 99–109.

Thingstad, T. F. and B. Pengerud. 1985. Fate and effect of allochthonous organic material in aquatic microbial ecosystems. An analysis based on chemostat theory. *Mar. Ecol.* 21: 47–62.

Thomas, H., and V. O. Sadras. 2001. The capture and gratuitious disposal of resources by plants. *Funct. Ecol.* 15: 3–12.

Thompson, B. D., A. D. Robson, and L. K. Abbott. 1986. Effects of phosphorous on the formation of mycorrhizas by *Gigaspora calospora* and *Glomus fasciculatum* in relation to root carbohydrates. *New Phytol.* 103: 751–765.

Thorpe, N. O. 1984. *Cell Biology*. Wiley, New York.

Tiessen, H., ed. 1995a. *Phosphorus Cycling in Terrestrial and Aquatic Ecosystems*. Wiley, New York.

Tiessen, H., ed. 1995b. *Phosphorus in the Global Environment*. SCOPE Synthesis Series Vol. 54. Wiley, New York.

Tilman, D. 1982. *Resource Competition and Community Structure*. Princeton University Press, Princeton, N.J.

Tilman, D. 1988. *Plant Strategies and the Dynamics and Structure of Plant Communities*. Princeton University Press, Princeton, N.J.

Tilman, D., J. Fargione, B. Wolff, C. D'Antonio, A. Dobson, R. W. Howarth, D. E. Schindler, W. H. Schlesinger, D. Simberloff, and D. L. Swackhamer. 2001. Forecasting agriculturally driven global environmental change. *Science* 292: 281–284.

Tilman, D., S. S. Kilham and P. Kilham. 1982. Phytoplankton community ecology: The role of limiting nutrients. *Annu. Rev. Ecol. Syst.* 13: 349–372.

Tilman, D., and D. Wedin. 1991. Plant traits and resource reduction for five grasses growing on a nitrogen gradient. *Ecology* 72: 685–700.

Tonkyn, D. W. and B. J. Cole. 1986. The statistical analysis of size ratios. *Am. Nat.* 128: 66–81.

Torres-Contreras, H., and F. Bozinovic. 1997. Food selection in an herbivorous rodent: Balancing nutrition with thermoregulation. *Ecology* 78: 2230–2237.

Turner, R. E., and N. N. Rabalais. 1991. Changes in Mississippi river water quality this century. *BioScience* 41: 140–147.

Turpin, D. H. 1986. Growth rate dependent optimum ratios in *Selenastrum minutum* (Chlorophyta): Implications for competition, coexistence and stability in phytoplankton communities. *J. Phycol.* 22: 94–102.

Tyrell, T. 1999. The relative influences of nitrogen and phosphorus on oceanic primary production. *Nature* 400: 525–531.

Tyrell, T., and C. S. Law. 1997. Low nitrate:phosphate ratios in the global ocean. *Nature* 387: 793–796.

Urabe, J. 1993a. N and P cycling coupled by grazers' activities: food quality and nutrient release by zooplankton. *Ecology* 74: 2337–2350.

Urabe, J. 1993b. Seston stoichiometry and nutrient deficiency in a shallow eutrophic pond. *Arch. Hydrobiol.* 126: 417–428.

Urabe, J. 1995. Direct and indirect effects of zooplankton on seston stoichiometry. *Ecoscience* 2: 286–296.

Urabe, J., J. Clasen, and R. W. Sterner. 1997. Phosphorus-limitation of *Daphnia* growth: Is it real? *Limnol. Oceanogr.* 42: 1436–1443.

Urabe, J., J. J. Elser, M. Kyle, T. Yoshida, T. Sekino, and Z. Kawabata. 2002a. Herbivorous animals can mitigate unfavorable ratios of energy and material supplies by enhancing nutrient recycling. *Ecol. Lett.* 5 in press.

Urabe, J., M. Kyle, W. Makino, T. Yoshida, T. Andersen, and J. J. Elser. 2002b. Reduced light increases herbivore production due to stoichiometric effects of light :nutrient balance. *Ecology* 83: 619–627.

Urabe, J., M. Nakanishi, and K. Kawabata. 1995. Contribution of metazoan plankton to the cycling of N and P in Lake Biwa. *Limnol. Oceanogr.* 40: 232–242.

Urabe, J., T. Sekino, K. Nozaki, A. Tsuji, C. Yoshimizu, M. Kagami, T. Koitabashi, T. Miyazaki, and M. Nakanishi. 1999. Light, nutrients and primary productivity in Lake Biwa: An evaluation of the current ecosystem situation. *Ecol. Res.* 14: 233–242.

Urabe, J., and R. W. Sterner. 1996. Regulation of herbivore growth by the balance of light and nutrients. *Proc. Natl. Acad. Sci. USA* 93: 8465–8469.

Urabe, J., and Y. Watanabe. 1992. Possibility of N or P limitation for planktonic cladocerans: An experimental test. *Limnol. Oceanogr.* 37: 244–251.

Urabe, J., and Y. Watanabe. 1993. Implications of sestonic elemental ratio in zooplankton ecology: Reply to the comment by Brett. *Limnol. Oceanogr.* 38: 1337–1340.

Vadstein, O. 2000. Heterotrophic, planktonic bacteria and cycling of phosphorus. Phosphorus requirements, competitive ability, and food web interactions. *Adv. Microb. Ecol.* 16: 115–167.

Valiela, I. 1984. *Marine Ecological Processes*. Springer, New York.

Van Cappellen, P., and E. D. Ingall. 1996. Redox stabilization of the atmosphere and oceans by phosphorus-limited marine productivity. *Science* 271: 493–496.

van de Koppel, J., J. Huisman, R. van der Wal, and H. Olff. 1996. Patterns of herbivory along a gradient of primary productivity: An empirical and theoretical analysis. *Ecology* 77: 736–745.

van der Heijden, M.G.A., T. Boller, A. Weimken, and I. A. Sanders. 1998. Different arbuscular mycorrhizal fungal species are potential determinants of plant community structure. *Ecology* 79: 2082–2091.

Van der Wal, R., H. Van Wijnen, and D. Bos. 2000. On facilitation between herbivores: How Brent Geese profit from brown hares. *Ecology* 81: 969–980.

Van Hook, R. I., M. G. Neilsen, and H. H. Shugart. 1980. Energy and nitrogen relations for a *Macrosiphum liriodendri* (Homoptera: Aphididae) population in an east Tennessee *Liriodendron tulipfera* stand. *Ecology* 61: 960–975.

Van Nieuwenhuyse, E. E., and J. R. Jones. 1996. Phosphorus-chlorophyll relationship in temperate streams and its variation with stream catchment area. *Can. J. Fish. Aquat. Sci.* 53: 99–105.

Vandermeer, J. 1980. Indirect mutualism: Variations on a theme by Steven Levine. *Am. Nat.* 116: 441–448.

Vanni, M. J. 1996. Nutrient transport and recycling by consumers in lake food webs: Implications for algal communities. Pages 81–91 in *Food Webs: Integration of Pattern and Process*, edited by G. Polis and K. Winemiller. Chapman and Hall, New York.

Vanni, M. J., and D. L. Findlay. 1990. Trophic cascades and phytoplankton community structure. *Ecology* 71: 921–937.

Vanni, M. J., A. S. Flecker, J. M. Hood, and J. L. Headworth. 2002. Stoichiometry of nutrient recycling by vertebrates in a tropical stream: Linking biodiversity and ecosystem function. *Ecol. Lett.* 5: 285–293.

Varma, A., B. W. Boesch, and B. O. Palsson. 1993. Biochemical production capabilities of *Escherichia coli. Biotechnol. Bioeng.* 42: 59–73.

Verhoeven, J.T.A., W. Koerselman, and A.F.M. Meuleman. 1996. Nitrogen- or phosphorus-limited growth in herbaceous, wet vegetation: Relations with atmospheric inputs and management regimes. *Trends Ecol. Evol.* 11: 494–497.

Vermeer, J. G., and F. Berendse. 1983. The relationship between nutrient availability, shoot biomass and species richness in grassland and wetland communities. *Vegetatio* 53: 121–126.

Villar-Argaiz, M., J. M. Medina-Sánchez, and P. Carrillo. 2002. Linking life history strategies and ontogeny in crustacean zooplankton: Implications for homeostasis. *Ecology* 83: 1899–1914.

Villar-Argaiz, M., J. M. Medina Sánchez, L. Cruz-Pizarro, and P. Carrillo. 2000. Life history implications of calanoid *Mixodiaptomus laciniatus* in C:N:P stoichiometry. *Verh. Int. Verein. Limnol.* 27: 527–531.

Vitousek, P. 1982. Nutrient cycling and nutrient use efficiency. *Am. Nat.* 119: 553–572.

Vitousek, P. M., J. D. Aber, R. W. Howarth, G. E. Likens, P. A. Matson, D. W. Schindler, W. H. Schlesinger, and D. Tilman. 1997a. Human alteration of the global nitrogen cycle: Sources and consequences. *Ecol. Applic.* 7: 737–705.

Vitousek, P. M., and H. Farrington. 1997. Nutrient limitation and soil development: Experimental test of a biogeochemical theory. *Biogeochemistry* 37: 63–75.

Vitousek, P. M. and R. W. Howarth. 1991. Nitrogen limitation on land and in the sea—how can it occur? *Biogeochemistry* 13: 87–115.

Vitousek, P. M., H. A. Mooney, J. Lubchenco, and J. M. Melillo. 1997b. Human domination of earth's ecosystems. *Science* 277: 494–499.

Vitousek, P. M., L. R. Walker, L. D. Whiteaker, and P. A. Matson. 1993. Nutrient limitations to plant growth during primary succession in Hawaii Volcanoes National Park. *Biogeochemistry* 23: 197–215.

Vogel, S. 1998. Academically correct biological science. *Am. Sci.* 86: 504–506.

Vogt, K. A., C. C. Grier, and D. J. Vogt. 1986. Production, turnover, and nutrient dynamics of above- and below-ground detritus of world forests. *Adv. Ecol. Res.* 15: 303–377.

Vogt, V. M., and R. Braun. 1976. Structure of ribosomal DNA in *Physarum polycephalum. J. Mol. Biol.* 106: 567–587.

Vogt, V. M., and R. Braun. 1977. The replication of ribosomal DNA in *Physarum polycephalum. Eur. J. Biochem.* 80: 557–587.

Vollenweider, R. A. 1965. Calculation models of photosynthesis-depth curves and some implications regarding day rate estimates in primary production measurements. *Mem. Ist. Ital. Idrobiol.* 18(suppl): 425–457.

Vollenweider, R. A. 1968. Scientific fundamentals of the eutrophication of lakes and flowing waters, with particular reference to nitrogen and phosphorus as factors in eutrophication. Organisation for Economic Cooperation and Development, Paris, Report No. DAS/CSI/68.27.

Vollenweider, R. A. 1970. Models for calculating integral photosynthesis and some implications regarding structural properties of the community metabolism of aquatic systems. Pages 455–472 in *Prediction and Measurement of Photosynthetic Productivity*, edited by I. Setlik. PUDOC, Wageningen.

Vollenweider, R. A. 1975. Input-output models with special reference to the phosphorus loading concept in limnology. *Swiss J. Hydrd.* 37: 53–84.

Vollenweider, R. A. 1976. Advances in defining critical loading levels for phosphrous in lake eutrophication. *Mem. Ist. Ital. Idrobiol.* 33: 53–83.

Vollenweider, R. A. 1985. Elemental and biochemical composition of plankton biomass: Some comments and explorations. *Arch. Hydrobiol.* 105: 11–29.

Vollenweider, R. A., and P. Dillon. 1974. The application of the phosphorus loading concept to eutrophication research. NRC Canada Rep. No. 13690.

Vrede, T., T. Andersen, and D. O. Hessen. 1999. Phosphorus distribution in three crustacean zooplankton species. *Limnol. Oceanogr.* 44: 225–229.

Walker, T. W., and A.F.R. Adams. 1958. Studies on soil organic matter, 1. Influence of phosphorus content of parent materials on accumulation of carbon, nitrogen, sulfur and organic phosphorus in grassland soils. *Soil Sci.* 85: 307–318.

Walker, T. W. and A.F.R. Adams. 1959. Studies on soil organic matter, 2. Influence of increased leaching at various stages of weathering on levels of carbon, nitrogen, sulfur and organic and total phosphorus. *Soil Sci.* 87: 1–10.

Walker, T. W. and J. K. Syers. 1976. The fate of phosphorus during pedogenesis. *Geoderma* 15: 1–19.

Wardle, D. A., G. M. Barker, K. I. Bonner, and K. S. Nicholson. 1998. Can comparative approaches based on plant ecophysiological traits predict the nature of biotic interactions and individual plant species effects in ecosystems? *J. Ecol.* 86: 405–406.

Wasser, J. S., R. G. Lawler, and D. C. Jackson. 1996. Nuclear magnetic resonance spectroscopy and its application in comparative physiology. *Physiol. Zool.* 69: 1–34.

Watanabe, Y., J.E.J. Martini, and H. Ohmoto. 2000. Geochemical evidence for terrestrial ecosystems 2.6 billion years ago. *Nature* 408: 574–578.

Watson, J. D. 1968. *The Double Helix: A Personal Account of the Discovery of the Structure of DNA*. Atheneum, New York.

Webster, J. R., and E. F. Benfield. 1986. Vascular plant breakdown in freshwater ecosystems. *Annu. Rev. Ecol. Syst.* 17: 567–594.

Wedin, D. 1995. Species, nitrogen and grassland dynamics: The constraints of stuff. Pages 253–262 in *Linking Species and Ecosystems*, edited by C. G. Jones and J. H. Lawton. Chapman and Hall, New York.

Wedin, D. A., and D. Tilman. 1990. Species effects on nitrogen cycling: A test with perennial grasses. *Oecologia* 84: 433–441.

Weers, P.M.M., and R. D. Gulati. 1997. Effect of the addition of polyunsaturated fatty acids to the diet on the growth and fecundity of *Daphnia galeata*. *Freshwater Biol.* 38: 721–730.

Welker, D. L., K. P. Hirth, and K. L. Williams. 1985. Inheritance of extrachromosomal ribosomal DNA during the asexual life cycle of *Dictyostelium discoideum*: Examination by use of DNA polymorphisms. *Mol. Cell. Biol.* 5: 273–280.

Werner, E. E., and J. F. Gilliam. 1984. The ontogenetic niche and species interactions in size-structured populations. *Annu. Rev. Ecol. Syst.* 15: 393–425.

Westheimer, F. H. 1987. Why nature chose phosphates. *Science* 235: 1173–1178.

Westheimer, F. H. 1992. The role of phosphorus in chemistry and biochemistry—an overview. *ACS* Symp. Ser. 486: 1–17.

White, M. M., and I. A. McLaren. 2000. Copepod development rates in relation to genome size and 18S rDNA copy number. *Genome* 43: 750–755.

White, T.C.R. 1993. *The Inadequate Environment: Nitrogen and the Abundance of Animals*. Springer-Verlag, Berlin.

Wiebe, H. H. 1978. The significance of plant vacuoles. *BioScience* 28: 327–331.

Wiegert, R. G., ed. 1976. *Ecological Energetics: Vol. 4, Benchmark Papers in Ecology*. Dowden, Hutchinson, and Ross, Stroudsburg, Pa.

Wilhelm, S. W. 1995. Ecology of iron-limited cyanobacteria: A review of physiological responses and implications for aquatic systems. *Aquat. Microb. Ecol.* 9: 295–303.

Williams, E. T., and R. C. Johnson. 1958. *Stoichiometry for Chemical Engineers*. McGraw-Hill, New York.

Williams, K. L. 1986. Extrachromosomal DNA elements in the nucleus of *Dictyostelium discoideum*. Pages 439–447 in *Extrachromosomal Elements in Lower Eukaryotes*, edited by R. B. Wickner. Plenum, New York.

Williams, R. F. 1955. Redistribution of mineral elements during development. *Annu. Rev. Plant Physiol*. 6: 25–42.

Williams, R.J.P. 1981. The natural selection of the elements. *Proc. R. Soc. London B Biol. Sci*. 213: 361–397.

Williams, R.J.P., and J.J.R. Fraústo da Silva. 1996. *The Natural Selection of the Chemical Elements: The Environment and Life's Chemistry*. Clarendon, Oxford.

Wilson, E. O. 1998. *Consilience: The Unity of Knowlege*. Knopf, New York.

Wittenberger, C. 1970. The energetic economy of the organism in animal evolution. *Acta Biotheor*. 19: 171–35.

Wittman, H. G., and B. Wittman-Leibold. 1974. Chemical structure of bacterial ribosomal proteins. Pages 115–140 in *Ribosomes*, edited by M. Nomura, A. Tissières, and P. Lengyel. Cold Spring Harbor Laboratory, Cold Spring Harbor, N.Y.

Woodwell, G. M., and F. T. Mackenzie. 1995. *Biotic Feedbacks in the Global Climatic System: Will the Warming Feed the Warming?* Oxford University Press, New York.

Wu, J., W. Sunda, E. A. Boyle, and D. M. Karl. 2000. Phosphate depletion in the western North Atlantic Ocean. *Science* 289: 759–762.

Wurtsbaugh, W. A. 1988. Iron, molybdenum, and phosphorus limitation of N_2 fixation maintains nitrogen deficiency of plankton in the Great Salt Lake drainage (Utah, USA). *Verh. Int. Verein. Limnol*. 23: 121–130.

Xie, L., and D.I.C. Wang. 1997. Integrated approaches to the design of media and feeding strategies for fed-batch cultures of animal cells. *Trends Biotechnol*. 15: 109–113.

Yodzis, P. 1989. *Introduction to Theoretical Ecology*. Harper and Row, New York.

Zak, D. R., K. S. Pregitzer, P. S. Curtis, J. A. Teeri, R. Fogel, and D. L. Randlett. 1994. Elevated atmospheric CO_2 and feedback between carbon and nitrogen cycles. *Plant Soil* 151: 105–117.

Zauke, G.-P., J. Bohlke, R. Zytkowwicz, P. Napiorkowski, and A. Gizinski. 1998. Trace metals in tripton, zooplankton, zoobenthos, reeds and sediments of selected lakes in north-central Poland. *Int. Rev. Gesamten Hydrobiol*. 83: 501–526.

Zeng, A.-P., W.-S. Hu, and W.-D. Deckwer. 1998. Variation of stoichiometric ratios and their correlation for monitoring and control of animal cell cultures. *Biotechnol. Prog*. 14: 434–441.

Index

absorption, 91, 191; of CO_2 in oceans, 359; of light, 52, 300. *See also* uptake
Acanthodiaptomus, 139
Acartia, 222–25
affinity strategy, 86
allocation, 289; algal, 111–15, 301; cellular, 38–39, 57, 80, 85–86, 132–33, 137, 377; the growth rate hypothesis and, 144–45, 160–168; leaf, 57, 72, 80, 103; to lignin, 56; plant, 88, 91, 96, 102, 131, 133, 279, 336, 372; to RNA, 104, 142, 151, 153, 172, 176–77; vertebrate, 170–74, 252, 254, 373
allometry. *See* body size
aluminum, 11–12, 76
amino acid, 7, 19, 50, 53, 59–60, 188, 218; N content in, 52, 59; polymerization rate, 162, 165; sequence, 62; storage of N by, 82–83, 93; in vacuoles, 70
ammonia (-ium), 19, 30, 49, 60, 99, 218, 248, 234, 250, 260, 295, 330, 334, 362
amphibian, 256–58; rDNA of, 156
amphipod. See *Diporeia*
Anabaena, 76, 103, 266
anemia, 189
annual plant, 88, 260, 278
anteater, 221
antler, 66, 169; formation, 211–12, 227; -phosphorus in forests, 319
apatite. *See* hydroxyapatite
aphid, 147, 211, 376
Archaea, 54
arctic, 321, 364; and the growth rate hypothesis, 147, 150, 215
argon, 11, 12
armored catfish, 169, 257
arthropod, 136, 183, 220, 221, 320; RNA in, 146
Asterionella, 104

atmosphere, 10, 47, 66, 323, 331, 354, 355–356; CO_2 in, 90, 100–101, 133, 359; O_2 in, 44, 333–335, 375
ATP, 47, 50, 52, 61–62, 71, 77, 114, 136, 142–44, 160–61, 164; C:N:P in, 61; in N fixation, 323; P turnover rate in, 77
angiosperm, 102, 105, 125–26

bacteria, 4, 158, 171, 173, 205, 209, 241, 317, 319–20, 324; and algae, 270–74; cell membranes, 69; cell wall, 58, 67–68; detrital breakdown, 208; gross growth efficiency in, 226; homeostasis in, 19–22, 195–96, 206–7; lipids in, 54–55; luxury consumption, 52, 82–83; methanogenic, 218; mineralization, 130, 249–52; polysaccharides in, 58; protein in, 60; respiration, 222; ribosomes in, 151–53; in seston, 120–21
bacterivore, 186, 209, 374
balance, defined, 15
balanced growth, 15, 80, 251; defined, 42; in DE biochemical allocation model, 162; in element linkage model, 338–39; in nutrient use efficiency, 342
bark, 56, 121, 123, 331; N content of, 58
bass, 174, 254, 293
bat. See *Myotis*
bean, 55; rDNA of, 156
beetle. See *Paropsis*, *Tenebrio*
beryllium, 11–12
Betula, 117–19
biological pump, 359
biotechnology, 188–89
birch, 342. See also *Betula*
bird, calcium needs in, 221, 241
bluegill, 174, 254
body size, 189, 210, 235; and gross growth efficiency, 187, 227; and organism nutrient content, 171–75, 177, 254; use in models, 24

Milton Keynes UK
Ingram Content Group UK Ltd.
UKHW020611031023
429847UK00005B/42